Energy Systems

Series Editor:
Panos M. Pardalos, University of Florida, USA

For further volumes:
http://www.springer.com/series/8368

Stamatina Th. Rassia
Panos M. Pardalos
Editors

Cities for Smart Environmental and Energy Futures

Impacts on Architecture and Technology

Springer

Editors
Stamatina Th. Rassia
ETH Zürich
Department of Architecture
Institute of Technology in Architecture
Zürich
Switzerland

Panos M. Pardalos
University of Florida
Department of Industrial and Systems Engineering
Gainesville, Florida
USA

ISSN 1867-8998 ISSN 1867-9005 (electronic)
ISBN 978-3-642-42971-2 ISBN 978-3-642-37661-0 (eBook)
DOI 10.1007/978-3-642-37661-0
Springer Heidelberg New York Dordrecht London

Printed on acid-free paper

Springer is part of Springer Science+Business Media (www.springer.com)

Preface

Urbanism and the design of *Cities for Smart Environmental and Energy Futures* form the basis of an ongoing effort upon which interdisciplinary architectural planning and philosophy as well as engineering design are developed.

As cities are aimed nowadays to be ever *smarter*, they are designed to offer competitive, remarkable and resourceful environments for the future. As such, cities may offer alternatives between energy strategies and optimization theories of mega-spaces in a territory or a landscape. Designing or redeveloping the city blueprint for today and for the future is and has been an open question in the architectural and engineering fields.

By the use of the words *smart*, *energy* and *future*, in connection with a city, it is desired to present a variety of viewpoints about the operational efficiency and the living conditions in a city. Design objectives lead to an "in-process" or to an entirely complete *smartness*.

Cities could be examined as environments of systemic change, as unified technological advancements, as historic sites, as high-rise or low-rise and formal or informal developments. They may appear branded, as zero-carbon, sustainable, eco, bionic, network, digital and ubiquitous. These form only a subset of the multitude of themes that exist or are referred to currently in the scientific community. Such themes, whether for complete or in-development process, coexist on a scientific platform that may offer the basis upon which new challenges could be addressed.

The chapters comprising this book are aiming to engage in further research and innovative practices. In this book, cities are examined from the macro to the micro level, from the sustainable city development to the human scale and further to the structural material properties and formation. Cities are presented to be invincible and utopic but also to have the magnetism of a digital dream. They are questioned further for their complex and conflicting ideologies, and they are analyzed in relation to disaster relief and indoor urbanism, sustainable remedies for building conservation and climate change as well as sustainable user mobility and accessibility.

In this process of interdisciplinary scientific enquiry, the present book offers a common framework where international experts present theories and new results, novel ideas, concepts and research findings, as these relate to the city of the past, of today and of tomorrow.

We would like to express our special thanks to all the authors of the chapters contributed in this volume.

Last but not least, we wish to acknowledge the superb assistance that the staff of Springer has provided during the preparation of this publication.

Zurich, Switzerland Stamatina Th. Rassia
Florida, USA Panos M. Pardalos

Contents

Contributors

Eveline Althaus ETH Wohnforum – ETH CASE Centre for Research on Architecture, Society & the Built Environment, Zürich, Switzerland, althaus@arch.ethz.ch

Zeynep Aygen Mimar Sinan Fine Arts University, Istanbul, Turkey, Aygen_zeynep@hotmail.com

Colin Beattie Curtin University, Curtin University Sustainability Policy Institute (CUSP), Bentley, Australia, colbeattie@gmail.com

Kobe Boussauw Ghent University, Geography Department, Krijgslaan 281, S8 9000, Gent, Belgium

Alfredo Brillembourg ETH Zurich, D–ARCH, Urban-Think Tank Chair, Zurich, Switzerland, brillembourg@arch.ethz.ch

Paul Budde BuddeComm, Bucketty, NSW, Australia, pbc@budde.com.au

Jessica Bunning Curtin University, Curtin University Sustainability Policy Institute (CUSP), Bentley, Australia, jessbunning@hotmail.com

Gousis Christos National Technical University of Athens, Athens, Greece, euchristos@gmail.com

Michael Contento ETH Zurich, D–ARCH, Urban-Think Tank Chair, Zurich, Switzerland, contento@arch.ethz.ch

Frank Duffy CBE, Architect, Independent Consultant, London, UK, frank@duffydesign.com

Bryan Eisenhower University of California, Department of Mechanical Engineering, Center for Energy Efficient Design, Santa Barbara, USA, bryane@engineering.ucsb.edu

Melanie Fessel Terreform ONE, Brooklyn, USA, fessel@terreform.org

Anthony Filipovitch Minnesota State University, Urban & Regional Studies Institute, Mankato, USA, Anthony.Filipovitch@mnsu.edu

Marie Antoinette Glaser ETH CASE, Centre for Research on Architecture, Society & the Built Environment, Zurich, Switzerland, glaser@arch.ethz.ch

Richard Hanley City University of New York, New York City College of Technology, New York, USA

Ioannis Tzouvadakis National Technical University of Athens, Athens, Greece, itzouvad@central.ntua.gr

Mitchell Joachim Terreform ONE, Brooklyn, USA

New York University, New York, USA, mj@terreform.org

Alex Kaiser Ordinary Ltd, London, UK, alex@anordinarywebsite.com

Gökhan Kılıç Izmir University of Economics, Izmir, Turkey, gokhankilic79@gmail.com

Hubert Klumpner ETH Zurich, D–ARCH, Urban-Think Tank Chair, Zurich, Switzerland, klumpner@arch.ethz.ch

Denia Kolokotsa Technical University of Crete, Environmental Engineering Department, Crete, Greece

Magnus Larsson Ordinary Ltd, London, UK, magnus@anordinarywebsite.com

Wei Lin Case Western Reserve University, Department of Electrical Engineering and Computer Science, Cleveland, OH, USA, linwei@case.edu

Xinghua Liu Shenzhen Graduate School, Harbin Institute of Technology, HIT Campus of Shenzhen Town, Xili, Shenzhen, China, leo.xinghua.liu@gmail.com

Peter Newman Curtin University, Curtin University Sustainability Policy Institute (CUSP), Bentley, Australia, P.Newman@curtin.edu.au

Fabiano Lemes de Oliveira University of Portsmouth, Portsmouth, UK, fabiano.lemes@port.ac.uk

Smita Rakshit Minnesota State University, Urban & Regional Studies Institute, Mankato, USA

Vanessa Rauland Curtin University, Curtin University Sustainability Policy Institute (CUSP), Bentley, Australia, V.Rauland@curtin.edu.au

Mark Raymond Architect, Belmont, Port of Spain, Trinidad and Tobago, mark@markraymond.com

Mattheos Santamouris National Kapodistrian University of Athens, Physics Department, Group Building Environmental Studies, Athens, Greece, msantam@phys.uoa.gr

Lindsey Sherman ETH Zurich, D–ARCH, Urban-Think Tank Chair, Zurich, Switzerland, sherman@arch.ethz.ch

Henriette Steiner ETH Zürich ETH Wohnforum – ETH CASE, D-ARCH, Zürich, Switzerland, steiner@arch.ethz.ch

Robert Stuart-Smith Kokkugia, London, UK, rob@kokkugia.com

Konstantina Vasilakopoulou Technical University of Crete, Environmental Engineering Department, Crete, Greece

Kristin Veel University of Copenhagen, Department of Arts and Cultural Studies, Copenhagen, Denmark, Veel@cantab.net

Frank Witlox Ghent University, Geography Department, Gent, Belgium, Frank.Witlox@ugent.be

A Time and a Place for Everything

Frank Duffy

Abstract Office buildings are used as an example of the unsatisfactory relationship between the architect's role of designing of the physical fabric of workplaces and the subsequent necessity of accommodating emerging – and almost certainly unexpected – patterns of beneficial use over their lifetime. The professionalisation of Facilities Management has so far failed to fill this gap. The concept of the "intelligent building" has further confused the issue because it is based on the unrealistic assumption that automated environmental control mechanisms will have the capacity to anticipate, and respond to, open–ended user expectations and ever changing patterns of occupancy. The very term is an excellent example of a "pathetic fallacy", the belief that lifeless objects, on their own, can anticipate change. The practical implication of this argument is that the software of space management must become at least as "intelligent" as the design of the hardware of the buildings themselves – perhaps much more so. Space management for many organisations is already transcending isolated, individual buildings and will have to expand through time and space to embrace the accommodation of the totality of ever shifting networks, connecting people, wherever they happen to be, in multiple organisations in many buildings – and places – of many different types.

1 Intelligent Buildings and Smart Cities

The adjectives, "smart" and "intelligent", applied to office buildings and sometimes even to entire cities, betray the anthropocentric underpinning of such claims. The dulcet, seductive whispers of advertising copywriters and similar self-promoters can be heard in the background. The terms "intelligent buildings" and "intelligent cities"

F. Duffy (✉)
CBE, Architect, Independent Consultant, London, UK
e-mail: frank@duffydesign.com

S.Th. Rassia and P.M. Pardalos (eds.), *Cities for Smart Environmental and Energy Futures*, Energy Systems, DOI 10.1007/978-3-642-37661-0_1,

are good examples of a "pathetic fallacy", i.e. the belief that lifeless objects have a will of their own. This cannot be the case however cleverly building systems are automated to respond to external factors such as climate and internal factors such as patterns of use: the capacity to respond to stimuli should not be confused with intelligent anticipation. In the end it is essentially the responsibility of owners and users themselves to make sure that they are smarter than their buildings and are using them purposefully to further their own and society's interests.

The truth is that buildings are predominantly inert – thankfully – and very commonly far from intelligent. They have to be led. They exist to be used. Unfortunately it's not easy to have a conversation with a building although some are undeniably capable of moving the heart. Certainly some buildings are badly designed but many, many more are very badly managed. It's certainly possible to have been the unlucky victim of an accident caused by careless design or poor construction or to have the bad luck to be struck down by, let's say, legionnaires disease. However, the vast majority of buildings don't bite back. What they are much more likely to be is lazy – i.e. far less efficient and effective than they ought to be, measured by various scales of utility, energy use and environmental performance over time.

What is even more critical than the initial impact and performance of the physical design of buildings and cities is the design of intelligent and efficient procedures and protocols to anticipate measure, and manage patterns of use during their entire lifetime. For example, when actual patterns of space use *are* measured, many buildings, especially offices, usually prove to be far from fully occupied most of the time[1] – which raises the question of the relationship between the design of the physical fabric of architecture and the 'design' of socially acceptable and ecologically sound patterns of use. As information technology develops, such changes in internal patterns of use as well as external seasonal and climatic factors can, to some extent, be anticipated and programmed into so called building information management (BIM) systems. Nevertheless, ultimate responsibility for beneficial use must remain firmly in the hands of not the developers, nor the designers, nor the contractors but the consumers, i.e. those who use the buildings for business purposes.

2 Why Measurement over Time Matters

The wider environmental performance of buildings and cities is far from stable, being subject to contextual changes, such as wars and civil disturbances that lead to fluctuations, to the price of such items as fuel – usually upwards. Technological innovations also affect the environmental performance of buildings, not always in

[1]Francis Duffy, David Craig, Nicola Gillen, *Design as a Research Tool,* Facilities, Vol. 29 Iss: 3/4, pp 97–113.

a good way. A case study example is the sad story[2] of the Olgyay brothers, both architects, exiled from Europe in the 1940s to come to work and teach at the School of Architecture at Princeton, who were once renowned for their highly intelligent and far sighted proposals for designing what would now be called "with", as opposed to "against", climate. Their work was based on meticulous and systematic studies of the siting and orientation of houses in particular to minimise over time the deleterious impact of direct sunlight and to maximise what we would now call the "environmentally friendly" benefits of natural cross ventilation. By the late 1960s, the Olgyays' work was neglected, their books forgotten and their environmental laboratory closed, direct results of the post war availability of cheap energy and of cheap packaged air conditioning units which by the 1970s in the United States had become practically universal – creating ironically enough in Princeton in Summer, as I well remember, the dominant background acoustic in every street and courtyard, drowning even the chirping of crickets. The extent to which packaged air conditioners were hugely wasteful of energy as well as irritatingly noisy only became apparent later, sadly too late to rescue the reputation of the far sighted Olgyay brothers in their lifetime.

Looking back the wider lessons from this local incident are that (1) it is far more ecologically sound to work with nature than against it (2) that technology develops rapidly but not always in a benign direction (3) that energy resources are by no means unlimited (4) that more intelligent design and technological solutions can be invented (5) that the successful adoption of such technological solutions depends upon user education and behavioural change and (6) by no means all contextual, political, economic, social and technological developments can be anticipated.

3 Information Technology and Intelligent Office Buildings

The use of term "Intelligent Building" in the UK can be traced back to the early Eighties[3] when it became apparent, particularly in the City of London, that what was then still called the computer was fundamentally changing the nature and processes of the what would come to be called the Financial Services industry. London, a city facing fierce competition from rival European banking centres such as Frankfurt and Paris, was at that time notorious for a stock of office buildings that were particularly poorly designed and underserviced by both North American and Northern European standards. Accommodating the environmental demands of the emerging technology of the computer in the midst of a major revolution in the "regulation" of the Financial Services industry made it imperative to construct in very short order an entirely

---

[2]Beautifully expressed in Victor Olygay, Aladar Olygay, *Design With Climate,* Princeton University Press, 1963.

[3]DEGW, Intelligent Building Studies, a series of multi client studies conducted over the Eighties and Nineties in the UK, Europe, the Far East and Latin America.

new stock of high performance office buildings for businesses, initially especially in London's burgeoning Financial Services sector, that had begun to operate globally on an integrated 24 h basis.[4]

North American standards of workplace design were initially adopted and then quickly modified and enhanced in order to ensure that the City of London could beat off competition from Paris and Frankfurt for the pole position in the time zone that lies between the Far Eastern and the North American financial markets. Initially the urgent overriding technological imperatives, particularly in London, were to provide (1) very large floor plates to accommodate financial trading, (2) substantial vertical and horizontal ducting for cabling and cooling and (3) the zoned air conditioning necessary to accommodate high levels of occupancy and concentrations of heat generating electronic equipment. On all of which the operational performance of the rapidly globalising Financial Services industry continues to depend.

The new types of office buildings that emerged were sometimes, if carelessly, described as "intelligent buildings", not least because, initially, in a very real and practical sense they were extensions of the computer systems and the electronic networks that they had begun to accommodate. The distinction between office buildings and the computer systems they housed was becoming blurred. This insight stimulated the ORBIT Studies[5] carried out between 1980 and 1990 by DEGW, first in the UK and subsequently in the USA and Japan, the purpose of which was to establish the technical performance specification for an emerging class of highly serviced and more thriftily managed office buildings. Such buildings were optimistically, and in retrospect erroneously, dubbed "intelligent buildings".

4 Facilities Management and the Development of the Internet

There is a critically important software dimension to the "intelligence" of buildings. By the early 1980s, more than a decade later than in the USA, the process was well underway in the UK of the professionalisation of the services of what were confidently expected by some, including me, to become the architectural equivalent of software. Facilities Management was the name given to procedures for managing how buildings could be best used over time to continue to achieve users' business and social purposes. This process of professionalisation, initiated in the United States, was enthusiastically aided and abetted in the UK by DEGW[6] which founded

[4]Francis Duffy and Alex Henney, *The Changing City,* Bulstrode Press, 1989.

[5]DEGW and BUS, The *Orbit Study,* Frank Duffy, 1983; *ORBIT 2 (North America)*, 1986, multi client studies, both privately published.

[6]DEGW was founded in London 1974 as an architectural practice specialising in workplace design. The practice no longer exists as an independent entity.

Facilities[7] at that time, initially published by the Architectural Press as a monthly, technical newsletter for Facilities Managers, very much based on the long standing technical sections of the *Architects Journal.* The expectation was that very specific skills and training, in effect the knowledge base of an emerging profession, had to be articulated and promulgated to help businesses and other clients make the most efficient, effective and expressive use of their increasingly complex building structures, servicing systems and interior working environments. In parallel great attention was given in these publications to the technological infrastructures upon which building performance increasingly depended to achieve organisation goals in a period of rapid developments on Information Technology. In effect the concept of software, so familiar in the world of the computer, had been quickly borrowed and extended to embrace the management over time of the physical working environment.

In more recent decades further technological developments have occurred – not simply the escape of the hardware of the computer from the computer room into the general office which had so rapidly accelerated the early process of professionalisation of Facilities Management, but an even more consequential series of temporal and spatial phenomena – the practical achievement and exploitation of instant connectivity and communication, everywhere and at any time, and in any order, through the creation, diffusion and the even more rapid universal popular acceptance of the new realities of the worldwide web and the internet. Time and space are no longer anything like the powerful constraints on communication and connectivity that had been taken for granted for centuries.

5 Redesigning the Relationship Between Supply and Demand

In the emerging world of instant and universal connectivity, the intellectual and operational challenges for architects, space planners and facilities managers, who together have the responsibility of articulating the demand side of the supply/demand equation, are fourfold: (1) to be parsimonious in justifying the value of place in an increasingly virtual universe; (2) to invent temporal and behavioural conventions that will respond to enhanced demands for instant accessibility and constant communication; (3) to value the modification and improvement of the existing built fabric as at least equally important to creating the new and (4) to invent the physical fabric that will be needed by emerging Twenty First Century cities and organisations. In this new world my conjecture is that less really will become more.

The supply side, which extends from the investors to the developers, through the real estate industry to the vast and incoherent construction "industry", faces even

[7] *Facilities,* 1983–91, originally a monthly newsletter edited by DEGW and published initially by the Architectural Press, London.

greater challenges: (1) to come to terms with "bigger" not necessarily being "better"; (2) to respond to user demands for more responsive and sophisticated services and (3) to concentrate not just on new construction but to give equal attention to ensuring the most effective and efficient use of what already exists.

In other words, the software of space management must become at least as "intelligent" as the design of the hardware of the buildings themselves – perhaps much more so. Space management for many organisations already is transcending isolated individual buildings and will have to expand through time and space to embrace the accommodation of the totality of ever shifting networks connecting people, wherever they happen to be, in multiple organisations in many buildings. The challenge is not so much the design of new buildings but of what the late Eric Trist[8] of the Tavistock Institute so presciently described 60 years ago as "open socio-technical systems". Trist's contribution, based on his work in the newly nationalised British coal mining industry, was to draw attention to the necessity of designing managerial, technological, social and remuneration systems in an integrated – and open ended – way.

6 An Alternative Model: Connecting Buildings and Their Use Over Time

The lesson is not that buildings are unimportant for business but rather the opposite: that measuring their utility depends on many other factors that transcend conventional architectural criteria and spill over existing organisational and temporal barriers. Unfortunately, over the Twentieth Century, procedures and processes have been developed for financing, designing and constructing buildings, especially office buildings, which are linear, fractured and largely feedback free. Far too often quick delivery is everything, moving on to the next project is the goal, blame is endemic, feedback is unwelcome – not least when there are lawyers on the prowl.

Facilities management potentially provides an alternative model, because, it already has an intellectual and practical perspective that embraces both multiple buildings and the efficient management of their use over time. Unfortunately the discipline is still too young and marginal to have been able to challenge the powerful, investment led supply chain that extends from capital accumulated in, let's say, Abu Dhabi and has been invested over the last two decades in vast new speculative office buildings, many of which accommodate the Financial Services sector in such centres as London and New York.

To create truly functional office buildings robust enough to accommodate rapidly changing office technologies and work patterns, we shall have to reverse that uni-directional supply chain, turning it into a feedback rich 'demand chain', i.e. creating

[8]Trist E.L. and K.W. Bamforth, *Some Social and Psychological Consequences of the Longwall Method of Coal Getting,* Human Resources, 4(1) 1951, pp 3–38.

a process less focussed on short term transactions and quick delivery and more related to the beneficial use of all buildings, through systematically connecting input with output over time. Then we would be able to use the term “intelligent buildings” truthfully without the letting agent’s and the copywriter’s condescending nod and wink.

Smart Cities of Tomorrow

Paul Budde

Abstract Tomorrow's digital cities will be the product of today's dreams. The world presently faces numerous challenges, many of which can only be overcome by changing the way we do things. We are now at a turning point and our future will be shaped by the way in which we respond to today's problems, and how successfully we can move into the digital world.

An important element of these developments lies in the creation of Smart Cities, consisting of cohesive and open telecommunication and software architecture, which will underpin the smart, citizen-centric applications that will abound. These applications will also be applied in areas of national interest and social need, such as e-health, smart energy and e-education. For this occur a trans-sector approach is needed, along with strong government leadership – a more horizontal system of collaboration which combines the knowledge and resources of different sectors.

The building blocks upon which smart cities will be created include smart and renewable energy; next-generation networks; smart buildings; smart transport; and, extremely important, smart government.

1 Digital Cities: Digital Dreams?

Half of the world's population are already city-dwellers, and the trend towards increased urbanization is accelerating rapidly. The future of the majority of the world's citizens is undeniably urban – 70 % will live in cities by 2050 – but exactly how that city of tomorrow will look, and how smart living will be implemented and experienced remains largely uncertain.

P. Budde (✉)
BuddeComm, 5385 George Downes Drive, Bucketty, NSW 2250, Australia
e-mail: pbc@budde.com.au

S.Th. Rassia and P.M. Pardalos (eds.), *Cities for Smart Environmental and Energy Futures*, Energy Systems, DOI 10.1007/978-3-642-37661-0_2,

We are already seeing some directions for the future emerging: public transport will become more efficient and predictable; cars will drive and park themselves; carbon emissions will be reduced as electric vehicles replace current models; and houses will be more energy-efficient. Tele-conferencing and video-conferencing is beginning to replace some business travel and e-health and home automation is offering higher quality smart health care and will eventually enable more elderly people to stay at home; smart governance will reduce costs whilst increasing safety and efficiency.

Eventually, perhaps, we will see self-cleaning carbon-neutral buildings tapping into an array of renewable energy sources and sending power back to the smart grid, and electrical appliances automatically switching off in response to demand. Smart cities will be green cities with e-generators routing power efficiently in response to demand.

The potential of smart applications is boundless – personal health applications fixed to clothes or beneath the skin sending back constant streams of data to medical centres and providing real-time alerts or diagnoses; biometric identity devices; always-on mobile access to social networks; and people-to-object digital connections. Smart education will empower the educationally disenfranchised through e-learning and m-education; devices and services will be targeted at the elderly and disabled to increase their inclusion.

The digital dreams of today will become a reality – not just in the privileged developed world but in the developing world also. Ubiquitous, universal broadband access can be achieved in a number of ways: through spectrum management to free up this increasingly scarce resource in a progressively data- and video-hungry mobile world, or the creation of a new investment model based on infrastructure-sharing; or next-generation wireless technologies. The early collaboration of all stakeholders from industry, government, regulators, urban planners, research companies and civic society alike is vital.

The city is essentially a form of cooperation. We can do nothing completely in isolation, but together we can do, make and accomplish anything.

These metropolises are being hit hard by the current stress, and so it would make sense to explore what city communities can do to survive, and even thrive, in the changing environment.

The infrastructure systems used in these cities to manage water, energy, food supply, transport, communication, economic and social structures are faltering. Cities are also the major polluters as they generate the vast bulk of CO_2 emissions.

At the same time, ever since cities started to emerge, they became the centre of knowledge, innovation and social and economic interaction. They contain a vast pool of human capital that can be tapped into for solutions.

Focusing the national search for solutions within cities will also create a grass-roots environment that could assist in addressing many of today's problems. And cities are in a better position than national governments to investigate trans-sector approaches – looking for more horizontal levels of collaboration by combining the knowledge of different sectors to address the shared problems we are experiencing.

The building of smart cities in a Greenfield environment will be difficult enough; managing the transformation of a Brownfield environment is a much more difficult problem, and unless the cities and their citizens become directly involved in working on solutions we see little hope for success.

And if we are going to use our cities as the spearhead in this battle we will need to make sure that they are operating effectively and efficiently – and at present this is often not the case.

2 Smart Cities Based on Three Pillars

Fascinating developments are already taking place towards smart city development. What is needed is for the concept and its many elements to be better communicated to the ordinary citizens. Very significant information and communication processes need to be put in place to support these communities as they work to transform themselves.

In essence, a smart city is built on three pillars:

1. To make quality of life an excellence hub, to deliver services tailored to the citizen;
2. To promote sustainable development through harmonised management of public services, which will increase productivity and generate savings on energy;
3. To work on economic development, so that the city remains an essential lever in the development of new services and the creation of innovative businesses and activities.

The smart city therefore relies on cohesive and open telecom and software architecture, which is the foundation for developing citizen-oriented applications. In particular, this includes:

- Access to superfast fixed and wireless broadband networks;
- A network based on machine-to-machine (M2M) sensors, data collection and storage/archive devices;
- Harmonised and open management of Big Data, and access via service platforms;
- A cohesive, citizen-centric information system.

Once these building blocks are in place it becomes easy to create and/or sustain applications that will make the most of the city's resources, assets and positioning.

It then becomes possible to develop a long-term view for the cohesive integration of the building blocks of the smart city's architecture (telecom and software infrastructure, data centres, network of sensors, etc.) and the short and medium-term sector-specific initiatives (e-health, e-transportation, etc.).

Cities need to be especially vigilant about integrating the following in their plans:

- The conviction that the smart city is geared to a close understanding of its residents and customising services;

- The understanding that the interplay between private and public initiatives, in the areas of both networks and applications, incorporate open innovation as a central tenet;
- The systematic review of those elements that are managed internally and those that are outsourced, and this for all layers of the architecture;
- The integration of new business models, which consider public information as an essential resource in drawing the maximum benefit from a city's actions.

3 Smart Cities Have Key Components

A smart city must address the following issues and may be made up of the following components:

3.1 Smart Energy

Many countries are in the grip of escalating energy prices and to a certain extent this is paralysing the debate regarding sustainability, renewable energy, energy efficiency, industry transformation and so on.

In analysing that situation it became clear that it is actually not the price of energy that is causing these increases. As a matter of fact, energy is still very cheap and in general we have plenty of it from many different sources – the traditional fossil fuels, uranium, geothermal etc. Furthermore there are sun and wind generation opportunities, so there is no reason to fear heavy cost increases due to a fundamental shortage of energy.

We also have to be realistic about the fact that India and China are opening up new coal-powered generators every week. Whatever the rest of the world is doing regarding clean energy is negated by this reality. An encouraging development is that these countries are moving towards using gas for new generators, and while that still has a significant CO_2 footprint it is much better than coal. China is also already the world leader in solar energy technology and their investments in Renewables overtake the investments made in many of the developed economies.

Of course, the million dollar question will be: "Do we have enough time? Is this relentless increase in the use of fossil fuels – for at least the next decade – going to affect our planet's climate?"

The scientists are very clear on this. They say it will have a devastating effect on the world as we know it. The frightening situation, however, is that it is unlikely that any significant reduction in CO_2 emission is going to happen as long as the developing countries are adding new fossil fuel generators to their energy infrastructure at the current rate. Just as similar developments were essential for building the western economies over the last 100 years, the same now applies to the developing economies. Their prosperity is dependent on access to cheap energy.

3.1.1 Renewable Energy

Everybody agrees that the best solution is to move as soon as possible towards renewable energy, despite the fact that these technologies have not yet matured and that it will take a long time to change over. But if we pursue this course, and if we have some luck with the climate change developments, we might still make the switch over to renewable in time. The truth, however, is that the scientists are very sceptical about our ability to respond in a sufficiently timely manner.

So what have smart grids to do with all of this?

> Smart grids form one of the bases of smart cities. There has been a global underinvestment in utility infrastructure. It is dumb, outdated and in desperate need of an upgrade. The latest development is the concept of smart grids – an intelligent IP overlay over the electricity grid, with sensors and other equipment. This will allow utilities too much better manage their network, limit electricity loss, prevent outages, load-shed and provide customers with in-house information and tools (smart meters) to better manage their own energy use. In addition, utilities will be able to reduce their carbon emissions, which will offer interesting opportunities on the carbon trading market. Paul Budde

Much of the reason for the recent price increases relates to the cost of the networks and systems needed to carry the energy in an efficient way to the users. Therefore from a cost perspective it is most certainly very relevant to look at the grid in relation to the price increases. But perhaps we should not do that in a linear way, extrapolating from the past to the future.

As more and more people are putting photovoltaic (PV) systems on their roofs we need to start looking at distributed energy systems, enabling streets and neighbourhoods to become energy producers. While some are describing this development as 'Off Grid', we would like to describe it differently. We do need to reposition the grid and review the investments in the grid in order to ensure that it becomes an interconnected system that can play a central role in distributed energy – like exchanges as we know them in telecommunications and internet services. Initially the distributed use of this system will be limited; however over time this becomes more significant, and eventually that could well become its main role.

What this means is that the value of the grid is not only the delivery of centrally produced energy (basically funded by a component of that energy price) but that its value becomes its intelligent network functions. This network will be shared by many applications such as:

- Managing distributed generation;
- Arbitrage of renewable energy;
- Managing electric vehicles;
- Extending network intelligence into homes and businesses;
- Providing data collection and data analyses in real time and making that (commercially) available to authorised users;
- Value-added network services in relation to billing, network management services for the new parties involved in energy generation and so on.

This means that the business model for grid operators will evolve from simply selling and delivering energy to the more sophisticated role of managing energy. Any new smart grid developments will need to be pursued with that future in mind.

There are now several countries that are putting smart infrastructure central to their energy policies, as they see this as the key to linking together the various policy and innovation developments. Germany, the Netherlands and the Scandinavian countries are among the thought leaders in this field and Germany has invested heavily to make this happen. In this country, the total cost of their energy investments amounts to around one-third of the country's Gross Domestic Product (GDP).

3.2 Next Generation Telecoms/National Broadband Networks

The second key component of a smart city involves telecoms infrastructure. Global telecom networks are undergoing extraordinary changes, with increasing investments in All-IP Next Generation Networks (NGNs) and National Broadband Networks (NBNs) based on fibre – in order to meet burgeoning consumer demand for high-bandwidth applications.

E-health, e-education, media and sustainability are also the key reasons developed nations need NGNs. IP is at the core of NGNs as it facilitates affordable triple play business models and seamlessly integrates voice, data and video. A proper inventory of national infrastructure assets is required if we want to establish an efficient and economically viable national broadband structure for these services. In the developing markets, next generation telecoms will primarily take the form of wireless NGNs (i.e. LTE/WiMAX).

The most critical element to the success of NBNs will be the infrastructure company that will run the network. They will have to make the critical architecture and design decisions for the open wholesale-only services which will form the basis of this new infrastructure for at least the next 25 years. Essential here will be for the network to facilitate the vision laid down by governments, which includes multiple use of the network by other sectors such as healthcare, education, energy, etc. At the same time the company will need to ensure that it remains an infrastructure company and doesn't become another telco.

Australia was the first country to get the national purpose vision right, thanks to government leadership. The USA soon followed and is now showing real leadership as well. The Netherlands, New Zealand and others are also are on the right track. Economic and trans-sector innovations are now key items on the political agenda of these countries. There is no silver bullet and each unique situation generates its own alternatives, which in turn inform others involved in similar national projects. The vision gives rise to the creation of social and economic strategies that need to be taken into account in the design and architecture of the infrastructure.

> Open networks also provide the possibility to create a multiplier effect. The infrastructure can be used for a range of smart community applications (healthcare, education, smart grids, media, etc.). Open networks are the next step in the evolution of telecoms infrastructure as it gives users full control of the services and applications that are made available over high-speed broadband infrastructure. Open networks also means a democratisation of the telecoms infrastructure. The topology and the architecture of the open network should be such that infrastructure, service and content providers can all offer higher quality and different 'premium' products and services. Paul Budde

Pragmatic solutions need to be developed to maximise the use of existing infrastructure and other resources. Un(der)served areas need to receive priority and local communities and councils can play a key role in this. Wireless broadband can play a major role as well. These early projects could also be an ideal testing ground for trans-sector applications.

3.3 Smart Government

The infrastructure above requires the governments around the world to modernise outdated infrastructure and implement new strategies and processes to meet the needs to the community. Governments taking a smart approach are heading in this direction which is referred to as Smart Government. It is an emerging area and a key element to the success of smart communities. One study by Pike Research estimated the global smart government technology market to be worth $706 million in 2011 and would grow to $2.5 billion by 2017. Between 2011 and 2017 cumulative investment in this area would total $4.8 billion worldwide. Pike Research identified North America to lead these developments during this time, with Europe and Asia Pacific not far behind.

In 2013 the future use and management of Big Data is becoming a key focus for governments. In addition, mobile services are being incorporated into service offerings and cloud computing is becoming integrated to network development. The impact of the Internet of Things (IoT) and M2M developments are also a key consideration for this sector.

3.4 Smart Buildings/Homes

In many cities around the world, high density living is the norm and attention is now turning towards making this style of living more sustainable. Technological innovations include water harvesting and re-use, solar collection and energy-efficient appliances including heating and cooling. Sustainable urban transport systems are also on the agenda for many governments.

Exhibit 1: Smart Homes

Smart homes such as 'habitat control' or 'intelligent home' type networks are equipped with devices that possess an amount of integrated intelligence required to manage and exchange data.

Though home automation systems being offered in the past were heavily marketed as devices that enhanced lifestyle, in recent times, security, energy and access control systems have gained increased prominence and usage. The market is also displaying signs of maturity as the demand for integrating home automation systems with the internet is on the rise. In fact, even the current demand for simple applications has prompted broadband service providers to include home networking products in their installation packages and even go to the extent of integrating them with the existing systems at home free of charge.

Smart home functions include: entertainment, communications, energy and climate control, security, alternative energy and energy neutral applications, lighting and robotics. Facilities for tele-working are another important element in smart home designs.

Increasingly the definition of smart homes is also beginning to include 'zero-energy'. These buildings have zero net energy emissions. Carbon emissions generated from on-site or off-site fossil fuel use are balanced by the amount of on-site renewable energy production. This can also include carbon emissions generated in the construction of the building and the embodied energy of the structure.

(Source: BuddeComm 2013)

3.5 Smart Transport

Smart transport systems or intelligent transport systems (ITS) encompass a range of wireless and wired communications-based information technologies that can be integrated into transportation infrastructure and vehicles.

Traditionally road network infrastructure has been built and then maintained, with little active management of how it is used. Investing in intelligent infrastructure allows for road managers to be more aware of how the network is being used, and how it can be used more efficiently.

Smart transport developments also include smart vehicles, which can communicate with each other and the transport network, and take action to improve safety and efficient operation. Smart vehicles include not only cars, but also trucks, trains, trams and buses.

Smart vehicle technologies are not something of the future. Almost all of us would be familiar with anti-lock brakes, where the vehicle thinks and then acts to brake more safely than a human can. Of course, there are even more advanced technologies already available, such as adaptive cruise control that can keep vehicles at a safe following distance, and dynamic navigation systems that receive data to calculate the best route home.

In 2013 there are many exciting ITS projects taking place around the world and Electric Vehicles, or Smart Cars, have become a reality with hundreds of thousands in operation. Recharging solutions are now under the spotlight, with developers attempting to find ways to power the cars based wireless and grid technology. Currently the infrastructure to facilitate this development is not in place. There are very few charging stations, and the electricity grids are incapable of handling any volume of smart cars. More coordinated smart infrastructure policies are required in order to move forward.

Exhibit 2: Learning from E-cars

It is interesting to note that the e-car is not new. In fact the first 30 years of the car industry was dominated by e-cars. The first e-cars began to emerge as early as 1897. They were mainly used as commercial vehicles, and taxis in particular ran on electricity until well into the 1920s. Some of the commercial e-cars lasted into the 1940s.

With the arrival of the combustion engine the reach of cars could be expanded, and one could drive into the country. There was already a well-established gasoline distribution network for household purposes (mainly stoves) and the first private cars were used purely for leisure, so a trip into the country was high on the list for those early drivers. At that time electricity was still a novelty outside the cities.

The residential mass market quickly discarded the e-car and soon after WWI most private vehicles were petrol-driven. After the oil crisis in the 1970s e-cars received renewed attention but the car and oil industries – while paying lip service to the development – never seriously threw their support behind them.

A very promising initiative was taken in California in 1990, which saw the arrival of modern well-designed e-cars. However, once again the American car and oil industries were able to undermine the project and eventually the last e-car was literally ‘destroyed’ in 2005.

In the meantime Japan continued its e-car development. Learning from the history of the e-car it is clear that there isn’t a one-size-fits-all solution and a balanced approach is essential to make sure this is not another false start.

(Source: BuddeComm)

4 Strategies for Smart Communities

The concept of smarter homes, communities and cities encompasses a range of policies and strategies that need to be developed parallel to each other.

Smart infrastructure is needed for communications, energy, transport, etc. The key element in 'smart' is its ability to gather data, analyse it and provide intelligent feedback. These developments are referred to as the Internet of Things (IoT) and comprise elements of M2M processing; real-time data and Big Data management.

Communications is an essential element in all smart activities. Currently all the silo-based sectors operate separate and independent infrastructure, which of course creates significant financial waste and a lack of interoperability.

Without sound government policies to actually direct the various sectors to share utilities-based infrastructure nothing 'smart' will happen in the near future. At BuddeComm we call this the trans-sector approach.

Leadership from the top is needed to direct the sectors to work together, share infrastructure and provide a seamless service for the customer. This involves:

- People – and empowering them to take control over, for example, the management of their energy use, e-health requirements, e-education, etc.; and
- Things – as in sensors and devices that collect data and process this in real time. The combined data from, for instance, the weather bureau, the transport authorities, energy companies, local authorities can then be analysed (also in real time) to supply instant information to users. This can be in relation to natural disasters, traffic jams, energy prices, energy efficiencies of buildings, personal healthcare information and so on.

Currently most of this information is collected and stored in silos, often on incompatible systems and most of the time not shared with others. It is highly unlikely that these sector-based silos will readily agree to work together. They will all take cover – rightly or wrongly – behind security, privacy, reliability and so on. Only a leader at the top (President, Prime Minister, Emir etc.) can provide the overall leadership that is needed to create a trans-sector policy that will bring together the various infrastructures and application policies.

Once the foundation is established, industry will jump on board, operating in accordance with clear policies, guidelines and regulations. Governments can promote this further with policies that will see some budget redirection from the various sectors (healthcare, education, energy, environment, transport) towards the development of e-services.

Through positive incentives private industry will develop strategies and come up with most of the investments needed for the building of these smart systems.

> Trans-sector thinking will be required to guide us through the next stage of human evolution. We are drawing attention here to the importance of looking across sectors to create synergy. BuddeComm has previously discussed at length the opportunities within the ICT industries of utilising new telecoms networks for e-health, e-education, smart grids, etc. A new way of thinking has emerged which also applies across infrastructure projects – looking at the potential synergies between the building of roads, sewerage systems, water and gas pipe networks as well as telecoms and electricity networks. Paul Budde

4.1 Greenfield Communities

The culmination of trans-sector thinking will be the evolution of smart connected communities.

Based upon the formation of a community based on FttH networks, a number of technologies and services can combine to create an enhanced-value opportunity for residents of these communities.

Smart homes that are connected to these networks can utilise a range of technologies such as:

- Smart grids, which will enable the management of renewable energy, including e-cars;
- Smart meters connected to smart grids, which will communicate intelligent data about power and energy usage;
- Environmental monitoring meters, which will allow for real-time collection of data via an in-home display to show potable water, non-potable water, power, gas and waste water usage;
- Community portals, which will provide a gateway to community-based information, government services, health services, education services, etc.;
- Home automation technology, which will allow control of energy and resource-saving devices such as photovoltaic cells, window shading, lighting, cooling, heating, garden irrigation, security monitoring, etc.;
- Home entertainment systems, which utilise internet-connected systems for audio and video;
- High-speed broadband capabilities that will support tele-working to allow workers to work from home using voice, data and video conferencing. This will significantly reduce carbon emissions normally associated with travel to a place of work.

To date the easiest path to incorporating these concepts into a smart community has been with the development of 'Greenfield' residential communities. These projects essentially start with no existing legacy utility infrastructure, and involve the construction of new dwellings which are capable of incorporating a range of new technologies.

While these Greenfield developments represent only a small percentage of the total population they provide a valuable opportunity to easily introduce FttH and to build services and capabilities into the community using this infrastructure within a relatively short space of time.

Therefore it is our recommendation to use Greenfield development sites as the test bed for new smart community and smart home concepts, which will accelerate the adoption of new technologies and will lead the way for other existing communities to follow suit.

5 Changes are Starting to Drive Action

Interestingly, we appear to be arriving at a turning point. People are now being confronted directly with environmental, energy, healthcare problems; they need no further convincing and are now starting to react out of a 'survival instinct'. It has become obvious that we need to do something about the problems ourselves.

In addition, the world's population is quickly growing from around seven billion to nine billion.

This obviously is causing stress in virtually all of our global systems. While it is essential that we address the causes of these problems, the reality is also that some of these issues, like over-population, which are causing the stress, are virtually unstoppable and we cannot alter this trend in the foreseeable future. While these problems need to be addressed in the longer-term, it is imperative that we focus on alleviating some of the stress now, so as to win some time to address the long-term problems.

So let us acknowledge that the world is facing a significant number of challenges. The key problem associated with these challenges is a lack of smart government policies based on integrated solutions that cross sector boundaries. Political leadership is needed to address these issues. Over the last few years citizens all round the world have indicated that they are ready for change. We have seen this in relation to climate change and the use of new and modern means of communication.

BuddeComm has argued that we can solve the challenges at hand, but we will have to do things differently. There is not a linear way forward – lateral solutions are needed. Over the last 60 years or so we have created a world of specialists who operate within silos. These silos need to be demolished and new horizontal structures established in which all sectors and disciplines work together.

Leadership from the top is needed if this is to be achieved. It is called the trans-sector approach, and (Information and Communication Technology) ICT is the glue needed to build more horizontal collaborative structures. Whether we are talking about smart cities, smart transport, smart grids, smart buildings or e-health – what is needed is useful data that can be analysed in real time, allowing people and/or machines to make instant decisions in relation to energy efficiency, traffic situations, weather activities, and personal health issues, as well as commercial decisions. These developments are referred to as machine-to-machine (M2M) or IoT and are the cornerstone of the smart developments of the future.

Rethinking Urban Landscapes: Self-Supported Infrastructure, Technology and Territory

Mitchell Joachim and Melanie Fessel

Abstract The tide of environmental decline is a multilayered dilemma in qualitative architectural research. To some extent, architectural directives have struggled to reverse this environmental decay. That is not to say the desire and aspiration to positively contribute to an ecological society is omitted from the design discourse. In fact just the opposite, an immense consortium, both formal and informal, of architectural thinkers are absolutely devoted to the task of sustainability. For at least this past decade the dominant meta-theme of a majority of architectural research has been the promise of sustainability. Architects have reflexively launched themselves into the center of the environmental polemic as both its source and resolution. This eco-crisis demands robust solutions on a considerable scale to deal with an imminent collapse. At this point, the radicalization of sustainability is widespread. As a principle it is the politicized mainstream agenda for most design procedures. How can you contend against platitudes like; 'save the planet'? The architectural responses range from fantastical feats of geo-engineering to low-flush toilets, all in service to assuage our fears. It's vital to concede that sustainability will happen in every shade of green. For if the mission fails, we will not be here to say otherwise. Therefore, we must shift beyond sustainability alone and its associative rhetoric. What are the latest comprehensive models within design research that can expedite a greener and grander shift? We need to prepare the next generation of innovators to be self-reliant in a world that requires regeneration. We also need to expand a discourse that continues an arc of humanitarian and technological assertions after sustainability is achieved.

M. Joachim (✉)
Terreform ONE, Brooklyn, NY, USA

New York University, New York, USA
e-mail: mj@terreform.org

M. Fessel
Terreform ONE, Brooklyn, NY, USA
e-mail: fessel@terreform.org

S.Th. Rassia and P.M. Pardalos (eds.), *Cities for Smart Environmental and Energy Futures*, Energy Systems, DOI 10.1007/978-3-642-37661-0_3,

Terreform ONE (Open Network Ecology) is a 501c3 non-profit experimental green design group that promotes new concepts for the advancement of cities. As a unique interdisciplinary laboratory for industrial design, synthetic biology, landscape ecology, art, architecture, and master planning, we work to explore the theoretical and practical framework of ecological design. Through innovative projects and community outreach efforts, we aim to integrate the ecological possibilities of New York City and inspire like-minded projects in urban centers worldwide.

1 Urbaneering Utopia, a New Profession for the Design of Cities

Who is the primary authority in the making of utopia or any extraordinary future city? An Urbaneer is a burgeoning discipline based on urban design that can negotiate the complex mix of technology, theory and practice that embraces the re-invention of the city to exceed the needs of the planet. Today, this nascent interdisciplinary field is in a state of radical development (Fig. 1).

Sparks of utopian reflection throughout human history have been indispensable in evolved societies.[1] Utopias, for the most part, are a necessary paradigm. Utopias display maximal solutions to existing real world problems. They tackle upheaval with orderly retribution. In nearly all variations, Utopias are deliberately excessive. They overshoot the answer to a crisis to accentuate the problem.

Society needs a psychological frame of reference.[2] It's helpful to depict utopia like a personal membership to a health club. What is the perfect picture of a physique? Why do we need to exercise? Aspirations to appear like a certain idyllic athlete or super model sustain our work-out objectives. Many of us understand we cannot be converted into the unattainable airbrushed automatons that decorate voguish magazines. Yet these portrayals provide a common measurement to reflect on our inadequacy. And in some instances, this false imagery makes us admire the precious imperfections and confines of reality. Instead of aims to be grandiose and ideal, the goal shifts to be earnest and good.[3]

One catalyst to reimage the good city is based on the establishment of a patently new field that trumps the outmoded agenda of urban design. Kevin Lynch originally conceived urban design as "City Design" in the late 1950s at MIT.[4] It has not been significantly upgraded to meet contemporary insights. Lynch couldn't have ever imagined more recent factors of change; Google, social networks, ubiquitous smart phones, climate dynamism, energy addictions, global economic calamities, and etc.

---

[1]Thomas More, Utopia, 1516.

[2]Harold Lasswell, Psychopathology and Politics, 1930.

[3]James C. Collins, Good to Great: Why Some Companies Make the Leap... and Others Don't, 2001.

[4]Kevin Lynch, The Image of the City, MIT Press, 1960.

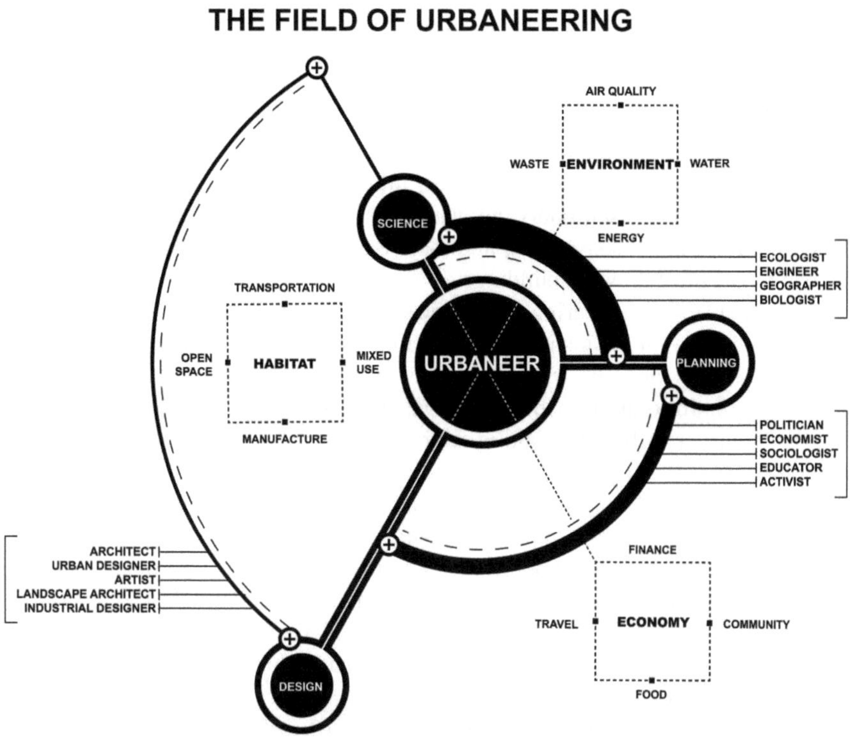

Fig. 1 The field of urbaneering. Diagram of urbaneering (Credits: Mitchell Joachim, Maria Aiolova, Melanie Fessel, Caleb Lowery)

Throughout the developed world, urban design is at an impasse, unable to mend the rift between theory and practice, and stuck in cryptic arguments, such as the one between "Landscape Urbanism" and "New Urbanism". Landscape Urbanists deploy concepts that favor the landscape over architecture in order to plan a city.[5] On the contrary New Urbanist schemes promote historical pedestrian centered neighborhood developments.[6] The dissimilarity between the two approaches is of interest to almost no public body beyond architects, planners and their ilk. Think of the neo-traditionalist town extension of Poundbury near Dorchester endorsed by New Urbanist; Prince Charles versus the high-tech interactive open space zone of Schouwburgplein in Rotterdam of the Landscape Urbanists; West 8. Both of these urban fractions have their merits yet fail to coalesce on a holistic idea of the future city. Moreover the public – every day citizens who occupy theses spaces – invests precious little time in understanding the minutiae. For them, it's an archaic clash between tradition and newness. Urbaneering embrace both and more.

[5]Waldheim, Mostafavi, Corner, Geuze et al., 1997.

[6]Calthorpe, Duany, Plater-Zyberk et al., 1993.

Urbaneering undertakes a diverse range of projects as a prescription for maximal design. It practices totalized schemes that rethink all scales of involvement from the doorknob to the democracy. Its projects can range from; new materials, transportation systems, open spaces, buildings, cities, and surrounding regions. Currently, a few Urbaneers have shaped phytoremediation ponds, living woody plant structures, rooftop farms, soft cars/buses, urban junkspace, and city-wide action plans. To inspire interdisciplinary innovation, Urbaneers encourage people to switch roles; architects must design cars, automotive engineers must devise eco-systems, and ecologists must draw up buildings.

At the core of Urbaneering is a variety of utopian agitation that dispels the defunct myths of modernism with equitable objectives. An Urbaneer replaces implausible rules and master planning with suggestive memes and polemical models. It is hard to argue with amorphous memes like; "city beautiful", "garden city," or "smart growth".[7] The public can rally themselves around these open-ended symbolic gestures and phrases. Since the meme is not fully explicit, the concept leaves room for broad cultural interpretations. It's almost exactly what communities yearn for; freedom to define their own urban spaces.

The Urbaneers aim is to support people to become part of an advanced intellectual initiative framed on the recalibration of the city. Projects such as; Canary Wharf, Potsdamer Platz, NYC Highline, Masdar UAE, and Tianjin China already demand fresh directives. The new profession of Urbaneering provides them in astonishing collaborative ways.

In the past Urban design has mostly been interdisciplinary, but it has not been revamped since its formal inception. Urbaneering will involve city design in a huge range of new ideas, including crowd-sourcing, DIY projects, localized energy, shared transport, e-government, high-throughput computation, biotechnology, and ecology. Urbaneers focus as much on cities' ecosystems and infrastructure – areas ripe for improvement – as well as on more conventional subjects such as buildings and parks.

What does Urbaneering look like? Form follows anything as long as no shape is unmotivated. Diagrams that combine ecology and urbanity called, "Ecograms" serve to prioritize design directives. Light and air are one of many causal factors that respond to context in such Ecograms. Equally, encapsulation of program is correctly orchestrated via any low energy embodied volumes or dense generic loft spaces. Highly malleable these spaces support growth, with elaboration and magnification of character at its limits. These are Ecograms of rampant plurality in which lifestyle is elective and fluid. Depending on the nerve with which the Ecogram pressures, the reinvention and re-privatization, of consequences could prove to be illimitable.

---

[7]Burnham, Daniel H., and Edward H. Bennett, Plan of Chicago, the Commercial Club, Chicago, 1909.

Howard, Ebenezer, Garden Cities of To-Morrow, 1902.

Calthorpe, Peter, and Doug, Kelbaugh (ed.), Pedestrian Pocket Book, 1989.

Beyond utilizing the Ecogram, Urbaneers have a set of proficiencies that merge previously disparate occupations. The discipline is home to almost any recombined professional activities, as long as they meet the constantly changing needs of urbanization. An excellent historical example of someone who now would be an Urbaneer is Frederick Law Olmsted (1822–1903), the nineteenth-century activist who combined journalism, social action and landscape architecture to a single political end.[8]

The next city needs a new breed of communicator. A person skilled in the art of cites beyond the typical utopists, planners, civil engineers, and architects in the present day. These fields need a multifaceted filter of reason to incorporate a profound knowledge of place. An Urbaneer posits the solutions to municipal problems that normally take multiple disciplines to solve. Urbaneers look to merge the edification and expertise needed to reform the city of today for the utopia of tomorrow.

2 Brooklyn Urbaneering

What is the key objective for ecological cities? A primary assertion for the next city is that all necessities are provided inside its accessible physical borders. In this intensified version all vital needs are supplied for its population. In this city, food, water, air-quality, energy, waste, mobility, and shelter are radically restructured to support life in every form. Infrastructure is celebrated as the new center (Figs. 2, 3, 4, 5, and 6).

The strategy includes the replacement of dilapidated structures with vertical agriculture and housing merged with road networks. Former streets become snaking arteries of livable spaces embedded with; renewable energy sources, soft cushion based vehicles for moving, and productive green rooms. The plan uses the former street grid as the foundation for up-to-the-minute networks. By reengineering the obsolete streets, we can install radically robust and ecologically active smart pathways. These operations are not just about a comprehensive model of tomorrow's city, but an initial platform for discourse. Urbaneers expect the future will necessitate marvelous dwellings coupled with a massive cyclical resource net.

3 Rapid Re(f)use, 3-D Fabricated Positive Waste Ecologies

Imagine our colossal municipal landfills as sensible resource sheds to build our future urban and peri-urban spaces. What kind of effort is required to reuse their bountiful contents? Now that the bulk of humanity has chosen to settle in urbanized areas, waste management needs a radical revision (Fig. 7).

[8]Beveridge, Charles E; Paul Rocheleau, Frederick Law Olmsted: Designing the American Landscape. New York, New York: Universe Publishing, 1998.

Fig. 2 Urbaneering Brooklyn 2110: City of the future. A spectacle of smart infrastructure highlights the New Downtown (Credits: Mitchell Joachim, Maria Aiolova, Melanie Fessel, Dan O'Connor, Celina Yee, Alpna Gupta, Sishir Varghese, Aaron Lim, Greg Mulholland, Derek Ziemer, Thilani Rajarathna, John Nelson, Natalie DeLuca)

For hundreds of years we designed cities to generate waste. Now it is time that we begin to design waste to regenerate our cities. What are the possibilities for urban environments after our aged infrastructure is recalibrated? How might urban intensification and waste mix? Terreform ONE's supposition is to reallocate resource streams to flow in a positive direction.[9] In this case, waste is not faintly recycled through infrastructural mechanisms but instead up cycled in perpetuity.

America is the lead creator of waste on the earth, making approximately 30 % of the world's trash and tossing out 0.8 US tons (0.72 t) per US citizen per year.[10] Ungracefully, our American value system is somewhat distressed. It seems value has devolved into rampant waste production: mega-products scaled for super-sized franchise brands, big-box retail, XXL jumbo paraphernalia and so on. The US mindset is thus encapsulating a joint race for ubiquity and instantaneity. Where does it all end up? Heather Rogers affirmed in her investigative book Gone Tomorrow that throwing things away is unsustainable.[11] The first step we must take is

[9]Mitchell Joachim, Maria Aiolova, Melanie Fessel, Philip Weller, Ian Slover, Emily Johnson, Landon Young, Cecil Howell, Andrea Michalski, Sofie Bamberg, Alex Colard, and Zachary Aders for Terreform ONE (Open Network Ecology), Ecological Design Group for Urban, Infrastructure, Building, Planning and Art.

[10]Environmental Protection Agency 2008 Report on the Environment: Highlights of National Trends; see www.epa.gov/roehd/pdf/roe_hd_layout_508.pdf

[11]Heather Rogers, Gone Tomorrow: The Hidden Life of Garbage, The New Press (New York) 2006, pp 54–67 and 104–32.

Fig. 3 Urbaneering Brooklyn 2110: City of the future. All necessities are provided inside the city boundaries (Credits: Mitchell Joachim, Maria Aiolova, Melanie Fessel, Dan O'Connor, Celina Yee, Alpna Gupta, Sishir Varghese, Aaron Lim, Greg Mulholland, Derek Ziemer, Thilani Rajarathna, John Nelson, Natalie DeLuca)

reduction – meaning a massive discontinuation of objects designed for obsolescence. Then we need a radical reuse plan. Our waste crisis is immense. What is our call to action?

One such dilemma lurks in New York. New York City is currently disposing of 36,200 US tons (32,840 t) of waste per day.[12] Previously, most of this discarded material ended up in Fresh Kills on Staten Island, before operations were blocked. Manhattan's inhabitants discard enough paper products to fill a volume the size of the Empire State Building every 2 weeks. Terreform ONE's Rapid Re(f)use and Homeway projects strive to capture, reduce and redesign New York's refuse infrastructure. The initiative supposes an extended city reconstituted from its own

[12]Steve Cohen, 'Wasted: New York City's Garbage Problem', New York Observer, 3 April 2008.

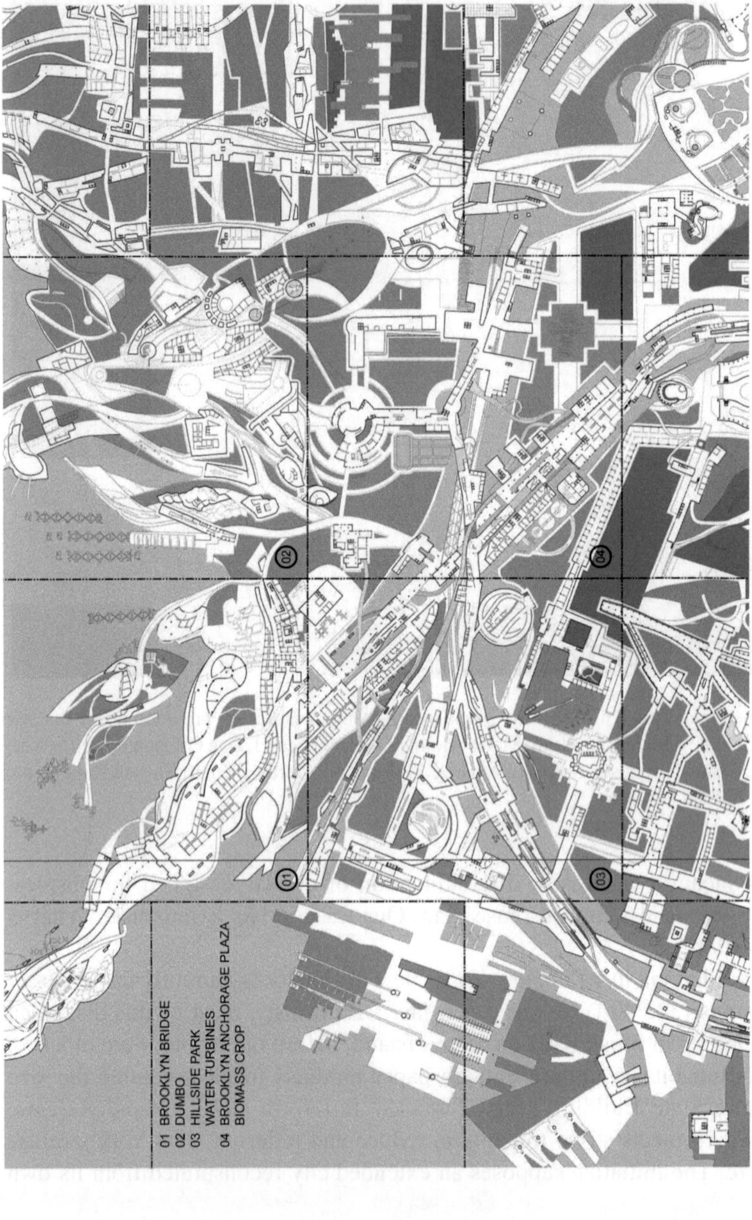

Fig. 4 Urbaneering Brooklyn 2110: City of the future. Site plan includes quadrants for energy, food, water, waste (Credits: Mitchell Joachim, Maria Aiolova, Melanie Fessel, Dan O'Connor, Celina Yee, Alpna Gupta, Sishir Varghese, Aaron Lim, Greg Mulholland, Derek Ziemer, Thilani Rajarathna, John Nelson, Natalie DeLuca)

Fig. 5 Urbaneering Brooklyn 2110: City of the future. View of Brooklyn Bridge into new downtown (Credits: Mitchell Joachim, Maria Aiolova, Melanie Fessel, Dan O'Connor, Celina Yee, Alpna Gupta, Sishir Varghese, Aaron Lim, Greg Mulholland, Derek Ziemer, Thilani Rajarathna, John Nelson, Natalie DeLuca)

junked materials. The concept remakes the city by utilizing all the trash entombed in the Fresh Kills landfill. Theoretically, the method should produce, at minimum, seven entirely new Manhattan Islands at full scale. New York City's premier landfill was started by Robert Moses and driven by apathetic workers and machines.[13] Now, guided by a prudent community with smart equipment, we must reshape it.

How could this work? Outsized automated 3-D printers could be modified to rapidly process trash and to complete the task within decades. These potential

[13]Parks Commissioner Robert Moses, NYC Proposal for Development at Fresh Kills, November 1951.

Fig. 6 Urbaneering Brooklyn 2110: City of the future. Green brain park space used for algae production (Credits: Mitchell Joachim, Maria Aiolova, Melanie Fessel, Dan O'Connor, Celina Yee, Alpna Gupta, Sishir Varghese, Aaron Lim, Greg Mulholland, Derek Ziemer, Thilani Rajarathna, John Nelson, Natalie DeLuca)

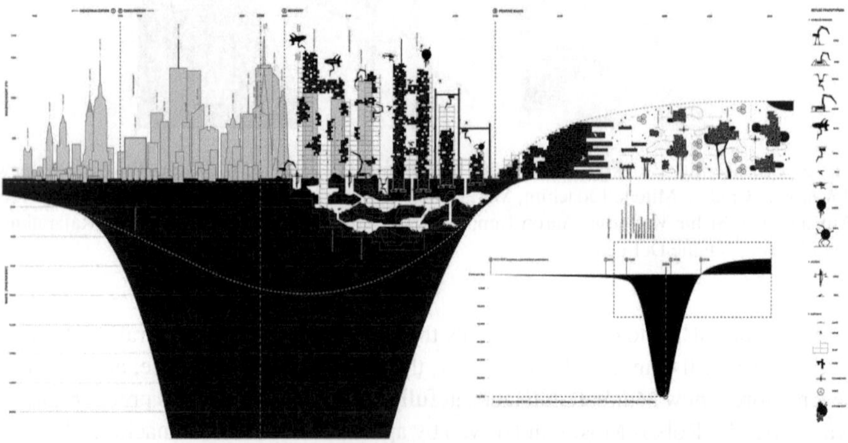

Fig. 7 RAPID RE(F)USE: Waste to resource city 2120. History and future of garbage in New York (Credits: Mitchell Joachim, Maria Aiolova, Melanie Fessel, Emily Johnson, Ian Slover, Philip Weller, Zachary Aders, Webb Allen, Niloufar Karimzadegan, Lauren Sarafan)

automatons would be entirely based on existing techniques commonly used in industrial waste compaction devices. To accomplish this job, nothing drastically new needs to be invented. Most technologies are intended to be off-the-shelf. Instead of machines that crush objects into cubes, compaction devices could benefit from adjustable jaws that would craft simple shapes into smart 'puzzle blocks' for assembly. The blocks of waste material could be predetermined, using computational geometries, in order to fit domes, archways, lattices, windows, or whatever patterns would be needed. Different materials could serve specified purposes: transparent plastic for fenestration, organic compounds for temporary decomposable scaffolds, metals for primary structures and so on. Eventually, the future city would make no distinction between waste and supply.

Admittedly, this meta-design theme is not entirely novel. At approximately the same time that Rapid R(e)fuse was initiated, the feature film WALL · E was conceptualised.[14] The film profoundly infused Terreform ONE's research agenda.

4 Excursion to Disneyland

Inspired by an equal interest in fictive productions of tomorrow such as Disney's Tomorrowland, Terreform ONE visited the Walt Disney Imagineering (WDI) headquarters in Glendale, Southern California. The group had prepared a presentation that would unpack a comprehensive view of its version of the future: a world free of carbon loading in the atmosphere and abundant in self-sufficient lifestyles. As architects invested in an ecological future vision, the team had meticulously crafted cities within the rubric of a socio-ecological domain – rethinking the design of entire systems, from doorknobs to democracies (Fig. 8).

When Ben Schwegler, mastermind and chief imagineer, pulled back the proverbial curtain to reveal WALL · E, the group was crestfallen. Disney had beaten them to it. WALL · E was perfect – almost: a tightly packaged, solar-powered, curious, obedient, evolved, robotic trash compaction and distribution device. His name is an acronym: Waste Allocation Load Lifter Earth Class. Left behind by mankind, he toils with trillions of tons of non-recycled inner-city trash. Not only is WALL · E a highly advanced rubbish manager, he also is a mechanized and inventive architect. He accomplishes his immense tasks while remaining completely adorable. Not easy to do.

WALL · E's life is a tale of an ultramodern trash compactor in love. Ceaselessly, he configures mountains of discarded material. Why pyramids of trash? WALL · E's daily perpetual feats seem almost futile. Disney omits exactly why he is programmed to pile refuse; and there is the shortcoming.

[14] Disney/Pixar Animation Studios, WALL · E, 2008.

Fig. 8 RAPID RE(F)USE: Waste to resource city 2120. Every hour NYC produces enough waste to fill the statue of liberty (Credits: Mitchell Joachim, Maria Aiolova, Melanie Fessel, Emily Johnson, Ian Slover, Philip Weller, Zachary Aders, Webb Allen, Niloufar Karimzadegan, Lauren Sarafan)

5 Future Waste and Past Cities

Collaborators at Terreform ONE were interested in exploring a deeper motivation for stacking refuse. Similar to the Disney film, what if the refuse was refabricated to become real urban spaces or buildings? If it is plausible to adapt current machinery, how much material is available? At first sight, any sanitary landfill may be viewed as an ample supply of building nutrients. Heavy industrial technologies to compact cars into lumber or to automatically sort out garbage are readily available. Other technologies, which would make possible the articulation of specific forms, are also available if scaled in larger sizes. 3-D printing has exhausting capabilities if adjusted to larger scales. This is where Terreform ONE's city began (Fig. 9).

The envisioned city would be derived from trash; not ordinary trash, but 'smart refuse'. A significant factor of the city composed from smart refuse is 'post-tuning'. Unitized devices would not immediately adapt. Integration into the city texture would be a learning process. In time, the responses would eventually become more attenuated to the needs of the urban dweller. This city is envisioned from trash, but each individual component would be enhanced with a modicum of CPU power. Brief durational events would endow these 'smart units' with experiences needed for their evolution.

The main objective for the city of Rapid R(e)fuse is to establish a smart, self-sufficient, perpetual-motion urbanism. It has been advocated that perpetual motion cannot exist. Perpetual motion defies the laws of thermodynamics and energy conservation, since it would necessitate a machine that produces more energy than

Fig. 9 RAPID RE(F)USE: Waste to resource city 2120. One-Day tower made of 24H of compacted waste in NYC (Credits: Mitchell Joachim, Maria Aiolova, Melanie Fessel, Emily Johnson, Ian Slover, Philip Weller, Zachary Aders, Webb Allen, Niloufar Karimzadegan, Lauren Sarafan)

it consumes. Cities, unlike machines, are similar to a complex ecology.[15] Ecology is capable of achieving a continuous harmonious state, or even further, a positive intensification. If ecological models are productively everlasting, urban models can logically follow.

Architects have ruminated over improbable instruments of physics since the Middle Ages. In the thirteenth century, evidence of the perpetuum mobile was uncovered in the sketchbooks of French architect Villard de Honnecourt.[16] What if the Rapid R(e)fuse city was like an instrument that produces more energy from renewable sources than the energy it consumes? In this case, 'nothing can be thrown away'. Every bit would be a vital piece of stored energy, poised to be reused in a cyclical nutrient stream.[17] Rapid R(e)fuse is imagined as a city without a tail pipe; a city that not only has zero impact, but a positive contribution towards the natural surroundings.

John Fitzgerald Kennedy once declared: 'Our problems are man-made, therefore they may be solved by man.'[18] The matter posed on the table is not only about solving our ecological issues, but also about returning to a system of perpetuity. This is the only possible future for a truly breathing, interconnected, metabolic urbanism. Cities have passed the age of industrialization and entered the age of recovery. After this great cleansing, we may transition into in a greater order: 'positive waste'. Here is an order that captures our socio-ecological needs: not utopia, but a place where everything is precious and nothing is disposed.

6 Envisioning Ecological Cities

How should urban design foresee new instrumentalist technologies for cities? For 150 years, the innovation of the elevator has done more to influence urban design than most urban designers. Elevator systems had incredible success in the creation of

---

[15] 'Cities are not machines and neither are they organisms, and perhaps resemble them even less – Rather then communities of non-thinking organisms undergoing inevitable phases until they reach a certain iron limit – cities are the product of beings capable of learning. Culture can stabilize or alter the habitat system, and it is not clear whether we wish it to be otherwise.' Kevin Lynch in Good City Form, MIT Press (Cambridge, MA), 1984, pp 26–27.

[16] Theodore Bowie, The Medieval Sketchbook of Villard de Honnecourt, Dover Publications (New York), 2006, pp 32–49.

[17] William McDonough, 'Waste Equals Food: Our Future and the Making of Things', in Judy Laddon, Tom Atlee and Larry Shook (eds), Awakening: The Upside of Y2K, Printed Word, 1998, pp. 5–57.

[18] John F. Kennedy. BrainyQuote.com, Xplore Inc, 2010. See http://webcache.googleusercontent.com/search

John F. Kennedy, speech at The American University, Washington, D.C., June 10, 1963. BrainyQuote.com. Xplore Inc, 2010. http://www.brainyquote.com/quotes/quotes/j/johnfkenn124671.html

Fig. 10 S.O.F.T. Mobility: Blimp bumper bus (Credits: Mitchell Joachim)

compact and greener cities. Imagine what the advent of the jet pack will do for cities. Urban design is greatly altered by such devices. For instance, automobiles have defined limits in cities for almost a century. Unlike the elevator, however, the car has arguably caused more problems than it has solved. Perhaps it is time for urban design to rethink technologies to fit cities, not constrain them. As a wide-ranging discipline, it can effortlessly illuminate the technological potentials for cities. Urban design will successfully situate itself through the production of future macro-scaled scenarios predicated on innovative devices (Figs. 10, 11, 12, and 13).

Physicist and polymath Freeman Dyson has said that the best way to comprehend our near urban future is to examine science fiction, not economic forecasts. In his experience, sci-fi is good for decades of technological fulfillment. Unfortunately, economic forecasts are only accurate within 5–10 years. Most of these predictive economic models are quantity based and find it difficult to extrapolate the qualifiers associated with creativity. Sci-fi is a phenomenal way to chronicle our plausible urban future that should not be dismissed by urban designers.

Dyson is certain that the urban era of information will soon transition into "the age of domesticated biotechnology."[19] In his novel Infinite in All Directions, he states: "Bio-tech offers us the chance to imitate nature's speed and flexibility." He envisions a realm of functional objects and art that humans will "grow" for personal use. According to a New York Times article on Dyson, "The Civil Heretic," he

[19]Freeman Dyson, Infinite in All Directions (New York: Harpercollins,1988).

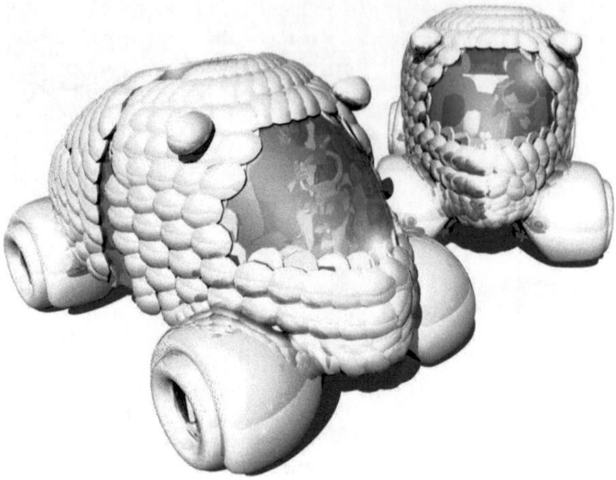

Fig. 11 S.O.F.T. Mobility: The Hug and Kiss Sheep car. Zero accident victims (Credits: Mitchell Joachim)

Fig. 12 S.O.F.T. Mobility: Stackable car. Urban stackable car with solar recharge port (Credits: Mitchell Joachim)

also believes that climate change is profoundly misstated. "He added the caveat that if CO_2 levels soared too high, they could be soothed by the mass cultivation of specially bred 'carbon-eating trees.'"[20] He is not concerned with predicting the future but rather with expressing the possibilities. These expressions are founded along societal desire lines as a kind of relevant optimism. Therefore Dyson measures the wants of civilization and advances our expectations.

At some level, urban design engages this position that promises a better tomorrow. Numerous practitioners and urbanists mildly suffer from this invariable

[20]Nicholas Dawidoff, "the civil Heretic," New York Times (March 25, 2009).

Fig. 13 S.O.F.T. Mobility: Homeway the great suburban exodus. Largest SUV's planned to achieve peak oil instantly (Credits: Mitchell Joachim. Credits: Mitchell Joachim, Maria Aiolova, Melanie Fessel, Philip Weller, Ian Slover, Landon Young, Cecil Howell, Andrea Michalski, Sofie Bamberg, Alex Colard, Zachary Aders)

search for direction and clairvoyance. Alex Krieger strongly asserts the broadly defined vocation is more of a scrupulous sensibility than an exclusive authority.[21] The profession is torn between many in- compatible agendas, weighty theories and oversimplified applications, ivory towers and new urbanism, developer brands and radical ecologies, and vernacular forms and futurology. One of my research group's chief directives is about shrewdly locating the intersection of technology and urbanism, especially under the rubric of ecology. Our projects range from highlighting the possible effects of self-sufficient cities to studying flocks of jet packs. These ideations keep us thriving as urban design researchers. It is our supposition that the prospective ecological city is about extreme solutions to an extreme predicament. Our future fundamentally depends on the immensity our solutions envision.

Envisioning is by definition a view or concept that evolves beyond existing boundaries. This notion of foresight may be interpreted in many different ways, each foregrounding particular ideations and processes describing the next event. Here in America, we need these radical new visions to assist in solving our current global calamity. As of now, the earth's cli- mate endures an unremitting state of trauma. We seek precise prescriptions that cover a wide scope to alter this massive dilemma. To paraphrase, John F. Kennedy said: "If man created problems, man can solve them." This future vision unfolds a truly breathing, interconnected metabolic urbanism. How does it reify from statistics to architectural form? What does the future look like for America's cities? How do technological devices affect these functions?

For a popular audience, the recent Disney sci-fi film Wall-E enabled society to anticipate one conceivable future. The film is set in a generic city that is completely buried in trash. Humans have abandoned life on earth for off-world dwelling, leaving being one lone solar-powered robot to clear the rubbish. Part of the message of the film is that technology alone can't solve humanity's "affluenza," yet the

---

[21] Alex Krieger and William S. Saunders, eds., Urban Design (Minneapolis: University of Minnesota Press, 2009).

film's powerful computer-generated visuals encourage us to confront our colossal wastefulness and rethink the city.

We foresee strategies for people to fit symbiotically into their natural surrounds. To achieve this, all things possible are considered. We design the scooters, cars, trains, and blimps, as well as the streets, parks, open spaces, cultural districts, civic centers, and business hubs that comprise the future metropolis. For centuries cities have been designed to accommodate the theater of our human desire. We have joined the ranks of those delivering a new sense of the city, one that privileges the play of nature over anthropocentric whims. We are constantly vying for a profound clairvoyant perspective. We desire to preview a likeness of our collective future yet untold.

Our foresight of ecological design is not only a philosophy that inspires visions of sustainability but also a focused scientific endeavor. The mission is to ascertain the consequences of fitting a project within our natural environment. Solutions are derived from numerous examples: living material habitats, climatic tall building clusters, and mobility technologies. These design iterations succeed as having activated ecology both as a productive symbol and an evolved artifact. Current research attempts to establish new forms of design knowledge and new processes of practice at the interface of design, computer science, structural engineering, and biology.

7 Extraterritorial Nexus: An Urban Genealogy

We no longer map territories, but territories map us. Humanity is inscribing its vast impacts on the Earth's surface; these new engraved territories have no boundaries, they portray who we are, what we have done and where we are moving (Figs. 14 and 15).

Within these global networked shapes and their interrelations they unveil our ever-changing cities and landscapes. These immeasurable human impressions are so complexly woven, that it is becoming increasingly difficult to maintain a distinction between nature, culture and the built environment. Fragmented exceptions defined as geographically autonomous zones of extraterritoriality disseminate these interacted figures. What are the unconditional opportunities that lie within these confined processes?

Utopian thought has been critical to our evolution and will be a necessary paradigm to envision an interconnection on extraterritoriality and ecology as a novel scenario of experimentation that challenges an innovative discourse beyond a conduit of sustainability. The investigations described by extraterritoriality nexus are centered on the comprehension of new approaches to the reification of the city. This may be best accomplished by further understanding these fragmented emerging ecologies of exception, and how they relate to architecture and urban design, as concise instruments in the development of society. In order to verify this vision in a socio-ecological realm, beyond a world of net-zero motivations as described

Fig. 14 Extraterritoriality Nexus: an urban genealogy. COMPOSITIONAL CONDITIONS; Speculation on unstable processes. An investigation on extraterritoriality emerges as a nexus for explorations of ecological and social disturbances as an active tool to re-invent and re-imagine our cities and vast networks today (Credits: Melanie Fessel)

by sustainability advocates, the aim is to structure an inquiry of mutable urban conditions as they pertain to global crisis and phenomena. There are a number of questions that come to mind addressing this predicament. How can we integrate and reinvent the already existing political mechanism of extraterritoriality as a tool of exception to address social and ecological disturbances that are prevailing our urban landscapes? What can we learn and extract from our past conceptions of cities and humanity? When confronted with any utter urban calamity from Port-au-Prince, to New Orleans, to Fukushima, what is the extent of architectures impact, if any? After the destruction of Fukushima, and many places like it, the bona fide intervention of architecture has been ineffectual. Events occur with such speed and complexity today that nothing remains certain. Large numbers live in a world where local economies and cultures are tightly bound into global ones through which effects ripple with enormous velocity and consequence.[22]

Natural and synthetic disturbances are dynamisms of immense force that set apart worlds. Governmental guidebooks and laws around the world map out detailed solutions for rescue operations in crisis scenarios. Hazard mitigation plans, military scenarios, crisis strategies, are instrumental directives that provide society with a safety mechanism to prevent and overcome the dangers created by forces of natural disasters, fire, mudslides, earthquakes, hurricanes, flooding, tsunamis as well as crime, violence, acts of terror, war, destruction. Planned in every detail it enables their creators to satiate society with a constant answer for unpredictable demands that relate to the influence of natural and societal instabilities on our built environment. [23]

Post the traumatic event of 9/11 the act of terror is still desensitizing our resilience to conquer new challenges. The effects of catastrophe on the built

[22]Saskia Sassen, Losing Control? Sovereignty in an Age of Globalization, 1996.

[23]County of Los Angeles, All-Hazard Mitigation Plan.

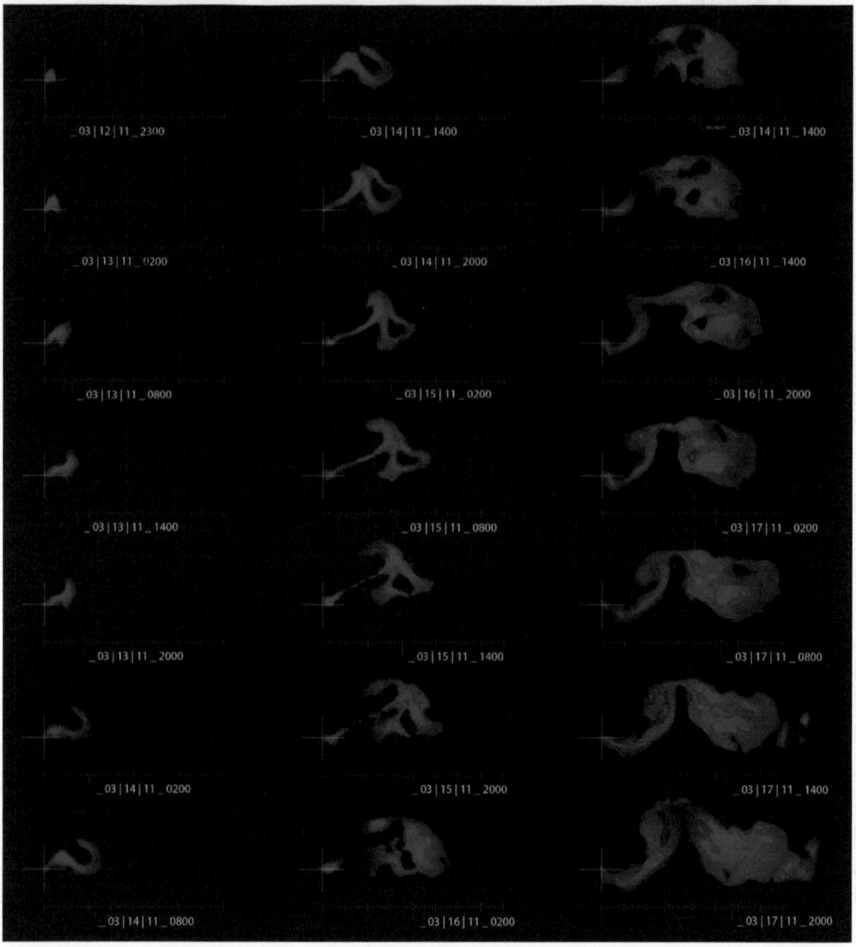

Fig. 15 Extraterritoriality Nexus: an urban genealogy. DIACHRONOUS LANDSCAPES; Radioactive Fallout. Recording of the movement of the radioactive cloud that was released by the fallout in Fukushima, Japan. Diachronous systems migrate geographically with their environment of formation (Credits: Melanie Fessel)

environment and its inhabitants continue to intrude every day life. New York City provides a significant locus to analyze the politics of shock and terror. A city of immigrants, many of whom are connected to other cities that have suffered catastrophe, New York citizens have a unique contribution to offer to the many urgent projects of reimagining cities around the world today. To address the devastating destruction by the events that took place on September 11, 2001, the call for entries to rebuild the World Trade Center site resulted in the world's largest architecture competition with the most amount of entries in history. The immediate urgency to restore and recover from such damaging incidents is implied.[24]

If we are going to study cities than we should also study that which is radically anti-city. Threats against our cities are measured in terms of armed gangs or moments of staged terrorist assaults as well as disturbances created by natural disasters, hovering over all of this is still the apocalyptic thought of something that could simply annihilate cities altogether. Nuclear catastrophes mark the only true long-term marker of human presence on earth. Weapons created by geology, minerals made all together unearthly, post-terrestrial, through anthropological intervention form a destructive power that turn them into a ubiquitous anti-landscape, something that no geography, built or natural, can resist.

The worst nuclear disaster to strike Japan since a single bomb fell over Nagasaki in 1945 occurred in the spring of 2011, at the Fukushima nuclear power plant following the epic tsunami.[25] The wide release of radiation, and fear of same, has forced the Japanese and others all over the world to reflect on what happened to the country in 1945, and the continuing threat of nuclear weapons and energy today. On August 6, 1945, the first atomic bomb is dropped from an American plane on the 245,000 residents of Hiroshima, Japan. Most of the city is destroyed and thousands of its inhabitants die. Some of its citizens survived and suffered the debilitating effects of terrible burns and radiation illness. The lives of six of those survivors are recounted in the days following the bombing. Hiroshima, John Hersey's journalistic masterpiece, tells what happened on that day.[26] More than six decades after the events in Hiroshima a new activism of survivors of the bombing is campaigning against nuclear power, which has provided most of their country's energy needs. Survivors, who are now called "hibakusha" have become the targets of politics and the peace movement. The radioactive fallout in Fukushima, Japan as the most recent nuclear disaster reminded everybody that there are no boundaries for threats. The fallout that was triggered by an earthquake that set off a tsunami heading towards the coast of Japan and destroying the nuclear power plant in Fukushima, was just the beginning of a keychain that will affect all future life on earth and beyond. Three elements – buildings, communications, and transportation – constituted the immediate effect of the disaster on the city. A further ominous phase of the disaster appeared after news of the first explosion at the Fukushima Daiichi nuclear power plant unwrapped. Power Cuts, product shortages, radiation and health warnings followed this information emphasized by uncertainties about the actual aftermath of this impact. Within a week the nuclear cloud arrived at the coast of California continuing its journey cross-country heading towards Europe and Asia. Carried by Aeolian forces and strong ocean currents the effects of the threat are immeasurable

[24]Columbia University, Engendering Archives Project of the Center for the Critical Analysis of Social Difference, Injured Cities: Urban Afterlives, October 2011.

[25]Fukushima Meltdown: The World's First Earthquake-Tsunami-Nuclear Disaster [Paperback] Takashi Hirose (Author), 2012.

[26]John Hersey, Hiroshima, 1989.

and its longevity is implied. This period of suffering, restraint and sobriety reminded of the struggles of the postwar years, and even further back to the years after the massive earthquake that destroyed Tokyo in 1923. Despite their difficulties, the reconstruction years were times of energy and aspiration, when a new city and a new nation were built.

Architecture was the immediate response to the threat of nuclear weapons by designing fallout shelters as a critical part of civil defense strategy of the 1950s and 1960s.[27] In an era of nuclear weapons the federal government, tasked with protecting American citizens and communities, relied on architectural expertise in order to survey, design, and build fallout shelters. During the height of the cold war architects and urban planners became instrumental to the importance and efficacy of both purpose-built and ad hoc fallout shelters, which granted them expert status. Architecture for civil defense planning in the United States was, ultimately, a failure due to a lack of federal funding, contradictions and ambiguities in fallout shelter design, and growing resistance to its political and cultural implications. Yet the partnership between architecture and civil defense influenced the perception and use of urban and suburban spaces. The result of this bunker architecture was a philosophy of building and urbanism that shifted focus from nuclear annihilation to urban unrest.[28]

The future will not be the same as the past but disruption on the one side sparks legacies of feedback on the other. Hip Hop, a cultural movement during the 1970s in New York City, in many ways comes out of a disruption – the disruption of a neighborhood and community in the Bronx through massive highway infrastructure that would soon segregate their neighborhood and leading newly formed gangs to resonate this impact through acoustic portraits of its interrupted built environment. Robert Moses was the lead figure for these massive urban renewal projects and practices in mid twentieth century New York City.[29] Facing the destruction of another local neighborhood in 1961, community activist Jane Jacobs organized grassroots efforts to block urban-renewal project for the Greenwich Village, which were instrumental in leading to the cancellation of the Lower Manhattan Expressway. [30]

Events all over the world identifiable as social, political, economical and ecological volatile disturbances are impacting the world's fragile ecosystem and ultimately influence how we reimagine cities around the world today. Comprehending unstable processes and environments is an attempt for a new and suggestive contemporary global order that recognizes a continuous spatial flow of extraterritorial zones and utilizes their exception as a progressive tool for envisioning our future. An

[27]David Monteyne, Fallout Shelter: Designing for Civil Defense in the Cold War (Architecture, Landscape and American Culture), 2011.

[28]U.S. Department of Defense, Office of Civil Defense, Fallout Protection: What to Know and Do About Nuclear Attack, 1961.

[29]Robert A. Caro, The Power Broker: Robert Moses and the Fall of New York, 1975.

[30]Jane Jacobs, The Death and Life of Great American Cities, 1992.

investigation on extraterritoriality emerges as a nexus for explorations of ecological and social disturbances as an active tool to re-invent and re-imagine our cities and vast networks today. The framework for this emergence defines our history, present and future: the impact to recognize these processes, as a positive impulse to regenerate new emergent landscapes of emergency, has to be created.

Extraterritorial places are positioned outside of the sovereignty and jurisdiction that surrounds them or are contiguous to them. International ownership treaties demarcate airports and ports, International Waters, International Seabed, Moon, Outer Space, International Zone, United Nations, Antarctica, Extraterrestrial Real Estate. These spaces aspire to be worlds within themselves, and provide vivid evidence of the weakness, resilience, or violence that these enclaves carry. Jurisdictionally ambiguous, they are infused with myths, desires, and symbolic capital.[31] The condition of extraterritoriality, which is defined by its segregation, transforms these hybrid spaces into spaces that attempt to create long envisioned utopian places and a radical new socio-ecological order. These new characteristics of our time could envision the power of new intense transformations and processes of global connectivity, where the cultural collision and segregation provide new techniques for a discourse on the political composition of our urban landscapes.[32]

8 The Necessity of All Scales: Planetary Design in the Age of Globality

Globalization has a final outcome; it's not unending. Without recourse, humanity is headed into the age of globality, the final state of the globalization process. Globality is the endgame, an all-inclusive terrestrial status. In this fully- connected world, populations will compete with everyone, everywhere, for everything at all times in all scales. Vast networked communication linkages will reverse developing world economies and cause developed nations to partially corrode if not yield. A conceivable state of planetary equilibrium ought to influence the major sectors of industry, commerce, resource management, infrastructure, technology, energy, and governance with profound transformations. What is design like in this saturated condition of globality? (Figs. 16, 17, 18, 19, 20, 21, and 22).

The American economist, Jeremy Rifkin infers that the modern age has been characterized by a "Promethean spirit", a restless energy that preys on speed records and rapid solutions, unmindful of the past, uncaring of the future, existing only for the moment and the quick fix.[33] The earthly rhythms that characterize

---

[31]Keller Easterling, Enduring Innocence: Global Architecture and Its Political Masquerades, 2007.

[32]Eyal Weizman, Anselm Franke, Thomas Keenan, Archipelago of Exception – Sovereignties of Extraterritoriality, CCCB – the Centre for Contemporary Culture in Barcelona, November 2005.

[33]Rifkin, Jeremy: The Age of Access: The New Culture of Hypercapitalism, Where All of Life is a Paid-For Experience, (Putnam Publishing Group) 2000.

Fig. 16 Super Docking Brooklyn Navy Yard. View of clean tech industrial water park (Credits: Terreform ONE + Planetary ONE, Mitchell Joachim, Maria Aiolova, Nurhan Gokturk, David Maestres, Jason Vigneri Beane. Design Team: Carlos Barrios, Alex Felson, Walter Meyer, Melanie Fessel, Zafirah Bacchus, Ivy Chan, Courtney Chin, Adrian De Silva, Julianne Geary, Francisco Gill, Shima Ghafouri, Jacqueline Hall, Kelly Kim, Florian Lorenz, Bart Mangold, Dustin Mattiza, Chema Perez, Alsira Raxhimi, Daniel Russoniello, Melody Song, Allison Shockley, Katherine Sullivan)

a more pastoral way of life have been shunted aside to make room for the fast track of an urbanized existence. Lost in a sea of perpetual technological transition, contemporaneous people find themselves increasingly alienated from the ecological choreography of the planet. Humanity expects jet airliners to collapse the space of a continent in a matter of hours. We have partly lost our sense of scale, time, and distance.

Scale also determines our profound connection to place. Yi Fu Tuan's environmental conjecture conveys knowledge and love of place with descriptions, maps and itineraries enabling people to appreciate their location as do aborigines or animals.[34] Such literature also enlivens the experience of everyday places with new facts and rhetorical devices that can recalibrate familiar environs to keep alive a sense of the undiscovered country of the nearby. These texts direct both official and intuitive knowledge toward "topophilia", the love of place. The intention is to expand our sense of the land, not to confine it to one size. We are in need of terrain propinquity, not dislocation.

Already the disasters at Chernobyl and Fukushima have demonstrated long reaching influences on worldwide environmental health and ultimately on global financial markets. Effects of scale become constantly transferred between the irreducible and the colossal. Small changes ramify into massive results and vice

[34]Tuan, Yi-Fu : Topophilia: A Study of Environmental Perception, Attitudes, and Values, (Prentice-Hall) 1974.

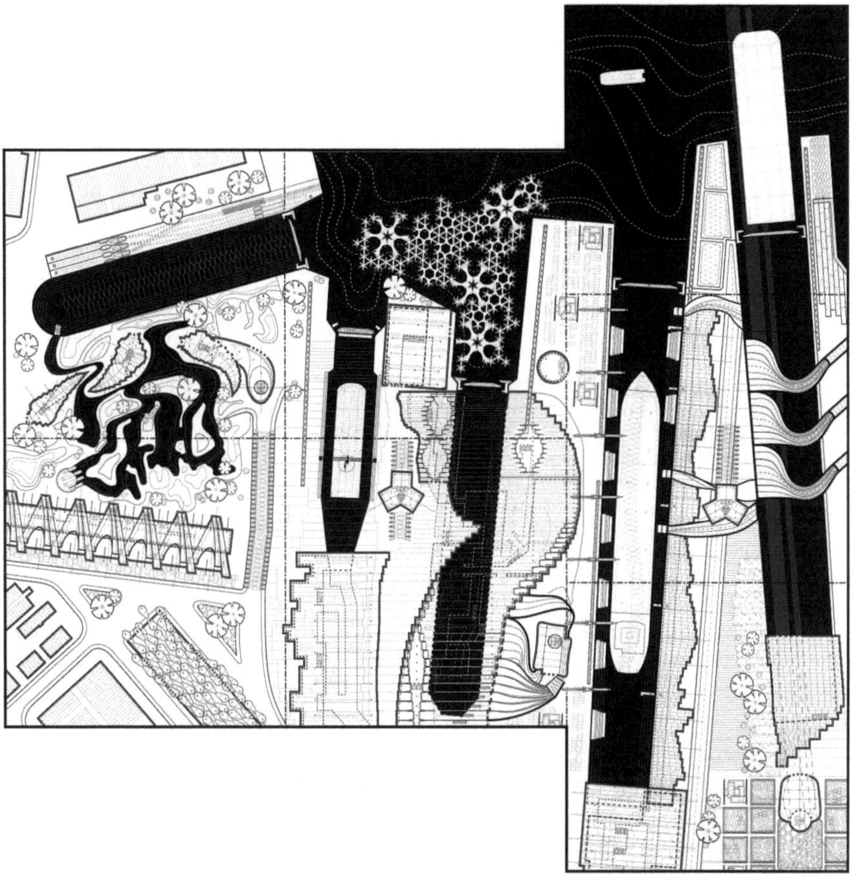

Fig. 17 Super Docking Brooklyn Navy Yard. Site with five retrofitted dry docks for green manufacturing (Credits: Terreform ONE + Planetary ONE, Mitchell Joachim, Maria Aiolova, Nurhan Gokturk, David Maestres, Jason Vigneri Beane. Design Team: Carlos Barrios, Alex Felson, Walter Meyer, Melanie Fessel, Zafirah Bacchus, Ivy Chan, Courtney Chin, Adrian De Silva, Julianne Geary, Francisco Gill, Shima Ghafouri, Jacqueline Hall, Kelly Kim, Florian Lorenz, Bart Mangold, Dustin Mattiza, Chema Perez, Alsira Raxhimi, Daniel Russoniello, Melody Song, Allison Shockley, Katherine Sullivan)

versa. Globality operates in all scales simultaneously without privileging one form. Its very nature implies measureless shifts in size. Ideas of thinking inside strict categories of scale are defunct and counter-intuitive. Charles and Ray Eames provided the perfect case. Illustrated in the "Powers of Ten", scale is conveniently defined in neat square-shaped frames.[35] The point of their animation is to bridge the different perceptions of scale cohesively. Their concept is to empower individuals

[35] Eames, Ray & Charles: Powers of Ten, American documentary film, 1968.

Fig. 18 Super Docking Brooklyn Navy Yard. Super sheds over docks merge Landscapes (Credits: Terreform ONE + Planetary ONE, Mitchell Joachim, Maria Aiolova, Nurhan Gokturk, David Maestres, Jason Vigneri Beane. Design Team: Carlos Barrios, Alex Felson, Walter Meyer, Melanie Fessel, Zafirah Bacchus, Ivy Chan, Courtney Chin, Adrian De Silva, Julianne Geary, Francisco Gill, Shima Ghafouri, Jacqueline Hall, Kelly Kim, Florian Lorenz, Bart Mangold, Dustin Mattiza, Chema Perez, Alsira Raxhimi, Daniel Russoniello, Melody Song, Allison Shockley, Katherine Sullivan)

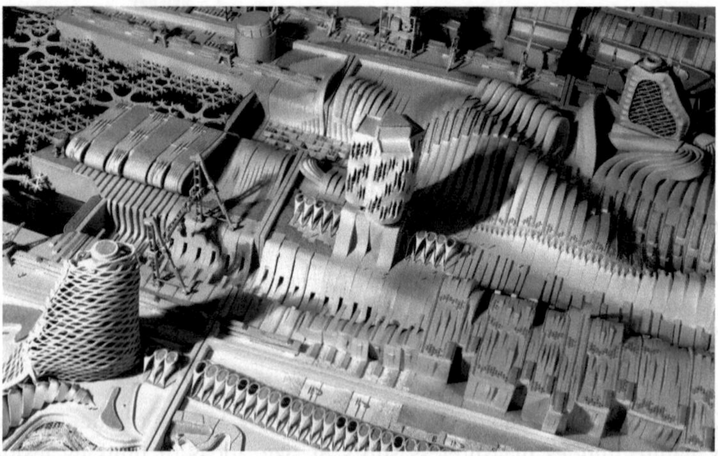

Fig. 19 Super Docking Brooklyn Navy Yard. Clean tech office towers (Credits: Terreform ONE + Planetary ONE, Mitchell Joachim, Maria Aiolova, Nurhan Gokturk, David Maestres, Jason Vigneri Beane. Design Team: Carlos Barrios, Alex Felson, Walter Meyer, Melanie Fessel, Zafirah Bacchus, Ivy Chan, Courtney Chin, Adrian De Silva, Julianne Geary, Francisco Gill, Shima Ghafouri, Jacqueline Hall, Kelly Kim, Florian Lorenz, Bart Mangold, Dustin Mattiza, Chema Perez, Alsira Raxhimi, Daniel Russoniello, Melody Song, Allison Shockley, Katherine Sullivan)

Fig. 20 Super Docking Brooklyn Navy Yard. Large scale 3D printer over dry docks (Credits: Terreform ONE + Planetary ONE, Mitchell Joachim, Maria Aiolova, Nurhan Gokturk, David Maestres, Jason Vigneri Beane. Design Team: Carlos Barrios, Alex Felson, Walter Meyer, Melanie Fessel, Zafirah Bacchus, Ivy Chan, Courtney Chin, Adrian De Silva, Julianne Geary, Francisco Gill, Shima Ghafouri, Jacqueline Hall, Kelly Kim, Florian Lorenz, Bart Mangold, Dustin Mattiza, Chema Perez, Alsira Raxhimi, Daniel Russoniello, Melody Song, Allison Shockley, Katherine Sullivan)

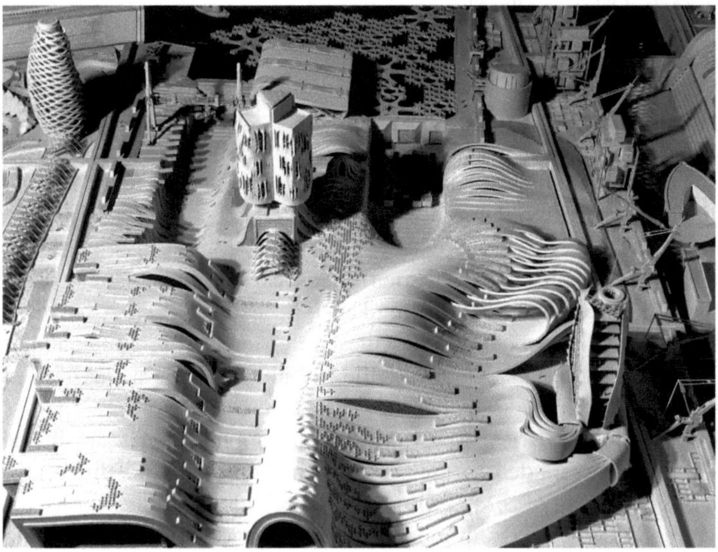

Fig. 21 Super Docking Brooklyn Navy Yard. Restorative ecology with phytoremediation systems in surfaces (Credits: Terreform ONE + Planetary ONE, Mitchell Joachim, Maria Aiolova, Nurhan Gokturk, David Maestres, Jason Vigneri Beane. Design Team: Carlos Barrios, Alex Felson, Walter Meyer, Melanie Fessel, Zafirah Bacchus, Ivy Chan, Courtney Chin, Adrian De Silva, Julianne Geary, Francisco Gill, Shima Ghafouri, Jacqueline Hall, Kelly Kim, Florian Lorenz, Bart Mangold, Dustin Mattiza, Chema Perez, Alsira Raxhimi, Daniel Russoniello, Melody Song, Allison Shockley, Katherine Sullivan)

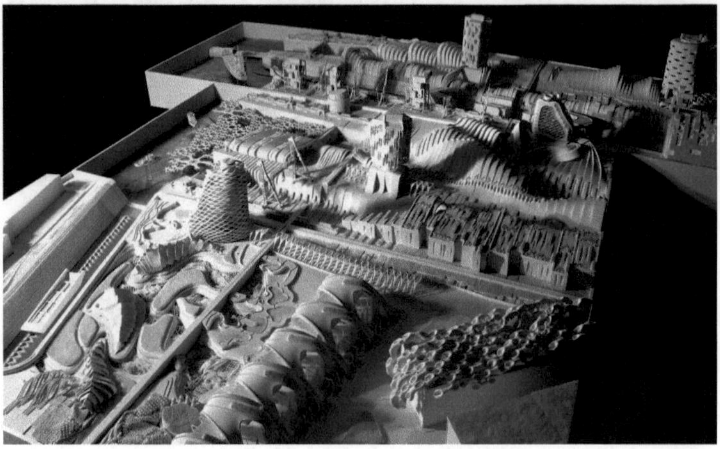

Fig. 22 Super Docking Brooklyn Navy Yard. Overall view with converted warehouses and sewer management (Credits: Terreform ONE + Planetary ONE, Mitchell Joachim, Maria Aiolova, Nurhan Gokturk, David Maestres, Jason Vigneri Beane. Design Team: Carlos Barrios, Alex Felson, Walter Meyer, Melanie Fessel, Zafirah Bacchus, Ivy Chan, Courtney Chin, Adrian De Silva, Julianne Geary, Francisco Gill, Shima Ghafouri, Jacqueline Hall, Kelly Kim, Florian Lorenz, Bart Mangold, Dustin Mattiza, Chema Perez, Alsira Raxhimi, Daniel Russoniello, Melody Song, Allison Shockley, Katherine Sullivan)

to visualize the ranges of observation melded as one. Unfortunately, a few viewers also interpret this to mean you actually should bracket places and things in specific scales. That's a common oversight of their "Powers of Ten" message. Nothing happens in only one frame of space/time. Framing can help to study a phenomenon at a particular moment but things always stir. Moreover, artificially binding a place to a numerical scale is to some extent random and arbitrary. What law declares measures must be in units divisible by ten? In his lectures, Jamer Hunt points this out nicely by making a reference to the film "This Is Spinal Tap". One scene in particular shows the actor, Christopher Haden-Guest referring to his guitar amplification equipment with a volume knob that goes to 11. Within the central idea of this farcical narrative, scale is portrayed as being truly capricious.

Rem Koolhaas and Bruce Mau intended an emphasis on scale in the book, S,M,L,XL, yet also alluded to the in-between thresholds and differentiation of projects.[36] A false read of S,M,L,XL is to suppose that the Office for Metropolitan Architecture's (OMA) works fit cleanly into categories measurable by an orderly unit. What exactly is designated a "small" project: a door hinge detail, an elevator, a wooden deck addition? More notably, what "small" project does not have significant impacts on an extra-large scale? Everything has consequent ramifications. Rem and

[36]Koolhaas, Rem, and Mau, Bruce: S,M,L,XL, New York (Monacelli Press) 1995.

group of course are fully aware, but may have failed to communicate its grandiose message in the title. Similar to Powers of Ten, it underscores an explicit simplification in degrees of change. Urbanism cannot be compiled into tidy categories defined by size. Design needs to break out of the emblematic question of size, especially as it relates to one planet. One size does not fit all; rather all sizes fit one.

Visualizing in one scale is markedly problematic. An analysis comparing the views of Johann Wolfgang von Goethe and Isaac Newton best demonstrates this assertion. Goethe pioneered a comprehensive lucid description of color within a perceptual human context.[37] He ardently conveyed that colors were defined by an inseparable relationship called Zur Farbenlehre (Theory of Colors). One color cannot be reduced to a single element. It requires a setting of other colors and circumstances to be fully perceived as an observable phenomenon by humans. He proclaimed that to identify blue you need to, at some level, recognize red, orange, yellow, and the entire spectrum. Newton, in opposition, treated color analytically and saw each color as discrete wavelengths in an optical spectrum.[38] Newton surmised it's absolutely possible to observe one color disconnected from its domain. Every color has a particular frequency and can be defined as such. While both thinkers are correct, the problem design professionals have with scale is identical. Scale can be Newtonian and viewable at one frame of unitized reference. However, scale is best perceived holistically in relationship to other scales in order to understand its true phenomenon.

Most archetypal designers tolerate a proclivity to divide concepts up into specific units. That is intended to help comprehend or visualize the problem better. However, this is inherently misguided. Design is a prescience that affects all disciplines and cannot be conceptualized as a bracketed or contained field. The condition of globality and scale asserts a restructuring of the design professions as we know them. This requires a new breed of designers who can speculate and produce at the nano-scale level up to feats of geo-engineering and be- yond. These thinkers can be referred to as planetary designers.

As a globalized meta-Pangea community, design is obliged to be pervasive. It simply cannot regulate itself to any one scale or project scope. If so, its relevance and instrumentality are greatly diminished. The principal operations of scale and systems that deploy it restrain and confuse the complicated reality of design problems. The more it's used the more designers fail to envision the whole picture. Computer aided design is part of this dilemma, both a solution and a quandary.

Most design software unwittingly forces designers into forms of measure. As soon as the file opens, designers are asked to define the units of measure, view ports, and scale. The same is not true when an individual picks up a pencil. Freedom to draw and therefore conceptualize without boundaries is practically inestimable. To paraphrase Frank Stella, "artists don't think in units".

---

[37] von Goethe, Johann Wolfgang: Theory of Colors, Boston (The MIT Press) 1970.

[38] Newton, Isaac: Opticks: Or a Treatise of the Reflections, Refractions, Inflections and Colours of Light-Based on the Fourth Edition London, 1730, Dover Publications, 2012.

The processing capacities of computers distort the implications of measure. Software allows for seemingly limitless flexibility. Operators can shift from the smallest possible detail to the largest components. It's possible to zoom endlessly outwards into entire regions. On one level, this is a tremendously effective visualization capacity. Consider the design of a rail spike and its connection to the track; zoom onto the tracks themselves and further out to all the trains on those tracks and advance past Penn Station and finally out of the entire New York City metropolitan region. This is an acutely impressive tool. However the software does not make any distinctions about the conventions of physics and natural forces that govern each zoomed layer. In many cases the current limitations of memory also fail to provide all the essential detail within each consecutive zone. Furthermore, the material and chemical behavior of the objects and places are not described in relationship to each other beyond geometric location. Admittedly, the current trend is to restructure computation to account for these missing characteristics. Envision a computer program that fully simulates the ecosystem of the earth and all the associative quantum mechanics. It would be a Jorge Luis Borges map of the world in such vast specificity that it would correlate to the exact size of the world at 1:1.[39]

Additional scenarios of scale in relation to the planet are depicted in The End of Nature by Bill McKibben.[40] He marshals the latest scientific evidence about the greenhouse effect, the depletion of the ozone layer and a harrowing array of other ecological ills, and unmistakably explains the frightening implications of the destruction cities have wrought on our planet. He questions ecological hysteria and reasonable scientific forecasts. Within either approach, The End of Nature has a philosophical position. McKibben declares confidently about the meaning of these changes, about the wretchedness of life where there is no escaping mankind. Although for centuries civilization has pillaged and polluted the earth, in the past those aggressions were relatively localized; now, with the globality shifts caused by greenhouse gases and ozone depletion, man and cities have altered the most elemental processes of life everywhere. Nature itself has been tainted, becoming the equivalent of a vast heated room. By turning nature into "an artifact" or by-product of economic development, society has lost some- thing of profound importance: nature as a quasi-celestial source of ultimate meaning and value. It is this loss that McKibben refers to as an apocalyptic calamity. The ending of nature is something independent of, larger than, and uncontrolled by man. In this nihilistic world every measurable space has been touched by human interference. Here even the right of all life to breathe has been impacted.

Multiple designers and planners have become concerned in recent years with revealing "truth windows" into nature to avoid its end. They highlight eco- logical processes in their designs so that the users of the environment may experience, comprehend, and appreciate the scales of those processes aesthetically. In practice,

---

[39]Borges, Jorge Luis: "A Universal History of Infamy", in: On Exactitude in Science, London (Penguin Books) 1975.

[40]McKibben, Bill: The End of Nature, (Anchor) 1999.

revelation of ecological process has meant everything from capturing storm water on the surface of the land before it drains away into sewers to planting a row of trees in an urban plaza where a creek once existed. In addition, the ecological processes that are revealed may themselves be truly natural, in the sense that they could continue to subsist without the management of society, or they may be deeply artificial, engineered systems that need relentless supervision if they are to persist in an urbanized context. Ultimately, the intention is make the scales of ecology visual and thereby expose an alleged spectacle of beauty otherwise unseen.

Our future in architecture recognizes that there is an immeasurable ecological quality that goes well beyond the borders of the building site. It's clear that ecological principles comprise a web of interconnected concepts, and a breathing, healthy metabolism has to fit inside. Thus, architecture in the future or architecture even now must be understood without a solitary scale. It must be planetary, the extent of which contains the outer edges of the atmosphere all the way into the deep regions of inner space.

The new sophisticated field of geo-engineering, for instance, exemplifies the lack of scale comprehension. In geo-engineering, exertion is made on the scope of an entire continent. Geo-engineers produce efforts equivalent to the Panama Canal as everyday feats. When looking at consequences and devices that can cause change at the regional level, we must look not at the region itself but the entire hemisphere, various megalopolises, as well as the smallest biological system. We must realize there is a kind of hubris, an unlimited bravery and power in conquering nature. Designing modifications at continental scale does infer accidents will be unmatchable.

The issue of scale seems rampant in the discussion and polemics of ecological cities. Scale is a pervasive term in engineering, architecture, urban studies, and design. It serves as a constant and definitive point of reference to help elucidate a given project although not without fault. We have separated our professional disciplines, project scopes, and programmatic language in terms of size, a redundant supposition in the age of globality. How designers can play a significant role in this expansive territory and live up to our proleptic merit is worthy of understanding further. Designers' responsibility is to re-tool the middle ground, the in-between, and the nexus points. In this case, infrastructure with applied innovative ecological directives becomes the penultimate goal, before completely reforming the world. Operations of infrastructure shape a broad range of circumstances. Infrastructure in all dimensions and extents is the actual frontier. Next to implemented technological, social, and ecological solutions from other disciplines, design innovations are rapidly being diminished if not surpassed by other competitive fields. We designers must take action and modify our stance on all scales and morphologies in order to have a positive effect on the global community. Our hypothesis is first and foremost based on one succinct predicate, the end of scale.

Invincible Cities for the Materiomic Age

Magnus Larsson and Alex Kaiser

Abstract Making novel buildings from novel materials equals a reenginering of the medium of architecture itself – a design optimisation of an entire creative field. The resulting schemes have less to do with high modernism than with synthetic biology.

The built environment is facing unprecedented opportunities brought about by recent scientific advances that open the laboratory doors to a future of material experimentation. In this chapter, the authors argue that humanity is on its way out of the Concrete Age, perhaps even out of the subsequent Timber Age, and moving rapidly towards the Materiomic Age.

The emergent field of materiomics is defined as the holistic study of material systems. Using this paradigm as a starting point, three speculative cities are presented, based on an architectural reading of Italo Calvino's 1972 novel *Le città invisibili.*

Positioned within a post-sustainable and post-hylomorphic model, the cities diverge completely in structural, programmatic, and architectural terms, and yet share a unique energy expressed through the audacious material attitude in which they have been steeped.

Advances within consilient fields such as synthetic biology, molecular chemistry, and bioengineering allow these conjectural urban prototypes to function as reconsiderations of the notion of materiality in architecture and urbanism. Positioned as Deleuzian "actualisations of the virtual," the Invincible Cities presented here push to a new level the idea that the city is mankind's greatest invention, through the simple guiding principle of material intelligence.

M. Larsson (✉) • A. Kaiser
Ordinary Ltd, London, UK
e-mail: magnus@anordinarywebsite.com; alex@anordinarywebsite.com

S.Th. Rassia and P.M. Pardalos (eds.), *Cities for Smart Environmental and Energy Futures*, Energy Systems, DOI 10.1007/978-3-642-37661-0_4,

1 A Novel Physicalist Urbanism

If you wish to make an apple pie from scratch, you must first invent the universe.[1]

In a famous and frequently quoted passage,[2] Estonian architect Louis I. Kahn proposed that architects should engage in conversations with their building materials:

> *When you are designing in brick, you must ask brick what it wants or what it can do. Brick will say,* I like an arch. *You say,* But arches are difficult to make, they cost more money. I think you could use concrete across your opening equally well. *But the brick says,* I know you're right, but if you ask me what I like, I like an arch.[3]

At a first glance, this animistic and spiritual vision of designing in accordance with material properties paradoxically seems to make a lot of sense. After all, no one would propose to construct, for instance, a large cantilever from non-reinforced concrete, which performs well in compression but has very low tensile strength. However, Kahn's statement carries at least two fallacies. The first is to do with the underlying conceptual logic: why ask the brick what it wants to be or what it can do, rather than ask the arch (if indeed an arch is what we have designed) which material it would like to be made of? Why engineer the design rather than the material? The second is to do with material science. Who cares what the brick wants? Its preferences are whatever we manipulate them to be.

This is the attitude at the heart of our speculative architectural investigations into what might be described as a novel physicalist urbanism.[4] Our position is based on an articulation of architecture as a principally material practice, not just the organisation of space but the organisation of matter in and into space. The world is composed of elements; putting these together in an ordered fashion allows us to create materials; ordering the materials in turn is the act of architecture; ordering architecture is urbanism. To create a work of architecture is to "reconstruct the

[1]Carl Sagan, "Cosmos". New York: Random House. 1980, p. 218.

[2]A passage later turned into Hollywood folklore as part of Woody Harrelson's lecture in a 1993 blockbuster movie: Adrian Lyne, Indecent Proposal (Paramount, 1993).

[3]Richard Saul Wurman, What Will Be Has Always Been: The Words of Louis I. Kahn (New York: Access Press & Rizzoli, 1986), p. 152. Kahn often repeated these statements, and they can be found in various sources. A recorded version of the architect delivering a similar passage to a meeting of students is available online: http://www.youtube.com/watch?v=2CYRSg-cjs4. The quote is slightly different: *If you think of brick, for instance, you say to brick,* What do you want Brick? *And brick says to you,* I like an arch. *And if you say to brick,* Look Brick, arches are expensive, and I can use a concrete lintel over you. What do you think of that, Brick? *Brick says,* I like an arch.

[4]Stoljar, Daniel, "Physicalism", The Stanford Encyclopedia of Philosophy (Fall 2009 Edition), Edward N. Zalta (ed.). Available online: http://plato.stanford.edu/archives/fall2009/entries/physicalism/

material as an enclosure".[5] In order to make an apple pie, you must first invent the universe; in order to design a city, you must first consult the periodic table. Just as the invention of the elevator (by Archimedes in 286 BC if we are to believe Vitruvius; a design importantly upgraded into the safety elevator by Elisha Otis in 1852) made the skyscraper possible,[6] so the invention of the microscope (by Hans and Zacharias Jansen around 1590) allowed our understanding of material properties to be informed by their internal realities.[7]

This new understanding of the interior structure of the fundamental elements making up the universe has given rise to an impressive body of scientific research, precious little of which has been applied within the conservative field of the built environment.[8] While humanity's relationship with the matter that makes up our environment – the stuff that surrounds us – has had a pivotal impact on our scientific and cultural evolution, to the point where anthropologists define historical epochs by the materials of their toolmaking (the Stone Age giving way to the Copper Age giving way to the Bronze Age giving way to the Iron Age), the architectural material palette is still predominantly composed of the usual suspects: concrete, glass, metals, brick, stone, and wood.[9]

---

[5]Peter Salter, TS Intuition and Process: Projects under the tutelage of Peter Salter (London: Architectural Association, 1989), p. 48.

[6]Cf. Mohammad Mehdi Masoumi & Soheila Naderinezhad, Elevator was Worked by Water and Water Pump. Research Journal of Applied Sciences, Engineering and Technology 4(24):5557–5560, 2012), p. 5557. For a dramatic account of Otis's first demonstration of his invention at the New York's World Fair in May 1854, and a discussion on the elevator's role in the evolution of the skyscraper, see Rem Koolhaas, Delirious New York (New York: The Monacelli Press, 1994), pp. 25–27 (originally published in 1978).

[7]The history of the microscope can be traced from the Jansens and possibly their contemporary, Hans Lippershey, via Galileo Galilei (who made a much better instrument in 1624), Anton van Leeuwenhoek, Robert Hooke, and Charles A. Spencer, through to today's rise of fluorescence microscopy in biology. Cf. John Gribbin, Science: A History 1543–2001 (London: Penguin, 2003) and Reginald S. Clay & Thomas H. Court, History of the Microscope (London: Chas Griffin & Co, 1932).

[8]Composed of a single proton paired with a single electron, hydrogen is the simplest and most abundant chemical element in the universe. By mass, however, iron is the most abundant element making up the Earth, while oxygen is the most common element in our planet's crust. Hydrogen makes up 75 % of normal matter (by mass) and more than 90 % by number of atoms. However, most of the mass of the universe is not in the form of chemical-element-type matter, but is postulated to occur as other forms of mass such as dark matter.

[9]The first pottery appeared around 29000BC. The oldest discovered bricks, originally made from shaped mud, dates to before 7500 BC. The history of concrete goes back some 7,000 years. Copper metallurgy was invented in the third millennium BC; bronze in the second millennium BC. Crude iron metallurgy was developed by the Hittites in the sixteenth century BC. The invention of steel is from the thirteenth century BC. Glass production began in ancient Near East in the tenth century BC. Porcelain was invented in Tang Dynasty China in the eighth century AD. Portland cement was invented by Joseph Aspdin in 1824, the first man-made plastic was created by Alexander Parkes and publicly demonstrated at the 1862 Great International Exhibition in London, stainless steel was invented by Harry Brearley in 1912, and nylon by Wallace Carothers in 1931.

The lack of technological information diffusion meant cultures reached different stages of material advancement at different times. In 1500 BC, Mesopotamia (current-day Iraq) was still in the Bronze Age, while people in Asia Minor (Turkey) were experimenting with Iron. Europeans, Palestinians, and Egyptians, meanwhile, were in the Copper and early Bronze Age. The Chinese, advanced in their development of bronze, had already melted iron, but the Spaniards and Portuguese, as well as those in the Americas, were in the Chalcolithic period (between the Stone Age and the Copper Age), with North Africa left in the late Stone Age.[10]

When Kahn tried to converse with a brick, however, that diffusion had already been in place for quite some time. Ernst Ruska had already co-invented the electron microscope for which he would later win the Nobel Prize.[11] Humanity knew how to view an object using electrons sped up in a vacuum until their wavelengths were extremely short, bringing the dimensions of visible objects down to the diameter of an atom. We had known for some 9,500 years how to make a brick, and we certainly had no reason not to challenge the makeup of its constituent parts. There was no need to remain passive with regards to the engineering of the brick itself. However, in a manner archetypal of the modernist generation he helped to define, this is exactly what Kahn proposed.[12]

2 Engineering the Medium of Architecture

We hold a different position. While we appreciate the advantages of allowing a material's properties to inform its formal extension, we see no reason not to challenge the status quo upheld by an anxious building industry comfortably making its living off of today's equilibrium rather than spearheading tomorrow's continual improvement at the intersection of material science and architecture. Japanese architect Kengo Kuma once wrote that if the act of designing were suddenly to be redefined as "selecting particles," we would be "thrown into a new world".[13] This redefinition now appears to describe rather well our contemporary condition and the

[10]Cf. Clive Ferguson, Historical Introduction to the Development of Material Science and Engineering as a Teaching Discipline (UK Centre for Materials Education). Available online: http://www.materials.ac.uk/pub/Materials-History-Intro.pdf

[11]The invention was made in 1931; Ruska won the Nobel Prize in Physics in 1986.

[12]Swedish architect Sigurd Lewerentz is another notable example: for his St Peter's Church at Klippan in Sweden, he refused to cut the standard bricks, Instead, he allowed the mortar joints to expand and contract freely. A rather active way of remaining passive with regards to the engineering of the brick. For a discussion, see Adam Caruso, Sigurd Lewerentz and a Material Basis for Form, OASE (Amsterdam, Netherlands: 1997) Issue 45/46, pp. 88–95. Available online: http://www.carusostjohn.com/text/sigurd-leverentz-and-a-material-basis-for-form/

[13]Kengo Kuma, The Relativity of Materials, JA The Japan Architect 38/Summer 2000, reprinted in Luigi Alini, Kengo Kuma – Works and Projects (Milan: Electa Architecture, 2005), pp. 222–223.

world we inhabit. We can make novel buildings from novel materials. As architects, we can design a version of the arch in the example above, and then engage with scientists to design the optimal material for its actualisation; we can engineer the very *medium* of architecture itself.

This is clearly an ambitious undertaking. All architects of ambition are embedded within the culture of their time, but they are also embedded within a material context that to an extent dictates the boundaries of their creative efforts. In 1891, the north half of the Monadnock Building at 53 West Jackson Boulevard in Chicago, Illinois, was built following designs by Burnham & Root. The drawings called for 17 stories (66 m – the tallest commercial building in the world at the time) of load-bearing masonry walls, braced by an interior frame of cast and wrought iron, which added support and counteracted wind loads. Two years later, the south half of the building had been added, designed by Holabird & Roche. Half of this addition was entirely steel framed. It reportedly cost 15 % less, weighed 15 % less, and had 15 % more rentable space than the northern half.[14]

The Monadnock Building can be viewed as the true starting point for architecture's Steel Age. In architecture before 1800, metals had played an auxiliary role. Wooden beams had been sheathed in iron for fire protection since at least 1793,[15] and Frank Lloyd Wright might have been right in calling the 1891 Wainwright Building in St Louis "the very first human expression of a tall steel office building as Architecture,"[16] but the surgically clear-cut division between the masonry and steel construction systems in the Monadnock was a symbolic rite of passage for the new material. It paved the way for countless steel-framed buildings to come, including icons such as the Seagram Building and the World Trade Center in New York, and the CCTV Headquarters in Beijing. Arguably, steel construction methods became the most important innovation in architecture since ancient times, allowing for far stronger and taller structures and less material expenditure than had been the case with buildings made from stone, brick, or wood.

3 The Perils of the Concrete Age

Being able to erect taller buildings on smaller budgets is not the only reason for conducting material research and speculating about alternative scenarios for material developments in the future. While one's personal environmentalist agenda

[14]Robert Bruegmann, The Architects and the City: Holabird & Roche of Chicago, 1880–1910. (Chicago: University of Chicago Press, 1997. ISBN 978-0-226-07695-9), p. 472.

[15]Fire was a major problem with the nineteenth-century timber framed mills – the example refers to the erection in 1792–1793 of a six-story cotton mill at Derby, designed by engineer William Strutt together with Richard Arkwright (inventor of the spinning jenny). Cf. James W. McAllister, Beauty & Revolution in Science (Ithaca: Cornell University Press, 1999), p. 143.

[16]Frank Lloyd Wright, The Tyranny of the Skyscraper in Frank Lloyd Wright, Modern Architecture: Being the Kahn Lectures for 1930 (Princeton: Princeton University Press, 1931) p. 85.

(or lack thereof) will assign a relative degree of alarmism to the current state of global affairs, measures that minimise the building industry's impact on the environment would seem to be at least politically and morally valuable in today's sustainability debate. However, despite ours appearing to be a favourable time for the evolution of greener building materials, and despite thousands of labs across the world having studied various biological and engineered synthetic materials at a molecular level, very little applied research seems to find its way into our buildings and cities. We appear to be largely stuck with an unchanging palette of concrete, glass, metals, brick, stone, and wood. A brief look at a substance that is at the same time stigmatised and admired – and arguably the most important building material today – might serve as an illustration of this.

Despite concrete having a history that stretches back some 7,000 years,[17] and despite having constituted a key event in the history of architecture during the Roman Empire (including the construction of the Pantheon in 200 AD),[18] it wasn't until 1903 and the world's first reinforced concrete skyscraper, the 15-story Ingalls Building in Cincinatti, Ohio, that the Concrete Age was a fact. The secret of concrete was essentially lost for 13 centuries until 1750, when the word was mentioned for the first time by Bernard Forêst de Bélidor as a description for a mortar (it should be added, however, that the Canal du Midi was built in concrete as early as 1670, and that concrete structures in Finland reportedly date back to the sixteenth century).[19] In 1756, British engineer John Smeaton pioneered the use of hydraulic lime in concrete, and in the early 1840s Portland cement was added to the palette of building materials. But with the Ingalls Building, considered a daring work of engineering in its day, the material world shifted once again. High-rises could now be built from concrete.

Today, the only thing humans consume more of than water (by volume and per capita) is concrete.[20] The annual production rate of hydraulic cement, the material that serves as a binder for the aggregate in concrete, is an extraordinary 3Gt (billion metric tonnes). In 2010, the cement output formed the basis for an annual global concrete production of around 23–30Gt, which at a global population of 6.85 billion people equals an annual production of roughly 1.6 cubic meters of concrete for each

[17]Reese Palley, Concrete: A Seven-Thousand-Year History (New York: Quantuck Lane Press, 2010).

[18]Roman concrete, *opus caementicium*, was made from quicklime, pozzolana, and an aggregate of pumice. Its widespread use freed Roman construction from the restrictions of stone and brick material, and allowed for the revolutionary new designs that heralded the arched, vaulted, and domed structures of the Roman Architectural Revolution (also known as the Concrete Revolution).

[19]Bernard Forêst de Bélidor, L'architecture hydraulique, ou l'art de conduire, d'élever et de ménager les eaux pour les différents besoins de la vie (Paris: Chez Firmin Didot, 1819; first published 1737–1753).

[20]World Business Council for Sustainable Development, The Cement Sustainability Initiative (Geneva/Washington: WBCSD, 2007). Available online: http://www.kankyo.metro.tokyo.jp/resource/attachement/Cement%20Sustainability%20Initiative.pdf

person on the planet.[21] To once again quote Kengo Kuma, concrete "rules the world, and not because it is beautiful but because it is a universal material".[22]

Kuma goes on to draw an interesting parallel between the surface-obsessed Concrete Age and the revolution in architectural design that has followed from progress within the field of contemporary computer graphics: "The 20th century developed as the age of poured-in-place concrete, which has produced a CG with modeling and texture mapping as its modes of operation. To us, who are used to architecture built out of concrete, this approach looks completely logical and natural. The system of construction and the means of representation seem to be marvelously in tune with one another. (... [But]) we are going to have to discover new materials that can replace concrete and use them to construct buildings, create cities and improve people's sensibility".[23]

Improving sensibilities is, however, not the only reason to search for ways out of the Concrete Age. Making cement for concrete involves burning fuel to heat pulverised limestone, clay, and sand to 1,450 °C – a process that generates a lot of carbon dioxide, as the naturally occurring calcium carbonate goes through its chemical transformation. The production of a single metric tonne of commonly used Portland cement releases 650–920 kg of CO_2. This energy-intensive industry accounts for around 5 % of global anthropogenic emissions of carbon dioxide (the equivalent of 60–130 kg of fuel oil and 110 kWh of electricity is typically required to produce one tonne of cement[24]), affecting climate change, air and water emissions, depletion of natural resources, as well as the health and safety of workers.[25] Cement plants are huge sources of man-made mercury as well as other toxic pollutants, and in stark contrast to renewable resources, a major part of the world's concrete production eventually ends up in landfills.

It is worth repeating that despite these well-documented shortcomings, each year we produce more than one and a half cubic meters of concrete *for every human being on this planet*. One alternative to this practice could be to open the door to a new material age in architecture (beyond not only the Concrete Age but also what we believe will be its successor, the Timber Age) – one that we will call the Materiomic Age. Certainly, in our practice, we already oscillate between designing for current scenarios for the Concrete and Timber Ages and exploring future possibilities through speculative schemes for the Materiomic Age. Using material aspects of our discipline as the starting point for our designs allows us to envision buildings and environments that promote radical shifts in spatial, structural, and sustainable thinking – an architecture that has less to do with high modernism

---

[21]Author correspondence with USGS Commodity Specialist for cement Hendrik G Van Oss, 1 August 2011.

[22]Kengo Kuma, A Return to Materials, in Luigi Alini, ibid., p. 15.

[23]Ibid., pp. 16–18.

[24]World Business Council for Sustainable Development, ibid.

[25]U.S. Geological Survey, Mineral Commodity Summaries, January 2012. Available online: http://minerals.usgs.gov/minerals/pubs/mcs/

than with synthetic biology.[26] To quote William Gibson, the future is already here, it's just not evenly distributed.[27]

4 Magical Transmutations: Consilient Design in the Materiomic Age

Defined as the holistic study of material systems, the emergent field of materiomics examines fundamental links between material processes, structures and properties from the nano to the macro scale, using systematic experimental, theoretical, and computational methods. The paradigm is closer to systems biology than to traditional material sciences (which is typically focused on mechanical properties such as strength, ductility, and hardness). Materiomics investigates both natural and synthetic materials, from complex biological systems and biomaterials through to non-biological systems. In an architectural context, materiomics can for instance be adapted and applied in areas such as rapid prototyping, lithography and formal/affectual experiments based on combinatorial chemistry. While several researchers have proposed the term independently with slightly different definitions, in an analogy to genomics (the study of an organism's entire genome), a material's *materiome* is the integrated view of its combined internal interactions across multiple scales.[28]

In their seminal book on biomateriomics (the holistic study of biological and bio-inspired materials and systems),[29] Cranford and Buehler point out a fundamental principle of materials science central to materiomics: "the inherent (and reciprocal) relation between a material's structure and material's function".[30] In a biological material, the simultaneous engineering of structure and function is naturally implemented through the process of growth. This production process differs from that of engineering materials in that it is highly adaptive and responsive, calling for a more comprehensive explanatory framework: the materiome.

---

[26]Polish geneticist Wacław Szybalski introduced the term "synthetic biology" already in 1974: "Up to now we are working on the descriptive phase of molecular biology. ... But the real challenge will start when we enter the synthetic biology phase of research in our field." Wacław Szybalski, In Vivo and in Vitro Initiation of Transcription, in A. Kohn and A. Shatkay (Eds.), Control of Gene Expression, pp. 23–24, and Discussion pp. 404–405 (Szybalski's concept of Synthetic Biology), 411–2, 415–7. (New York: Plenum Press, 1974), p. 405.

[27]American-Canadian writer William Gibson is reported to have first said this in an interview on Fresh Air, NPR (31 August 1993). Gibson repeated the quote, prefacing it with the words "As I've said many times ...," in The Science in Science Fiction on Talk of the Nation, NPR (30 November 1999, time code 11:55).

[28]S.W. Cranford, M.J. Buehler, Biomateriomics, Springer Series in Materials Science 165, 27 DOI 10.1007/978-94-007-1611-7_2 (Dordrecht: Springer Science + Business Media, 2012), p. 33.

[29]Ibid.

[30]Ibid., p. 22.

From an architectural point of view, the idea of the materiome is an interesting concept in that it opens up for design speculations on human-scaled neoteric structures that exploit the physical, chemical, biological and mechanical characteristics and properties of novel and complex building materials, a possibility that is rapidly becoming a realistic possibility in the wake of recent advances within emergent areas such as bioengineering and nanotechnology.

Key to this development is the notion of *consilience*, the unification (or literally "jumping together") of knowledge. Coined by William Whewell in 1840, it is a term that has come to denote the cross-pollination of ideas from different fields of expertise. All but forgotten by the broader public, the term was resurrected by humanist biologist Edward O. Wilson in his 1998 book Consilience: The Unity of Knowledge.[31] Wilson's overarching aim is to tie the fundamental importance of biology and genetics to social sciences, the arts, ethics, and religion. He promotes collaboration between the arts and the sciences, but perhaps jumps too quickly to reassurances that in time "all will be scientifically named, the explanations will be joined and all the facts will prove consilient. Now, with science and the arts combined, we have it all".[32]

If to some readers such a passage indicates the musings of a scientist trying too hard to rationalise and understand the world of art using his brain rather than his soul and mind, it should be remembered that Wilson is aware of the potentially presumptuous nature of his undertaking: "I admit that the confidence of natural scientists often seems overweening. Science offers the boldest metaphysics of the age. It is a thoroughly human construct, driven by the faith that if we dream, press to discover, explain, and dream again, thereby plunging repeatedly into new terrain, the world will somehow come clearer and we will grasp the true strangeness of the universe. And the strangeness will all prove to be connected and make sense."[33]

And yet, aren't those "boldest metaphysics" precisely the promise of architecture in the Materiomic Age? While inherently multidisciplinary ("borrowing from fundamental physics and chemistry at the atomistic scale, integrating biological mechanisms at the molecular and cellular level, traversing hierarchical scales, and linking a material's structure and mechanical properties with its natural requirements and functionalities"[34]), materiomics is currently positioned within a materials science paradigm, geared towards providing laboratory-based scientists with the means to invent and control the supermaterials of tomorrow, from mechanomutable materials through to tissue engineering. From our point of view, this is too narrow an application: we envisage consilient design teams consisting of members from a wider range of disciplines, from biologists and material scientists via architects and urbanists to psychologists and artists, designing the buildings and cities of tomorrow

---

[31]Edward O Wilson, Consilience: The Unity of Knowledge (New York: Alfred A. Knopf, 1998).

[32]Ibid., p. 258.

[33]Ibid., p. 12.

[34]Steven Cranford & Markus J Buehler, Materiomics: biological protein materials, from nano to macro, Nanotechnology, Science and Applications 2010:3 127–148, p. 133.

literally from the ground up – if by "ground" we mean the basic elements our planet is composed of.

Giving designers access to material science labs might be a good strategy for the development of new materials. As Chris Lefteri has pointed out, of the three materials on Time Magazine's list of 50 Best New Inventions from 2010 (Sugru, Fabrican, and BioCouture), all were invented by designers rather than scientists.[35] Furthermore, an expansion of the gradient of specialist knowledge within the field of materiomics seems to be in line with the application of multiscale investigative methods that are used to "probe material properties across a multitude of length scales" in order to "establish cross-scale interactions and mechanisms" that "supplement material characterization and properties at a single scale level".[36] The materiome is not only a description of a particular material system's components, structure, and properties, but also an exploration of, for instance, the spatial relationships and communicative interactions within that system.

Multiscale interactions, hierarchical structuring, chemical interactions, multifunctionality, and structural–property relations are all central to the field of materiomics, but why stop there? The time has come for materiomics to branch out even further than the historical collective of physics, biology, chemistry, and materials science into an applied context – our built environment and the implications it has locally for its inhabitants and globally for the entire ecology of the world. Surely, being able to trace and control the effects and affects created by the manipulation of a material at the nano scale all the way up to the scale of the urban landscape would be the ultimate implementation of what Roland Barthes called "the magical operation par excellence: the transmutation of matter".[37] In order to do that, the study of materiomics needs to be widened to encompass the spatial and tactile potentials of materials and how they affect agents interacting with them.

In our vision for the Materiomic Age, bio-inspired material research and development yields novel biological and synthetic materials that are applied at the architectural and urban scale. Knowing that multiscale interactions make some materials susceptible to having their macroscopic characteristics be controlled by their microscopic properties, we engineer the part in order to inform and control the whole. The more we understand about microscopic behavious, the better we can engineer macroscopic properties: the sensorial, tactile, and aesthetic properties of our materials – those applied physicalities that science often neglects. In tandem with such aesthetic concerns, biological materials in particular also give us the opportunity to achieve one of the holy grails of material science in architecture: applications that are not just renewable and therefore carbon neutral, but go beyond

[35]Chris Lefteri, Tangible vs Intangible, article in Innovation (Vol. 30, No. 1, 2011; Library of Congress Catalog No. 82–640971; ISSN No. 0731–2334; USPS 0016–067.), p. 29. Available online: http://idsa.org/images/pdfs/InnSpring2011_Tangible.pdf

[36]Cranford & Buehler, ibid., p. 132.

[37]Roland Barthes, Plastic, in Roland Barthes, Mythologies, transl. Anette Lavers (New York, The Noonday Press, 1991), p. 97.

the mere reduction of net carbon emissions through the harnessing of natural processes as part of their creation, essentially turning biological agents (such as bacteria), chemical procedures (such as crystallisation), raw energy (such as solar power), or basic forces (such as tidal phenomena) into manufacturing catalysts that allow the resulting materials to potentially perform as carbon-positive building blocks. In our view, this materiality-based position is a necessary first step on humanity's crucial journey towards a future post-sustainable world.

5 Matter Forms the Post Sustainable

As Swiss architect Andrea Deplazes has elegantly described it, using an analogy taken from the world of pottery, the paradox of architecture is that it reaches its goal – the creation of *space* – only through the production of its antithesis, the creation of a *non-space*, a limiting material membrane: the wall. "Architecture obtains its *memoria*, its spatial power and its character, from this material." The membrane has a thickness: as Deplazes points out, in architecture the line and two-dimensional area do not exist; even a micro-thin layer of paint is three dimensional and thus material. Construction is not only a question of technique or technology, but also of *techne*, the urge to create, which is the starting point for a trajectory that allows us to grasp construction *first* in a material sense and *only then* intellectually.[38]

This tactile approach is mirrored in the practice of another Swiss architect, Peter Zumthor, who has pointed out that materials are not poetic in themselves, but that they can assume a poetic quality in the context of an architectural project if the architect is able to "generate a meaningful situation for them". "Sense emerges," writes Zumthor, "when I succeed in bringing out the specific meanings of certain materials in my buildings, meanings that can only be perceived in just this way in this one building".[39] Furthermore, these meanings are not representations of ideas, but conceptual building blocks in their own right: "Rooms may owe their existence to an idea but, in the end, they consist of physical matter, of material that often obeys no idea at all and only wants to exercise its own rights".[40] For Zumthor, material properties and effects are intimately connected to the resulting experience: "The reality of architecture is the concrete body in which forms, volumes, and spaces come into being. There are no ideas except in things."[41]

No ideas except in things. And yet, as architect Neri Oxman points out in an incisive paper, at least since the early writings of modernist theoreticians such

[38]Andrea Deplazes, The Importance of the Material, in Andrea Deplazes (ed.), Constructing Architecture – Materials Processes Structures – A Handbook (Basel, Switzerland: Birkhäuser, 2005), p. 19.

[39]Peter Zumthor, Thinking Architeture (Basel, Switzerland: Birkhäuser, 1998), p. 11.

[40]Peter Zumthor, Therme Vals, (Zurich, Scheidegger & Spies, 2007), p. 136.

[41]Peter Zumthor, Ibid., 1998, p. 34.

as Sigfried Gideon, theories of form (as opposed to theories of material and matter) have dominated the discourse of architectural modernism. However, "design experimentalists such as Gaudí, Fuller, Prouvé, and Otto" carried out material research, and, according to Oxman, we currently "appear to be in a period in which the contribution of material experimentation is emerging as a new locus of theory and research in architecture".[42]

Oxman seeks to extend this material experimentation through notions of performance ("a multi-functional approach to design whereby matter is distributed where needed responding to its structural, environmental and indeed, social performance").[43] This is a highly fascinating topic within the realm of architectural materiomics, in particular within the context of digital design processes, but in our practice, we are even more interested in the matter/meaning connections reflected in the quotes by the two Swiss architects above, and how they might be used as tools for the investigation of "evolving theoretical positions on the perceived or potential role of materials in contemporary culture".[44]

One of those roles is arguably the concurrent evolution and potential intersection of novel materials and unprecedented urban conditions. The growing urbanization of the planet, visible in the fact that by the middle of this century some 70 % percent of the world population will live in cities,[45] situates the urban question as a key issue for the global environment. As architects, we need to engage with questions arising from current analyses of a world in which about 3.4 billion people are concentrated on roughly 1 % of the earth's surface, consuming 75 % of the world's energy use and producing 80 % of our total CO_2,[46] and predictions that in 2050 the world's population will increase to 9.3 billion and the urban population will double to 6.8 billion.[47] Alarming as this may sound to some, our metropolitan areas are the very places where increased density can allow us to live extraordinarily energy-efficient

[42]Neri Oxman, Oublier Domino – On the Evolution of Architectural Theory from Spatial to Performance-based Programming (conference proceedings, First International Conference on Critical Digital: What Matter(s)? Harvard University Graduate School of Design, 18–19 April 2008), p. 393.

[43]Ibid., p. 394.

[44]Liat Margolis, Encoding – Digital & Analoge Taxonavigation, in Thomas Schröpfer, Material Design – Informing Architecture by Materiality (Basel, Switzerland: Birkhäuser,), p. 152.

[45]The United Nations, World Urbanization Prospects: The 2007 Revision (United Nations, 2008). Available online: http://www.un.org/esa/population/publications/wup2007/2007WUP_ExecSum_web.pdf

[46]Wuppertal Institute for Climate, Environment and Energy GmbH/Siemens AG, Sustainable Urban Infrastructure – Munich Edition: Paths Toward a Carbon-Free Future (Munich: Siemens, 2009). Available online: www.siemens.com/munichstudy.

[47]UN Population Division, World Population to reach 10 billion by 2100 if Fertility in all Countries Converges to Replacement Level (Population Division, UN Department of Economic and Social Affairs, 2011). Available online: http://esa.un.org/unpd/wpp/Other-Information/Press_Release_WPP2010.pdf

lives; as David Owen observed in an article in The New Yorker, "Eighty-two per cent of Manhattan residents travel to work by public transit, by bicycle, or on foot".[48]

While that would typically be regarded as a "sustainable" urbanist argument, we hesitate to use that term, as we are rather seeking a *post-sustainable* condition. Humanity clearly needs to move beyond practices that merely do not increase the harm already done; we need to readjust our environmental strategies to focus on continuous improvement rather than mere sustainability. It could be argued that in 1987, when the Brundtland Commission of the United Nations declared that "sustainable development is development that meets the needs of the present without compromising the ability of future generations to meet their own needs,"[49] it laid the foundation for a focus on the capacity to *endure* rather than the capacity to *excel.*

The projects presented in this chapter are based on a set of material principles that attempt to overcome this perceived lack of ambition. Based on advances within consilient fields such as synthetic biology, molecular chemistry, and bioengineering, we seek to reconsider the notion of materiality in architecture and urbanism. While the building industry specialises in turning naturally occuring processes such as sedimentation, solidification, and crystallisation into engineered and/or synthetically produced products (usually protected under intellectual property laws), our aim is to investigate the possibility of turning such procedures into conceptual and constructional strategies, effectively allowing the material to come into existence from a set of architecturally controlled parameters, rather than vice versa.

As such schemes eventually reach economies of scale, the subsequent paradigm shift – a move from the constraints of pre-engineered materials and elements dictating the limits of our built environment towards a production of architectural form through the controlled in-situ growth of materials – would signal the emergence of an architecture that is not just environmentally responsive, but environmentally produced using easily transferable local techniques (that is, literally fabricated from its immediate surroundings). This constitutes a conscious break with the part of contemporary architectural discourse that investigates the nature of architecture; we are currently more interested in the architecture of nature. Not the observation of nature's forms so much as the constructional and material possibilities inherent within the processes that produce those forms.

The city is the obvious place to start, not the least because the metropolis is where change is most easily implemented. To quote Edward Glaeser, urban density "creates a constant flow of new information that comes from observing others' successes and failures... Cities make it easier to watch and listen and learn.

---

[48]David Owen, Green Manhattan: Everywhere Should be More like New York (The New Yorker, October 18, 2004), p. 111.

[49]World Commission on Environment and Development, Our Common Future (Oxford, Oxford University Press, 1987). Available online: http://www.un-documents.net/wced-ocf.htm

Because the essential characteristic of humanity is our ability to learn from each other, cities make us more human."[50]

Cities are architectural membranes between civilised society and primal nature. These membranes have a thickness, within which as architects we find a great potential to make a difference. Around half of all the non-renewable resources that mankind consumes are used in the construction of the built environment, making it one of the least sustainable industries in the world.[51] Cities are long lived: their impacts will stretch into the lives of many generations to come, into a future of unknown resources, pollution, climatic conditions, and political circumstances. We believe the way to confront the challenges associated with designing post-sustainable cities for the future is to begin with an understanding of the materials that they will be made of – to accept and embrace the fact that matter matters – and use this as a starting point from which to build up an intellectual understanding of their urban logics, formal organisation, and performative strengths. If architects and urbanists are allowed to break out of the established material distribution of our urban conglomerates and introduce bold alternatives to the prevailing state of affairs, these prospective post-sustainable conurbations might even become *invincible cities.*

6 Invincible Cities: A Calvino Addendum

Invincible Cities is our term for the post-sustainable cities of the future: the city as open-ended organism of the Materiomic Age rather than sealed machine of the Industrial Age. These cities are invincible because they make a difference even before commonly cited parameters of the urban success story – economic growth, housing supply, incidences of poverty and crime, traffic congestion, communication networks, quality of schools and cultural amenities, and so on – are taken into account. They are invincible not only because they offer the power of proximity (facilitating the transmission of ideas) and accelerate knowledge transfer, or even because they are invariably the healthiest, greenest, economically richest, and culturally most exciting places to live (all of this is true for most urban cores in the developed world), but because they push to an entirely new level the idea that the city is mankind's greatest invention, using one simple guiding principle: material intelligence.

Zero carbon homes are being widely touted as the be all and end all solution to humanity's environmental predicament; if we can only reduce our carbon emissions

[50]Edward Glaeser, Triumph of the City: How Our Greatest Invention Makes Us Richer, Smarter, Greener, Healthier, and Happier (Macmillan, 2011), p. 247.

[51]WD Re-Thinking Ltd, Briefing Notes 33: The Impacts of Construction and the Built Environment (Willmott Dixon, 2010). Available online: http://www.willmottdixongroup.co.uk/assets/b/r/briefing-note-33-impacts-of-construction-2.pdf

to zero, the thinking goes, we will be safe from harm. It has been pointed out, however, that while the world target is to cut industry's carbon emissions in half by 2050, the global demand for materials is set to double by that year. To meet our emissions target, then, we have to achieve a fourfold reduction in emissions per unit of material used.[52]

Set against this background, it is plain to see that playing the zero-sum game is not good enough. Rather than bland repetitions of failed experiments – the concrete islands and closed-off high-rise environments of the Modernist era[53] – we need compelling alternative visions for how to actively reduce our cities' environmental footprints while offering novel urban ways of life starting with a responsible attitude to the materiality of the city fabric itself; rather than more thoughtless Concrete Age exploitation of natural resources, we need a mindful exploration of carbon-positive materials for the Materiomic Age. We don't need more "passive houses," but rather more active houses; the building as power plant, the city as mitigation strategy for a warming globe. When it comes to the materials our cities are made of, we need to go far beyond existing typologies that no longer provide adequate support structures to the contemporary condition. We need radical social and political reinvention through a profound reconsideration of the materiality of the metropolis. And we need to build our cities on material ideas rather than match materials to our urbanist visions: "Instead of using materials within a given context, architecture increasingly entails the invention or construction of contexts specifically designed to support the performance of a given set of materials".[54]

This insight, however, does not necessarily provide enough guiding principles to plan and design an entire city. As all creative agents, architects need to conquer our fear of the blank page, the initial horror of the unmarked canvas, that dreaded *tabula rasa* of a project before its boundaries have been outlined, its parameters defined, its program, form, construction, and location fixed. Profusion stifles creation; rules and restrictions need to be invented and applied in order to plot a trajectory across the infinite invisible grid of unlimited design possibilities. As architect Thom Mayne explains, a certain material resistance is key to the act of architecture: "I have to start with material, and I'm going to work with that material. It's going to begin within a more normative logic, and I'm going to attack it. I'm going to keep attacking it and questioning it and challenging it, and it's going to move through reiterative acts".[55]

---

[52]Julian Allwood & Jonathan Cullen, Sustainable Materials – with Both Eyes Open: Future Buildings, Vehicles, Products and Equipment – Made Efficiently and Made with Less New Material (Cambridge, UIT Cambridge, 2011).

[53]For a popular account of these phenomena, see J. G. Ballard's Concrete Island (Jonathan Cape, 1974) and High Rise (Jonathan Cape, 1975).

[54]Sheila Kennedy, Material Presence: The Return of the Real, in Sheila Kennedy and Christoph Grunenberg, KVA: Material Misuse (Architecture Landscape Urbanism 4, London: AA Publications, 2001), p. 10.

[55]"If I were a Lebbeus Woods, I'd die." Thom Mayne interviewed by Magnus Larsson on 8 December 2009. Kultureflash, 18 February 2010. Available online: www.kultureflash.net/interviewDetail.aspx?interview=2-Mayne-Thom

So where might we find this generative friction, this material resistance necessary to begin designing our materiality-based post-sustainable Invincible Cities for the Materiomic Age?

In his famous 1947 comparative essay on the compositional rules underpinning Palladio's Villa Foscari (the Malcontenta) and Le Corbusier's 1927 Villa for M. de Monzie at Garches, architecture critic Colin Rowe points out that both architects, despite being separated in time by hundreds of years, share a strategy of "elaborations in detail" of a dominant scheme, complicated by "imposition upon (a) subsidiary system".[56] Rowe could have been referring to the present project, an attempt to develop materiality as a starting point from which to envision new urban realities (the detailed elaboration) while imposing and expanding on ideas from a famous work of literature (the subsidiary system).

This coalescence of two seemingly disparate conceptual trajectories is a classic creative device: John Lennon famously composed The Beatles' 1969 song *Because* by playing the chord progression of Beethoven's *Moonlight Sonata* backwards; George Perec wrote his lipogrammatic 1969 novel *La Disparation* without using the letter "e"; Matthew Barney physically restrains his body while attempting to make drawings for his *Drawing Restraint* series.[57] Our architectural strategy is one of monadic chains of association resulting from this conceptual imposition, this violent "intrusion of one order into another".[58]

We find such a frictive intrusion in a well-known work of fiction by Italian author Italo Calvino, who, in the early 1970s, was convinced, just as we are today, that "urban and social renewal was urgently needed, and that a prerequisite for the same was a humanistic inquiry into the nature of the city itself".[59] In his 1972 novel *Le città invisibili* (published in English as *Invisible Cities* in 1974), Calvino tells the story of the (invented) Venetian explorer Marco Polo describing 55 cities to the aging (fictitious) Tartar emperor Kublai Khan, accounts supposedly meant to show the state of the old man's expanding and vast empire. It is a tale of tales, storytelling about storytelling, ideas of cities turned into fanciful theories and words turned into fabulous cities.[60]

---

[56]Colin Rowe, The Mathematics of the Ideal Villa – Palladio and Le Corbusier compared, Architectural Review, March 1947, p. 103.

[57]John Lennon tells the story of Imagine and the Moonlight Sonata in David Sheff, All We Are Saying (New York: St. Martin's Press, 2000), p. 191. (Ludwig van Beethoven's 1801 Piano Sonata No. 14 in C-sharp minor "Quasi una fantasia" (Op. 27, No. 2) is popularly known as the "Moonlight Sonata".) George Perec, La Disparation (Paris: Gallimard, 1969) is available in an English translation by Gilbert Adair that also omits the letter "e": George Perec, A Void (New York, The Harwill Press, 1995). Matthew Barney's Drawing Restraint (1987-) series is documented online: http://www.drawingrestraint.net

[58]"Any relationship between a building and its users is one of violence, for any use means the intrusion of a human body into a given space, the intrusion of one order into another." Bernard Tschumi, Architecture and Disjunction (Cambridge: MIT Press, 2001), p.122.

[59]Letizia Modena, Italo Calvino's Architecture of Lightness – The Utopian Imagination in an Age of Urban Crisis (New York: Routledge, 2011), p. 18.

[60]Italo Calvino, Invisible Cities (London: Vintage Classics, paperback edition, 1997).

Essentially a fantastical re-write of the actual Polo's already-fantastic thirteenth-century book *The Travels*,[61] Calvino's seventh novel is filled to the brim with imaginative and innovative accounts of unorthodox metropolitan realities and unconventional urban organisations. "To see a city," Calvino once explained, "it is not enough to keep your eyes open. You must first discard everything that prevents you from seeing it – all your inherited ideas and preconceived images"[62] – and it is this kind of inspired ingenuity that has made his book a minor classic in the architectural and urbanist canons.

Offering an alternative view of what the city might become and how it might be used, Calvino's invisible cities – all named after women: Raissa, Irene, Phyillis, Chloe . . . – are structured within an elusive pattern: Cities and memory, Cities and desire, Cities and signs, Thin Cities, Trading Cities, Cities and eyes, Cities and names, Cities and the dead, Cities and the sky, Continuous Cities, Hidden Cities. One invented metropolis after another passes by, but in the end, the only city Marco Polo/Italo Calvino has woven stories about is Venice, or rather Venices: several cities folded into one, reflected and refracted inside of each other, becoming invisible. As writer Jeanette Winterson puts it, we read Calvino reading Venice.[63]

But we also read Calvino reading contemporary architectural and urbanist theory. Because of its author's understanding of urban issues and creative approach to the imaginative potentialities of the metropolis, Invisible Cities has been used by architects and artists to visualise what cities could be and what future conurbations might become. The book offers an alternative approach to thinking about how cities are formed and how they function. Maybe this could be used in a novel sense, one that goes beyond the semantics and poetry of the original to arrive at cities for a post-sustainable era: carbon-negative cities, urban nodes that excel rather than endure. Invincible cities.

This possibility becomes the connective tissue between our questioning of current urbanist thought and our quest for a redefinition of the role materials might play in the future design of cities. Materiality is what makes architecture architectural, and yet in a historical context, materiality has held a secondary status to form in the architectural discourse, while playing even less of a role in discussions of urbanist theories and realities. We are not interested in narrative readings of Calvino, but rather seek inspiration and conceptual potential in the instrumental and organisational (as opposed to iconographic or metaphorical) qualities of the novel. Through extrapolation and inference, Calvino's metaphorical descriptions of invisible cities are thus transmuted into our material descriptions of invincible cities.

---

[61]Marco Polo, The Travels of Marco Polo (London, Penguin Classics, 2004).

[62]As discussed in John Welsh, Erasing the Invisible Cities: Italo Calvino and the Violence of Representation (Berkely Electronic Press, 2008), p. 5. Available online: http://repository.upenn.edu/cgi/viewcontent.cgi?article=1008&context=wproml

[63]Jeanette Winterson, Invisible Cities (Building Design, July 2001). Available online: http://www.jeanettewinterson.com/pages/journalism_01/journalism_01_item.asp?journalism_01ID=171&journalism_01_Category=Building%20Design

This kaleidoscopic reading, through which widely disparate ideas are refracted into new patterns that give rise to novel conceptual trajectories, has a parallel in the monadology of German mathematician and philosopher Gottfried Wilhelm von Leibniz, in which "each monad is different yet the same in its difference, giving rise to the most variety with the greatest coherence".[64] Similar ideas seem to be echoed in Calvino's book, in which he writes (about the city of Pyrrha) that it was "one of the many cities where I had never arrived, that I conjured up, through its name: Euphrasia, Odile, Margara, Getullia. Pyrrha had its place among them, different from each of them, and like each of them, unmistakable to the mind's eye".[65]

Analogously, we have designed a series of cities that are as internally different as possible, while remaining firmly placed within our self-imposed paradigm of exploring new material possibilities in the fantastic vein of Calvino's dialogues between Marco Polo and Kublai Khan. For the three cities presented here – designs taken from an upcoming study of 55 speculative urban schemes – we have used three different material frameworks, three different formal typologies, and three different organisational layouts and circulatory principles. The cities diverge completely in structural, programmatic, and architectural terms, and yet share a unique energy expressed through the audacious material attitude in which they have been steeped. They are different yet the same in their difference.

Cities have always created energy. Create energy is what cities do. Yet in a world that has experienced continuous population growth since the end of the Bubonic Plague,[66] with current projections still showing a continued increase of population to upwards of 10.5 billion by the year 2050,[67] this flow of energy is accompanied by environmental concerns. The three invincible cities presented in this chapter offer alternative visions of future metropolitan centres put forward as if they were realities, much in the way that Calvino's characters discuss non-existing urban agglomerations as if they were real.

In one of those discussions, perhaps the most poetically compressed passage in the book, the Khan explains that he has constructed in his mind "a model city from which all possible cities can be deducted" and that "contains everything corresponding to the norm". This, he argues, means he only needs to "foresee the exceptions to the norm and calculate the most probable combinations". To this

[64]The quote is from Jeffrey Kipnis, Form's Second Coming, in Bernard Tschumi & Irene Cheng (eds.), The State of Architecture at the Beginning of the 21st Century (New York: The Monacelli Press, 2003), p. 59. For more on Leibniz's monadology, cf. Anthony Savile, Routledge Philosophy guidebook to Leibniz and the Monadology (London: Routledge, 2000).

[65]Italo Calvino, ibid., p. 83.

[66]Jean-Noël Biraben, An Essay Concerning Mankind's Evolution, Population, Selected Papers, Vol. 4 (1980), pp. 1–13. Original paper in French: Jean-Noël Biraben, 1979, "Essai sur l'évolution du nombre des hommes", Population, Vol. 34 (no. 1), pp. 13–25.

[67]Population Division of the Department of Economic and Social Affairs of the United Nations Secretariat, World Population Prospects: The 2008 Revision (2009). Available online: http://www.un.org/esa/population/publications/popnews/Newsltr_87.pdf

Marco Polo replies that he has also thought of such a model city, but one "made only of exceptions, exclusions, incongruities, contradictions":

> If such a city is the most improbable, by reducing the number of elements, we increase the probability that the city really exists. So I have only to subtract exceptions from my model, and in whatever direction I proceed, I will arrive at one of the cities which, always as an exception, exist. But I cannot force my operation beyond a certain limit: I would achieve cities too probable to be real.[68]

We might be forcing the present operation beyond such a limit. Perhaps our Invincible Cities are too probable to be real. Or perhaps – if again we consider for a moment the world's absurd annual concrete production as outlined above – they are too consequential in their promotion of radical reconsiderations of our cities' material building blocks, too bored with the status quo so eagerly defended by industry and politicians, too real to be probable.

7 Sandra

> Each city receives its form from the desert it opposes; and so the camel-driver and the sailor see Despina, a border city between two deserts.
>
> (...)
>
> In colored miniatures the atlas depicts inhabited places of unusual form: an oasis hidden in a fold of the desert from which only palm crests peer out is surely Nefta...[69]

Standing at the highest point of Sandra, the sandstone city, one can see the division line created by the crest of the sand dune into which, transverse to the primary wind direction, the city has been built. Its terraces follow precisely the contour of the surrounding dunescape, a gigantic hillside labyrinth of shaded walkways in between buildings made from bacteria and sand, stretching out into the desert horizon.

Sandra, defender of man, like her older sister Alexandria, is a link between countries and cultures, a cosmopolitan nucleus in an arid climate. The people of Sandra, inhabiting what shouldn't be inhabitable, defy the desert precisely by not deserting their city. Instead they live on in buildings sculpted by the wind, walk through souks carved by the fluctuating movements of grains of sand, lean against walls cast by millions upon millions of microscopic construction workers.

Solidified only in parts, Sandra is a city in constant motion: no planner knows in advance exactly the topography or program of the next stretch of its development, or indeed even where it will be situated. The future of Sandra is blowing in the wind (Fig. 1).

A single grain of sand is almost nothing – a splinter of rock, topographic memorabilia, the residue of a microcosmic event. But put myriad quartz grains together and they turn into a fluid material capable of being transformed and solidified into mesmerising landscapes, vast deserts, solid structures, and, ultimately, flourishing

---

[68] Italo Calvino, ibid., p. 61.

[69] Italo Calvino, ibid., p. 15 and 124.

Fig. 1 The terraces of Sandra follow the contour of the surrounding dunescape, a gigantic hillside labyrinth of shaded walkways in between buildings made from bacteria and sand

cities. In the desert, fluctuating masses of shifting sands express emergent properties; millions of tons of grains move "inexorably in regular formation, over the surface of the country, growing, retaining their shape, even breeding, in a manner which, by its grotesque imitation of life, is vaguely disturbing to an imaginative mind".[70]

The bacterial biomass on this planet exceeds that of all plants and animals combined. And yet we only really have a scientific understanding of about a fraction of the approximately five nonillion (5×10^{30}) bacteria on Earth. Some are incredibly sturdy: the Apollo 12 mission found specimens of *Streptococcus mitis* that had survived for 31 months of massive temperature swings in the waterless vacuum of the Moon's atmosphere.[71]

Now, add bacteria of the *Bacillus pasteurii* strain to loose grains of sand and in a short amount of time they will solidify into bacterial sandstone. A microscopic process allows the awe-inspiring desert landscape, these immense masses of sand, to be manipulated into habitable structures through the strategically controlled local solidification method known as microbially induced carbonate

[70] R. A. Bagnold, The Physics of Blown Sand and Desert Dunes (New York: Dover, 1941), p. xix.

[71] Jones E (1995) Apollo 12 Lunar Surface Journal, Surveyor Crater and Surveyor III, Corrected Transcript and Commentary: NASA, Viewed 13 April 2011 (search for "little bacteria"), http://www.hq.nasa.gov/office/pao/History/alsj/a12/a12.surveyor.html

precipitation (MICP).[72] While this is a naturally occuring phenomenon, an architectural exploitation of the process as a design and construction method – the controlled "growing" of sandstone structures that lies at the heart of Sandra – is a novel architectural speculation with staggering potentials.[73]

The individual grain of sand is an elementary particle within the material history of architecture. Take a poetic look at a city: what you will see is a conglomeration of sand pile buildings, frozen grains stacked high into the sky, an urban version of our deserts' dunescapes – streets and avenues lined with *castles made of sand.* Without sand there would be no brick, no concrete, no glass. Sand is ubiquitous and renewable: one *billion* grains of sand come into existence around the world *every second* through a cyclic process that sees entire mountain ranges weather and release tiny splinters. Some of those fragments lithify into a clastic sedimentary rock, a sandstone. As that sandstone weathers, new grains break free. The majority of quartz sand grains are derived from the disintegration of older sandstones; perhaps half of all grains of sand have been through six cycles. A typical mountain will be lowered by a few millimetres every year.[74]

That is a lot of sand. Dry areas cover more than one-third of Earth's land surface, and desertification – "the diminution or destruction of the biological potential of the land" – is a major threat on all continents, affecting more than 100 countries in the world. Some estimates suggest that the livelihoods of 850 million people are at risk, spread out across 35 % of the Earth's land surface.[75]

Sandra is a direct response to this challenge: a very narrow and roughly 6,000 km long pan-African city that would support the idea of a "Green Wall for the Sahara". This scheme, a barrier of greenery that is being planted to halt the shifting sands, was first proposed by former Nigerian president Olusegun Obasanjo in 2005. The initiative originally called for 23 African countries to come together in order to plant trees across a 15 km wide stretch south of the Sahara in order to stop the dunes

[72]Cf. Le Metayer-Levrel G, Castanier S, Orial G, Loubiere JF and Perthuisot JP (1999) Applications of bacterial carbonatogenesis to the protection and regeneration of limestones in buildings and historic patrimony. Sedimentary Geology 126: 25–34. Nemati M and Voordouw G (2003) Modification of porous media permeability, using calcium carbonate produced enzymatically in situ. Enzyme and Microbial Technology 33: 635. DeJong JT, Fritzges MB, Nüsslein K (2006) Microbially Induced Cementation to Control Sand Response to Undrained Shear. Journal of Geotechnical and Geoenvironmental Engineering, Vol. 132, No. 11, November 1, 2006. Whiffin VS, van Paassen LA, Harkes MP (2007) Microbial carbonate precipitation as a soil improvement technique. Geomicrobiology Journal, Volume 24, Issue 5, pp. 417–423.

[73]Sandra is an updated version of the previous *Dune* project. For a more in-depth explanation of this scheme, cf. Larsson M (2011) Dune – Arenaceous Anti-Desertification Architecture in Badescu V and Cathcart R (eds.), Macro-engineering Seawater in Unique Environments, Environmental Science and Engineering, DOI: 10.1007/978-3-642-14779-1_20 (Berlin Heidelberg: Springer-Verlag, 2011).

[74]Welland M (2009) Sand – The Never-Ending Story: University of California Press, Berkeley and Los Angeles, California.

[75]Grainger A (1990) The Threatening Desert – Controlling Desertification: Earthscan, London.

Fig. 2 Sandra is a direct response to the challenge of desertification: the roughly 6,000 km long pan-African city holds a barrier of greenery that halts the shifting sands

from migrating. While the initial ambitions have now been scaled down, we use the framework of the Green Wall as a context for our planning of Sandra (Fig. 2).

The mountain-to-sand cycle described above gives us a glimpse of one of nature's rudimentary rules: that erosion follows aggregation follows erosion. Sandra is a city that utilises aeolian forces to define its exterior while harnessing the power of biocementation processes to sculpt its interiors. A sensitive physical intervention in the landscape, it is a study in *aggrerosion*, a unique design strategy based on the accumulation and reduction of granular materials.

As a building material, grains of sand can be employed across a gradient of conditions: granular mass, solid stone, dynamic medium carried by aeolian forces, compressive membrane, and so on. The granularity allows for both structural and constructional inventions: the sand can for instance be used both as primary ingredient of the finished structure and as its formwork during construction. If we solidify parts of a dune to hold it in place, the non-solid parts will continue moving, making it possible to plan for aeolian forces to excavate the structure for us. Conceptually, we can read the desert dunes as readymade buildings; the only architectural intervention needed is to solidify the sand wherever we need solid surfaces, and then excavate the sand we don't need – or have the wind excavate it for us.

Using the existing local sand as our base material, we can sculpt arches, caves, tunnels, and other patterns straight into the dune, allowing saltation (a type of particle transport) to propel the grains in place, and then, once the sand has aggregated into a beneficial form, use MICP to solidify it. The spatial pockets created within the resulting solid rock structure would help retain scarce water and mineral resources, while also functioning as programmable spaces – a habitable wall straddling an entire continent, binding villages, people, countries, and cultures together.

While the background to this project has already been described at length elsewhere, it is perhaps prudent to note briefly how it came into being. In 2007, professor Jason DeJong at the University of California at Davis proposed a sustainable alternative to the often toxic chemicals that are injected into soils to make them withstand the phenomenon of *liquefaction*, which often causes more damage during an earthquake than its tremors: use *Bacillus pasteurii* to cause calcite (the most stable polymorph of calcium carbonate, $CaCO_3$) to precipitate, a process that glues the grains together and turns loose sand into solid rock. "Starting from a sand pile, you turn it back into sandstone," DeJong explained, before making clear that there are no toxicity problems, that the treatment could be applied after the construction of a building, and that the structure of the soil doesn't change as the void spaces between the grains are filled in.[76]

Inject sand with cultures of *Bacillus pasteurii*, feed them well and provide them with oxygen, and they will solidify loose sand into sandstone. DeJong and his colleagues experimented with sterilized sand and bacteria, and were able to control and monitor nutrients, oxygen levels, and other variables to determine exactly how the bacteria hardened their sand specimens. The microbes were added in a liquid growth medium amended with urea and a dissolved calcium source (nutrients are required by microorganisms for cellular material – carbon and minerals – as well as for an energy source). Subsequent cementation treatments were passed through the specimen to test the increase in cementation level of the sand particle matrix. The results for the MICP-cemented specimens were compared with those of gypsum-cemented equivalents, and were assessed by measuring the shear wave velocity with bender elements.

The MICP-treated specimens exhibited a noncollapse strain softening shear behaviour, with a higher initial shear stiffness and ultimate shear capacity than untreated loose specimens – a behaviour similar to that of the gypsum-cemented specimens, which represent typical cemented sand behavior. SEM microscopy formation of a cemented sand matrix with a concentration of precipitated calcite forming bonds at particle-particle contacts, while x-ray compositional mapping confirmed that the observed cement bonds were comprised of calcite.

For MICP to be effective, the microorganism must be capable of CO_2 production paralleled by a pH rise in the surrounding environment to an alkaline level that induces precipitation of calcium carbonate. This forces calcium and carbonate dissolved in the water to combine and form crystals of calcium carbonate – calcite – the same natural cement that binds together sandstone, as well as manmade concrete. Aerobic microorganisms capable of consuming urea as an energy source (such

[76]The Engineer (2007) Editorial, Bacteria Help Protect from Quakes: The Engineer, 23 February 2007. Available online: http://www.theengineer.co.uk/news/bacteria-help-protect-from-quakes/298382.article

as *Bacillus pasteurii*) are particularly good candidates because they provide two sources of CO_2 – respiration by the cell and decomposition of urea. Furthermore, cells of *Bacillus pasteurii* do not aggregate, which ensures a high cell-surface-to-volume ratio, an essential condition for efficient cementation initiation.[77]

How long would the microbial part of the construction process take? According to DeJong's research, factors critical to the success of the microbial treatment include pH, oxygen supply, metabolic status, concentrations of microbes, ionic calcium in the biological and nutrient treatment flushes, and the timed sequence of injections. While the treatment time is dependent on numerous factors such as microbial concentration, reaction kinetics, and soil characteristics, DeJong found the maximum shear wave velocity versus time value to be 1,700 min, and the relative amount of cementation resulting from each treatment to reach its peak towards the fifth or sixth injection. This would indicate that an initial sandstone surface could be created within the dune in approximately 28 h, and that the structure could reach its optimal structural strength in a single week ($1{,}700 \times 6 = 10{,}200$ min) (DeJong et al. 2006). In private conversation with the author, however, DeJong has indicated that variations are likely to occur, and that "the more patient you can be, the better".[78]

The invincible city of Sandra is constructed using modified injection piles of the kind that is normally used for grouting processes. The piles are pushed through the dune, after which a first layer of bacteria is distributed through the piles to solidify an initial surface within the dune. They are then pulled up, creating habitable spaces along the way, with the loose sand acting as a mould before being excavated either by hand or by the wind to create individual buildings. This method would be in line with one particular system that has been developed specifically for the purpose of using calcium carbonate as a laboratory cementation agent: the Calcite In-Situ Precipitation System, or CIPS. This involves injecting a proprietary chemical solution that causes the precipitation of calcite crystals within the pore fluid and on the surfaces of constituent sand grains.[79] Studies using CIPS have shown that the solidification level and cementation rate can be altered by using different chemical formulations or multiple solution flushes.[80]

The novel process of engineered architectural lithification that gives rise to Sandra establishes a radical shift in structural thinking, away from both pre-fabricated and in-situ construction, towards the localised biocementation of granular materials. Once we are able to control the solidification of sand into sandstone – the addition and subtraction of densifications mentioned above – we can investigate the possibilities of a completely monolithic architecture based not on components that are attached to each other in order to create a structural system, but on the binary

---

[77]DeJong et al., ibid.

[78]DeJong JT (2008), private correspondence with the author, 12 March 2008.

[79]Ismail MA, Joer HA, Randolph MF, Kucharski E (1999a) Cementation of porous materials using calcite precipitation. University of Western Australia Geomechanics Group, Geotech. Rep. G1422.

[80]Ismail MA, Joer HA, Randolph MF, Kucharski E (1999b) CIPS, a novel cementing technique for soils. University of Western Australia Geomechanics Group, Geotech. Rep. G1406.

Fig. 3 The novel process of engineered architectural lithification used to construct Sandra explores a radical shift in structural thinking, towards the localised biocementation of granular materials

densification of aggregate matter. The sand is either turned into stone, or it is not (Fig. 3).

While we have yet to explore this potential, future *programming* of the bacteria could potentially add another dimension to the strategy of using the existing sand dunes as granular readymade structures, working within the material volume itself, controlling the bacteria to augment the loose sand into solid structures. This strategy has been imaginatively compared to "a kind of *infection of the earth* ... a vast 3D printer made of bacteria (that) crawls undetectably through the deserts of the world, printing new landscapes into existence over the course of 10,000 years".[81]

In a mineral world like ours, all design is fundamentally about aggregation and erosion. Giving shape is a matter of deciding where elements will aggregate and matters densify, and where they will erode or stay as voids. Architecture at its most fundamental is an act that follows simple laws of arithmetic: we add something to contract a space, we subtract something to expand a space. At its core, architecture is about the manipulation of landscape, and Sandra the first bacterial sandstone city based on such aggrerosion strategies.

[81]Geoff Manaugh, Sand/Stone (BLDGBLOG, 19 April 2009). Available online: http://bldgblog.blogspot.com/2009/04/sandstone.html

8 Arboria

> After a seven days' march through woodland, the traveler directed towards Baucis cannot see the city and yet he has arrived. The slender stilts that rise from the ground at a great distance from one another and are lost above the clouds support the city. You climb them with ladders. On the ground the inhabitants rarely show themselves: having already everything they need up there, they prefer not to come down. Nothing of the city touches the earth except those long flamingo legs on which it rests and, when the days are sunny, a pierced, angular shadow that falls on the foliage.
>
> There are three hypotheses about the inhabitants of Baucis: that they hate the earth; that they respect it so much they avoid all contact; that they love it as it was before they existed and with spyglasses and telescopes aimed downwards they never tire of examining it, leaf by leaf, stone by stone, ant by ant, contemplating with fascination their own absence.[82]

> Walking through Arboria is walking high above the ground within a gigantic, undulating timber lattice, through the infrastructural core of the city itself: along its passageways and sewage systems, its shiny avenues and dark underbelly. Hovering above the landscape beneath, emulating precisely its topographic contours, Arboria is an augmentation of the terrain it covers, a carbon copy of the ground below, raised and turned into a thin trellis membrane that is at once protection and foundation.
>
> This vast megalopolis celebrates difference over repetition: despite the democratic equality in the footprints of all plots, each living capsule extends in a unique fashion above and below the gridshell, expressing its inhabitant's social, financial, and intellectual standing.
>
> Having already everything they need inside of their timber mesh, the people of Arboria rarely venture down to the ground. They prefer life at the level of the treetops or above, looking down at the leafs and stones and ants below, contemplating with fascination their elevated position in the world (Fig. 4).

An invincible city needs to be more than just city. It needs to add rather than subtract, it needs to not simply refrain from doing damage but constitute an actual environmental improvement. It needs to continuously adapt and evolve, constantly look so far into the future as to prepare for scenarios that would be impossible to foresee today, and use as few resources as possible to do as much good as it can. In colloquial terms, the objective is to get the "biggest bang for the buck". This is where timber enters the picture. A building made from wood is a carbon dioxide store, which is why building an entire city (for as many inhabitants as possible) from timber can be viewed as a form of active climate protection. If more builders were to use engineered wood, we would be putting less waste in landfills while prices would come down as supply increased to meet demand. Arboria is an aggressively provocative essay in file-to-factory principles translated into an environmentally conscious megastructure made from the most environmentally-friendly standardised building material on Earth: wood.

In an interesting 2011 essay, Hani Buri and Yves Weinand discuss how different technological ages have influenced the tectonics of timber structures, from the artisanal "wooden age" through to our present "digital age". Having established that

---

[82]Italo Calvino, ibid., p. 69.

Fig. 4 The vast megalopolis of Arboria celebrates difference over repetition: despite the democratic equality in the footprints of all plots, each living capsule is unique

the "ability to control machines with the help of a computer code eliminates the need for serial production," the authors go on to explain that the easy machinability of wood "makes it an ideal material for digitally controlled processing portals," which has led to timber "taking on the status of a high-tech material".[83]

This age-old high-tech material can now be shaped using a combination of computer-aided manufacturing (CAM) and computer numerically controlled (CNC) technologies linked in real time to a continually updated 3D model of the scheme, that might even incorporate live-fed contextual data from the site in question. Such recent technological advances enable the materialisation of architectural forms straight from a digital file. As the necessary equipment, machinery, infrastructure, and tools are becoming available to a larger part of the building industry, the manufacturing and fabrication of mass-customised elements according to specific instructions derived from modelling software is becoming increasingly popular.

Connect a computer to a CNC router, and a lot of the prevailing rules for how a building (and by extension a city) should be designed and constructed can be left behind. This was the starting point for Arboria, a plug-in timber gridshell city designed to contain a minimal amount of basic elements: glulam beams make up the main structural lattice, cross-laminated timber sheets add structural support, and surfaces (engineered wood boards) come together to form the plugged-in living capsules. Nearly all of these components, however, differ from each other, as they are interlinked through our file-to-factory system, which updates the underlying 3D

[83]Hani Buri & Yves Weinand, The Tectonics of Timber Architecture in the Digital Age, in Hermann Kaufmann and Winfried Nerdinger, Building with Timber – Paths into the Future (Pretzel, Munich, 2011), pp. 56–63.

Fig. 5 The construction system that supports Arboria collate several convergent data streams into finalised design files that are fed straight to the factories: a highly agile production process

model with new data at regular intervals, producing a perfect representation of the entire city at any one moment in time. The machines in the factory don't know, and don't care, whether the boards they are instructed to manufacture are to fill the function of a wall or a floor. As Kas Oosterhuis describes it, "Architecturally, convergence sees the beauty in the purest solution, the new modern: one building, one detail".[84]

While the construction system that supports Arboria doesn't go quite so far as to define a single, parametrically controlled detail with which we are able to assemble the entire city, it does collate several convergent data streams into finalised design files that are fed straight to the factories, which in turn manufacture the parts in accordance with a predefined time schedule. This makes for an extraordinarily agile production process, where each new building element is immediately manufactured as and when it is needed, and new parts get pushed to new positions within the work schedule if for some reason the construction program changes (Fig. 5).

It also completely does away with the need for conventional, two-dimensional drawings. As the design that goes into the centralised three-dimensional project model is converted into instructions that are sent straight to the factory, nothing gets lost in translation. An annotation system makes assembly a matter of simply lining up one marked-up panel against the next, and if a builder needs to double check its placement, he or she can always bring up the latest version of the 3D model on a computer or tablet. The file-to-factory process makes Alberti's orthographic projections and ideas of identicality redundant: plans and sections become useless descriptions.[85] This enhances efficiency and speeds up the building process. As Mark Burry has pointed out, significant economic benefits can also be derived

[84]Kas Oosterhuis, Cockpit Building in an AcousticBarrier, paper (2004). Available online: http://www.oosterhuis.nl/quickstart/fileadmin/Projects/142%20Cockpit/02_Papers/040831-Hessing-Cockpit_paper.pdf

[85]For an excellent discussion on Alberti and identicality, cf. Mario Carpa, The Alphabet and the Algorithm (Cambridge, MA: MIT Press, Writing Architecture series, 2011).

from "automating routines and coupling them with emerging digital fabrication technologies, as time is saved at the front-end and new file-to-factory protocols can be taken advantage of".[86]

This kind of precise and immediate procedural route from 3D model to manufactured element is particularly appealing when it comes to structures made from engineered wood, as accuracy is of the essence. The design of a building that is to be constructed from elements that are pre-cut off site must be finished when it leaves the architect's office. This suggests a production strategy based on mass customisation and file-to-factory processes. Such new manufacturing methods give rise to new construction strategies, new structural ideas, new buildings, new ways of using those buildings: new architectures and urbanisms.

Urbanising literally means "burying," but in the design of Arboria, we were not interested in burying the agricultural or suburban landscape. Rather, we wanted to celebrate the untouched ground datum as a counterpoint to the necessary further densification of our urban nodes. By having the structure undulate across the topography, touching the ground in as few places as possible, we can potentially achieve a dichotomy between landscape and urbanism allowing the project to reverse today's migration patterns, bringing the city to the people living in less densified areas rather tham bringing people to the already-dense urban core. Arboria redefines the notion of sprawl, turning it into a progressive strategy for linking the infrastructure of the city to the rural outskirts of our urban conglomerations, extending the metropolitan condition (Fig. 6).

Arboria is not just a single city, but rather a set of organisational and constructional principles that can be used to produce a range of mirror cities in different parts of the world. For each new iteration, the design process begins with a topographic analysis of the site in question. The act of measuring its geography turns the territorial continuum into an architectural element or system. This study is then carefully indexed and turned into a topographical network representation of the ground datum, which is raised above ground in a reversal of the typically predatory consumption of land by many urbanisation processes. Shielding the ground below from the forces of nature while offering shadow and even irrigation and/or rainwater distribution possibilities, this matrix canopy initiates a symbiotic relationship between the elevated city grid and the terrain beneath.

Adding to and extending this initial lattice structure calls for further analyses. The system becomes a continously moving factory, trawling the suburban and rural landscape to measure and survey it in preparation for the next stretch of the scheme. The gridshell diagram of the site topography next becomes the underlying array that controls the computation of a set of operational rules, yielding the living capsules through mass customisation scripts, habitation units that are then "stacked" as they get hung within the structural matrix. The relationship between the local conditions of the grid and the various instantiations of living units could be viewed as analogous

[86]Mark Burry, Scripting Cultures: Architectural Design and Programming (Architectural Design Primer, John Wiley & Sons, 2011).

Fig. 6 Arboria celebrates the untouched ground datum as a counterpoint to the further densification of urban nodes. The structure touches the ground in as few places as possible

to how, in genetics, a single genotype might produce a differentiated population of phenotypes in response to diverse environmental conditions.

Arboria is designed to be a perpetual work in progress based on the file-to-factory paradigm described above, an operation that is based on the material logic of wood, and that makes it possible to manufacture individualised building components in a highly economical manner that maximises efficiency while minimising waste. In a provocative 2004 essay, David Schaller pointed out that "a typical coffee business uses 0.2 % of the coffee bean to produce a cup of coffee. This means 99.8 % of the coffee bush becomes 'waste'". According to Schaller, there is "an abundance to nature that, in our ignorance and even arrogance, we are only beginning to fathom… Our microbiologists, botanists, biologists, mycologists, wood chemists and geneticists are only now scratching the surface of this great diversity and plenty. What we don't understand, we can't possibly explain, value, or protect."[87]

The choice of timber for a structure at this scale is a proposal that could perhaps open up one line of inquiry into this unharvested natural abundance. The Earth contains about one trillion tonnes of wood, which grows at a rate of ten billion tonnes per year.[88] As an abundant, carbon-neutral or potentially even carbon-negative renewable resource, timber is the only readily available building material on this

[87] David Schaller, Beyond Sustainability: From Scarcity to Abundance (BioInspire 13, 2004). Available online: http://bioinspired.sinet.ca/content/february-2004-beyond-sustainability-scarcity-abundance

[88] Nimz H, Schmitt U, Schwab E, Wittmann O, Wolf F (2005) ''Wood'' in Ullmann's encyclopedia of industrial chemistry. Wiley-VCH, Weinheim.

planet that stores carbon dioxide, minimises or eliminates its carbon footprint, and hence is truly post sustainable. The use of prefabricated, mass-customised living capsules further reduces construction times and minimises waste, thus decreasing the building's ecological footprint.

Despite its scale, the scheme is not as dense as a regular city. Instead of density, we get intensity: the idea of individual buildings holding collective housing gives way for a collective building holding individual housing. The lattice frame itself becomes programmable, accommodating aerial walkways, terraces, transportation routes, and services. Non-standard architectures call for non-standard ideas about how to use strategies such as mass customisation and prefabrication in the manufacturing process, though, crucially, it does not necessarily call for non-standard building materials or building processes. While its expression, program, manufacturing process, and design approach are all atypical, Arboria remains a city that can be constructed by everyday builders from perfectly normal engineered wood.

Arboria is a plug-in city in the shape of a gigantic horisontal skyscraper, a huge canopy weather-sealed by living units, an oversized tribute to that great British tradition, the Pavilion in the Park. It is impossible in this context not to mention Jurgen Mayer H's Metropol Parasol, a giant latticed timber canopy that forms part of the architect's redevelopment of the Plaza de la Encarnacíon in Seville, Spain. This scheme, finalised and opened to the public in April 2011, includes an archaeological museum, a farmers market, an elevated plaza, an aerial walkway, as well as bars and restaurants, all contained beneath and within the parasol structure made from engineered wood. Within the enormous timber lattice of Arboria – a wooden gridshell reminiscent of Frei Otto's 1975 Multihalle Mannheim structure, Buro Happold's 2006 Savill Building, or maybe Barkow Leibinger's Company restaurant/auditorium in Ditzingen[89] – prefabricated, pod-like living units are inserted and hung from the gridded framework. Those living units could equally well accommodate alternative programs: public squares, orchards, swimming pools, shops, libraries, restaurants, bicycle workshops ... In the shade of the structure, we plant seedlings that grow over time into a stretch of forest. Trees are cut down to create this habitable canopy for trees to grow under.

Arboria embodies the ideals of the Japanese Metabolists – never a group to hold back in terms of scale. Their proposals for future cities were inspired by the technological fervour of the 1960s, resulting in large-scale, flexible, and expandable structures that evoked the processes of organic growth.[90] That basic unit of Metabolism, iterated to perfection by Kisho Kurokawa, the prefabricated capsule, is important also to this scheme. Kurokawa's 1960 Neo-Tokyo Plan and grid system on stilts, Agricultural City, as well as his capsule-within-space frame 1969 Odakyu Drive-in Restaurant, and of course his famous 1972 Nagakin Capsule Tower, are

---

[89] "All of which are presented in Hermann Kaufmann and Winfried Nerdinger, ibid."

[90] Zhongjie Lin, Kenzo Tange and the Metabolist movement: urban utopias of modern Japan (Routledge, 2010).

all somehow related to the present project, as are other Metabolist proposals such as Arata Isozaki's 1960 *Clusters in the Air* project and perhaps the 1961 *Disaster Prevention City* by Kiyonori Kikutake.[91]

Living "off the grid" is a fashionable way of referring to a self-sufficient lifestyle that doesn't rely on public utilities (such as the main electricity transmission grid). The invincible city of Arboria promotes the perhaps less trendy alternative of a communal lifestyle *within the grid.* Over and above the obvious environmental credentials – a carbon dioxide store that uses the relatively low density and very high degree of stability (greater, in relation to its mass, than that of steel) of wood to soar high above the ground – the plug-in gridshell is a programmatically interesting alternative to rural or suburban lifestyles, which, as we have seen, are rarely as energy efficient as their urban counterparts. The systemic, adaptive variation and continuous differentiation arising from the dynamic design system, which anchors the city to the site as it continues to shoot through the landscape (in a way perhaps reminiscent of Superstudio's 1969 *The Continuous Monument* scheme[92]) offers an interesting contrast between input data and output result, the model/city dichotomy responding to external conditions in accordance with its own predefined logic.

9 Arachnia

> Now I will tell how Octavia, the spiderweb city, is made. There is a precipice between two steep mountains: the city is over the void, bound to the two crests with ropes and chains and catwalks (...) This is the foundation of the city: a net which serves as passage and as support. All the rest, instead of rising up, is hung below (...) Suspended over the abyss, the life of Octavia's inhabitants is less uncertain than in other cities. They know the net will last only so long.[93]

Suspended above the fault zone, the buildings, streets, and squares that make up the city of Arachnia quiver as the tremblors fracture the Earth's crust below. Safely relocated before the earthquake hit the region, the inhabitants watch the seismologic spectacle from afar, as they have done so many times before.

This is the reality of life in Arachnia, the spiderweb city, where once daring planners made a decision hailed as simultaneously rational and audaciously bold: to adapt the metropolis to the area's frequent tectonic displacements by lifting the entire city off the ground and having genetically modified spiders spin a supporting web around its dangling living units.

Serving as passage and support, the spider silk also protects the city from the violent vibrations of the recurring quakes. The people only ever touch the ground when the quake alarm sounds and they climb down from their pendent city to watch as the next set of seismic waves rocks their homes... (Fig. 7)

---

[91]For an unparalleled excursion into the world of the Metabolists, see Rem Koolhaas and Hans Ulrich Obrist, Project Japan – Metabolism Talks... (Taschen, 2011).

[92]Peter Lang & William Menking, Superstudio: Life without Objects (Skira, 2003).

[93]Italo Calvino, ibid., p. 67.

Fig. 7 Arachnia, the spiderweb city, is designed to be suspended above the fault zone of an earthquake-prone area

Spider silk is a biopolymer with fascinating properties: nearly as elastic as rubber on a weight-to-weight basis, thinner than human hair, lighter than cotton, and about five times stronger than steel of the same density. The silk of spiders first evolved

400 million years ago and orb webs emerged 180 million years ago.[94] Most of the 40,000 spider species on our planet produce silken threads of incredible extensibility and toughness. Spider silks are not only typically two to three times tougher than synthetic fibres like nylon or kevlar, but also antimicrobial, hypoallergenic, and completely biodegradable.[95] It is also made from renewable raw materials (substances that can be grown again), and spiders themselves often recycle the silk proteins by eating some of the web they have spun in order to recoup some of the energy used in spinning. An oft-repeated example of the material's relative strength is that a pencil-thick strand of spider silk could stop a Boeing 747 in mid flight (while mechanically true, it has been pointed out that this strand would need to be 30 km in length, and that it would stretch another 9 km before bringing a plane travelling at 320 km/h to a stop[96]).

Spiders have used protein-based nanomaterials with the ability to self-assemble into fibers and sheets for more than 450 million years, and humans have been attracted to spider silk since ancient times. Fairly intense study of spider silk was carried out prior to the second world war and in the late 1950s. However, it is only in the past decade or so that an understanding has emerged of the underlying reasons behind the unique mechanical properties if spider silk.[97] And it is only in the past few years that genetical engineering has allowed for spider silk production using artificial manufacturing methods.

This development has inspired artists, designers, and researchers to consider the possibility of using natural and/or genetically engineered spider silk in their work. As Goslin et al. points out, there are two paths to turn our knowledge of the molecular biology and network structure of spider silk into a structural material through genetic engineering: a "purely synthetic approach" where genetic sequence motifs are designed and spun into fibers, or an approach where we "build upon the variation created through millions of years of evolution to develop an understanding of structural designs that already exist in nature and have been tested in the day-to-day function of spiders".[98]

One exceptional project falling into the first category is bioartist Jalila Essaïdi's *2.6 g 329 m/s* scheme, positioned at the intersection of art and science, which

---

[94] L. H. Lin, D. T. Edmonds, F. Vollrath, Structural engineering of an orb-spider's web. Nature (London), Jan 12, 1995. v. 373 (6510), pp. 146–148.

[95] R V Lewis, Spider Silk: Ancient Ideas for New Biomaterials (Chemical Review, 2006, 106 (9), 3762–3774, DOI 10.1021/cr010194g).

[96] Cf. S.W. Cranford, M.J. Buehler, Biomateriomics (Springer Series in Materials Science 165, 27 DOI 10.1007/978-94-007-1611-7_2) (Dordrecht: Springer Science + Business Media, 2012), p. 48.

[97] L Römer & T Scheibel, The Elaborate Structure of Spider Silk (Prion 2:4, 154–161, October/November/December 2008). Available online: http://www.ncbi.nlm.nih.gov/pmc/articles/PMC2658765/pdf/prion0204_0154.pdf

[98] J. M. Gosline, P. A. Guerette, C. S. Ortlepp & K. N. Savage, The Mechanical Design of Spider Silks: From Fibroin Sequence to Mechanical Function (The Journal of Experimental Biology 202, 3295–3303 (1999)), p. 3302.

explores "the limits of acceptable safety improvements, by embedding a bulletproof matrix of spider-silk in human skin".[99] Essaïdi imagined replacing keratin, the protein responsible for the toughness of human skin, with spider silk protein, to create a bulletproof human being. By implanting spider silk produced by transgenic goats in human skin, Essaïdi enhanced its properties towards the point where it might be capable of protecting the agent from the weight and velocity of a .22 calibre Long Rifle bullet (the 2.6 g 329 m/s of the project title, analogous to the capacity of a Type 1 bulletproof vest).[100] The experiment has been hailed as the first in vitro-generated bulletproof skin prototype, though the corresponding tests carried out at the Forensic Genomics Consortium Netherlands/Dutch Forensic Institute used less gunpowder and only half the speed of a normal bullet.[101]

This transgenic approach – genetic engineering between organisms – may sound like science fiction, but is actually hard science: a result of recent developments in biotechnology that have tempted researchers to transfer spider silk genes to different recombinant protein production hosts such as the bacterium *Escherichia coli*. Alternatives to bacterial expression have also been tested with transgenic bacterial, yeast, plant, insect and mammalian cells. Spider silk fragments have been produced by baby hamsters, bovine species, goats, and silkworms, as well as plants such as tobacco and potato.[102] Biosupramolecular chemistry has been used to fabricate silk using microfluidic approaches, which brings us even further towards the large-scale production of recombinant spider silk proteins needed for the construction of a city like Arachnia.[103] In October 2012, one company made official news that they have achieved a large-scale production validation after a technical test campaign for the production of recombinant spider silk proteins – potentially the beginning of artificial spider silk production at the industrial scale.[104]

Science writer Philip Ball presciently proposed as early as 2002 a fantastic-sounding scenario in which "a team of engineers arrives to build a bridge across a gorge, but they have no lorries filled with materials. Instead, they eat a hearty breakfast and then begin to pull sticky cables out of their bottoms. Leaping from one side of the chasm to the other, they create a criss-crossing web of cables, each one stronger than steel. If they make a mistake, putting a cable in the wrong place,

[99]http://jalilaessaidi.com

[100]The project is described in detail, with essays by 14 authors, in Jalila Essaïdi (ed.), Bulletproof skin, Exploring Boundaries by Piercing Barriers (ISBN: 978-90-819957-0-2).

[101]Ibid., p. 95.

[102]Martin Humenik, Andrew M. Smith, Thomas Scheibel, Recombinant Spider Silks—Biopolymers with Potential for Future Applications (Polymers 2011, 3, 640–661; doi:10.3390/polym3010640), pp. 645–646.

[103]European Science Foundation (2008, February 25). Bacteria Can Be Made To Spin Spider Silk Through Understanding Of Big Molecules. ScienceDaily. Available online: http://www.sciencedaily.com/releases/2008/02/080221100539.htm

[104]AMSilk Achieves Large-scale Production Validation for Spidersilk® and Hires New CBO. (AMSilk, press release, 30 October 2012) Available online: http://www.amsilk.com/en/news/article-view/article/amsilk-achieves-large-scale-production-validation.html

they simply eat it up and start again." He then proceeded to note that "because we are not spiders (and not Spiderman either), we'll never be able to build bridges this way."[105]

While this is rather likely to be true for the foreseeable future (though Jalila Essaïdi's embedding of spider silk in human skin inevitably raises the question of the possibility of genetic couplings of spider spinnerets – the tube through which the spider rapidly pulls a soft protein/water mixture into a long silk fibre – with the human engineers Ball describe), the jury is out on whether in the near future, despite not being spiders, we might be able to build structures at the architectural scale using renewable spider silk protein materials.

The formal language, mechanical properties and construction process of spiderwebs have of course inspired artists, architects and engineers at least from the cable-net roofs of Frei Otto's pavilions for the 1972 Olympic Games in Munich, which were clearly inspired by the web of the *Cyrtophora citricola* (the Tropica Tent-Web Spider),[106] through to Neri Oxman proposing in 2012 that "in a way the spider is a kind of multi-material 3D-printer. The spider itself is a kind of printing machine, only instead of printing plastics, it prints with silk".[107] In this context, it would be ignorant at best not to mention Tomás Saraceno's fabulous project *14 Billions (Working Title)*, a work of art that used 36,400 meters of thin nylon together with 210 m of tape and 14,000 hooks to create a three-dimensional model of a laser-scanned spiderweb at the Bonniers Konsthall in Stockholm in 2010.[108] In 2011/2012 the same artist increased the scale even further with his *Cloud City* installation at Berlin's Hamburger Bahnhof, and yet again in 2012/2013 with *On Space Time Foam* at Hangar Bicocca di Milano.[109]

In our vision for the invincible city of Arachnia, the urban construction is not carried out by either tobacco plants or goats, but by actual spiders, genetically modified and reprogrammed to follow instructions. It is interesting to note that very few attempts seem to have been made to farm spiders, apparently mainly due to the territorialism and cannibalism among them.[110] While it is not yet possible to modify spiders in this way, it doesn't take a great leap of the imagination to

[105]Philip Ball, Smart Stuff – A booklet to accompany The Royal Institution Christmas Lectures 2002. Lecture 1: The Spider That Spun a Suspension Bridge. Available online: http://www.philipball.co.uk/index.php?option=com_content&view=article&id=81:smart-stuff-ri-christmas-lectures&catid=18:materials&Itemid=20

[106]Philip Steadman, The Evolution of Designs: Biological Analogy in Architecture and the Applied Arts (Abingdon: Taylor & Francis, 1979), p. 261.

[107]Neri Oxman, Printing 3D Buildings: Five tenets of a new kind of architecture, 30-min profile on CNN's The Next List, 7 December 2012. An excerpt is available online: http://whatsnext.blogs.cnn.com/2012/12/07/printing-3d-buildings-five-tenets-of-a-new-kind-of-architecture/

[108]Tomás Saraceno, 14 Billions (Working Title) (Milan: Skira Editore S.p.A., 2011).

[109]Cf. http://www.tomassaraceno.com

[110]Adam Rutherford, Synthetic biology and the rise of the 'spider-goats' (interview with professor Randolph Lewis, The Observer, Saturday 14 January 2012). Available online: http://www.guardian.co.uk/science/2012/jan/14/synthetic-biology-spider-goat-genetics

envisage a future in which it will become feasible: just combine the famous images of trees cocooned in spiderweb following the widespread and prolonged flooding in Pakistan in 2010[111] and the spiderweb blanketing of a field following flash floods in Australia's Wagga Wagga in 2012,[112] with the latest progress in genetic programming as carried out by researchers such as Christopher Voigt[113] – would it be too far a stretch to think that it will become possible to program not only bacteria, but also spiders into construction agents? (Fig. 8).

This would be the foundation for a new breed of buildings and cities that are proposed as examples of adaptive architecture for earthquake-prone areas. In organisational and architectural terms, one of the most interesting consequences of building with spiderweb is the formal and structural possibilities inherent in a *suspended* design that does away with any notion of ground datum. This *elastic city* would essentially change its internal structure and behaviour in order to allow an earthquake to dislocate the individual living units in a semi-controlled way, without compromising the structural integrity of the tension members. While cities are normally envisaged as two-dimensional layouts aligning with the ground topology, Arachnia can be organised three-dimensionally, with buildings floating on top of each other in a manner similar to that of Yona Friedman's famous *Spatial City*,[114] opening up for a plethora of interesting design challenges that we plan to return to in future papers.

The spider silk buildings of Arachnia would be true marvels of design, biotectonic replications of the idea of structural and material symbiosis evidenced in the near-perfect relationship between the spider's web and the spider silk it is spun from. Enlarge a spider to the size of a human being, and it wouldn't be uncommon for the web it weaves to be in the region of 150 m high. Of particular interest are the silks from spiders that produce orb webs which are used to catch aerial prey. The silk in these webs needs to be capable of capturing and holding the spider's flying prey, which requires interplay of various silks with different properties. Female orb-weaving spiders can produce up to seven different silks that have a range of properties.[115] Now imagine normally-sized but reprogrammed spiders working at the same scale, tirelessly using each silk-weaving gland at a time to create flagelliform, tubuliform, aggregate, piriform, aciniform, auxiliary spiral

[111]Russell Watkins in National Geographic, Pictures: Trees Cocooned in Webs After Flood. 7 March 2012. Available online: http://news.nationalgeographic.com/news/2011/03/pictures/110331-pakistan-flood-spider-trees-webs/

[112]Daniel Munoz (Reuters) in National Geographic, Spiderwebs Blanket Countryside After Australian Floods (Pictures). Available online: http://news.nationalgeographic.com/news/2012/03/pictures/120307-spiderwebs-australia-floods-wagga-wagga-world-science/

[113]Tamsir A, Tabor J J, Voigt C A, Robust multicellular computing using genetically encoded NOR gates and chemical 'wires' (Nature 469, 212–215 (13 January 2011) doi:10.1038/nature09565).

[114]Yona Friedman, Mobile Architecture: 10 Principles of Spatial Urbanism (Principles Mobile Architecture, 1959).

[115]Martin Humenik, Andrew M. Smith, Thomas Scheibel, ibid., p. 641.

Fig. 8 Genetically modified spiders are reprogrammed to become the construction agents that build the invincible city of Arachnia for us: the spider becomes a biological spinning machine

and dragline motifs, overlaying them into walls and floors and ceilings as per the architects' models and instructions (Fig. 9).

As are all our invincible cities, this is a project in progress: planned future work includes building a scale model of suspended living units and having different

Fig. 9 The spiders would work at the urban scale, weaving flagelliform, tubuliform, aggregate, piriform, aciniform, auxiliary spiral and dragline motifs around the city's living units

species of spiders spin webs around them. This will of course be carried out in collaboration with arachnologists, and will form the basis for the production of a digital test environment in which we can program spider agents to weave patterns according to given instructions. Having a second spider construct a second web on top of an initial web spun by a different spider – experiments already begun by Tomás Saraceno with *Cyrtophora* and *Tegenaria* species[116] – would also be an interesting starting point for future development.

10 Beyond Hylomorphism: Sensational Architecture

If for some reason we can't find a laboratory in which to carry out this reasearch, we can always set up shop in a garage. The most famous garage in the world is probably located at 11161 Crist Drive, Los Altos, California. It sits next to a house in the spare bedroom of which an exuberant start-up duo built their fifty first Apple 1 computers in 1976, before running out of space. A few months later, Steve Jobs and Steve Wozniak moved their combined headquarters and manufacturing plant into the garage next door, which belonged to Jobs's parents.[117]

The garage turned into the eye of the storm that was to become the microcomputer revolution. In the late 1970s and early 1980s, predictions had it that home computers would revolutionise daily life through "kitchen computer" databases, disk-based encyclopedias, interactive television, and smart home technology.[118] However, the computers were not powerful enough to support the revolution, and so it took another decade for the technology to mature. This delay mirrored that of other technologies that have been introduced to an unprepared public: early motorists were shouted at to "Get a horse!," and television didn't find its way out of the research lab for decades before regular public broadcasts begun.

Today, the biohacking culture growing within DIY synthetic biology circles seems to be at the same entrepreneurial level that spurred on those early pioneers in their quest to build the first home computers, an industry "still nascent, but about to explode".[119] It has been pointed out that biohacking has the same momentum as the mid-1970s computer scene, with a grassroots biohacking community developing rapidly and transforming biology from a purely professional activity to something hobbyists, artists, and architects can do in their garages and studios. A combination of the development of standard biological parts (genetic sequences that enable a

[116]Conversations between Tomás Saraceno (TS), Marion Ackermann (MA), Daniel Birnbaum (DB), Hans Ulrich Obrist (HUO) and Udo Kittelmann (UK) in Tomás Saraceno. Cloud Cities, DISTANZ Verlag Berlin, 2011, 256 Seiten, deu/eng, etwa 180 Farbabbildungen, ISBN: 978-3-942405-37-9.

[117]The address changed to 2066 when the land was annexed from the county to the city in late 1983. The coordinates are N37°20′, W122°04′.

[118]For an account of the microcomputer revolution, cf. Roy A Allan, A History of the Personal Computer: The People and the Technology (London, Ontario, Canada: Allan Publishing, 2001). Available online: http://books.google.ca/books?id=FLabRYnGrOcC&pg=PA14&dq=history+of+the+microcomputer+revolution&lr=#v=onepage&q=history%20of%20the%20microcomputer%20revolution&f=false

[119]Mike Loukides, Biohacking: The next great wave of innovation, O'Reilly Radar (website article, 10 October, 2012). Available online: http://radar.oreilly.com/2012/10/biohacking.html

cell to perform certain standardised tasks, allowing for an abstract programming at a higher level than DNA, the equivalent of binary code) and lowering prices turn the synbio revolution into an echo of its microcomputer precedenct, with equally ambitious startups beginning to see the light of day.[120]

Just as the next great wave of commercial start-up entrepreneurs are likely to be biologists rather than programmers, we remain convinced that the most interesting aspects of our contemporary architectural discourse have less to do with traditional architecture than with biology. Architecture takes its cues from prevailing scientific and cultural movements, and at the moment, we're working within a veritable explosion of new biology. This naturally shapes today's (or at least tomorrow's) architecture much as the explosion of data and computers shaped architecture in the era following the microcomputer revolution. However, the reason why there is less risk than ever for this revolution to get delayed is exactly the efforts of trailblazers such as Jobs and Wozniak: we now have the necessary infrastructure, computing power and means of communication at hand to facilitate a radical worldwide exchange of information.

French architect Le Corbusier defined his houses as machines for living in.[121] Futurist inventor Ray Kurzweil named one of his books *The Age of Intelligent Machines*.[122] Leading theorist, author, and founding editor of Wired magazine, Kevin Kelly, pronounced the future of machines to be biology.[123] Once we learn how to create novel materials from computer-controlled materiomic processes, architects can start designing *intelligent biological machines for living in.* Habitable biological machines – organisms – constructed perhaps from biomaterials that are assembled in accordance with chemical signal inputs received by microbe agents from a human designer; a spectacular design strategy falling somewhere between *in silico* and *in vivo* – sculpting with biological/binary code in real time on the microbial scale.

While futurist inventor Ray Kurzweil has noted that utilising full-scale nanoengineering to construct macroscale objects at the molecular scale is "still considered a middle to late 2020s technology",[124] that is no reason not to begin designing these new biological buildings/machines, the materiomic architecture facilitated by a globally distributed research effort into new areas of biology. Scientists have already managed to create basic circuit components in the bacterium *Escherichia coli,* and are now "trying to make the cell understand where it is and what it should

[120] Ibid.

[121] Corbusier L (1927) Towards a new architecture. Architectural Press, London, p. 10 (French original Vers Une Architecture, 1923).

[122] Kurzweil R (1992) The age of intelligent machines. The MIT Press, Boston.

[123] Kelly K (1995) Out of control – the new biology of machines, social systems, and the economic world. Basic Books, New York. http://www.kk.org/outofcontrol/contents.php

[124] Kurzweil R (2010) How my predictions are fairing. http://www.kurzweilai.net/how-my-predictions-are-faring-an-update-by-ray-kurzweil

be doing based on its understanding of the world,"[125] a breakthrough that would allow them to be programmed on an abstract level to perform complex, coordinated tasks such as forming and solidifying a particular shape inside of an aggregate mass.

The implication for architecture is breathtaking. Once the materials we use become biologically programmable, the very nature of the discipline is bound to change irrevocably: while interdisciplinary research has already begun to yield *metamaterials* (artificial materials engineered to have properties that go beyond those found in nature), the use of computational processes to control cellular compositions gives us the power to precisely configure the built environment, vastly expanding our capacity to use, for instance, programmable bacteria as building elements, or programmable spiders as construction workers. Soon this new generation of synthetic biologists will be able to effortlessly program cells, and when that happens, a whole new realm of architecture opens up, based on "the design rules by which simple logic can be harnessed to produce diverse and complex calculations by rewiring communication between cells".[126]

Every major architectural revolution is also a material revolution. Ever since the first Stone Age man or woman took shelter in a cave, architecture has advanced through instances of radical adaption to new circumstances through the adoption of new materials, or vice versa.[127] The breakthrough of early modernism might suffice as an obvious example: when glass, iron, steel, and reinforced concrete were first being explored in the nineteenth century, they had not been historically considered "architectural" materials. There were no precedents for these new materials, and that excited the modernists: breaking with architectural tradition "facilitated the uninhibited exploration of the new materials and methods".[128] We imagine they had the same giddy feeling of getting ready to explore territories for which there are no maps as we enjoy at the present moment in time.

What are the philosophical implications of this material revolution? In a stunningly succinct reading, Brian Massumi uses the example of a woodworker choosing the right piece of wood when making a table to explicate Deleuze and Guattari's ideas of signs being qualities that envelop potential capacities to be affected or to affect (release a force). "The presence of the sign is a contraction of time. It is simultaneously an indicator of a future potential and a symptom of a past. (...) The wood's individual and phylogenetic past exists as traces in the grain, and its future as qualities to be exploited. On a first, tentative level, meaning

---

[125]Christopher Voigt quoted in Katherine Bourzac, Software for Programming Microbes – A simpler way to modify microbes could help produce biofuels and drugs efficiently (article in MIT Technology Review, January 5, 2011). Available online: http://www.technologyreview.com/news/422283/software-for-programming-microbes/

[126]Tamsir A, Tabor JJ, Voigt CA (2011) Robust multicellular computing using genetically encoded NOR gates and chemical 'wires'. Nature 469(7329):212–215. doi: 10.1038/nature09565.

[127]A moment described by Umberto Eco in Function and Sign: The Semiotics of Architecture, in Geoffrey Broadbent, Richard Bunt & Charles Jencks (eds.), Signs, Symbols and Architecture (Chichester: John Wiley, 1980), pp. 12–13.

[128]Brent C. Brolin, The Failure of Modern Architecture (London: Studio Vista, 1976), p. 14.

is precisely that: a network of enveloped material processes."[129] Massumi then goes on to modify a quote from Deleuze: "A thing has as many meanings as there are forces capable of seizing it."[130]

Fittingly, the term Aristotle used for matter – *hyle* – also means wood (it is also occasionally used for other materials). In the *Metaphysics*, Aristotle understands substance as a compound of form and matter. This is his theory of hylomorphism, in which change is analysed as a material transformation: matter is what undergoes a change of form.[131]

As Katie Lloyd Thomas has pointed out, Aristotle is careful to exemplify his theory with a plastic material – bronze – that supports his argument, and argues that the hylomorphic concept of matter is "not as easily conflated with materials as it would appear".[132] Despite this, she notes, "architecture has this concept at its base and sets up a discourse in which form must be realized in matter, with the material being seen as merely interchangeable – just one instance of matter rather than another".[133]

This is precisely the attitude we set out to challenge with the present project. While the hylomorphic position "understands materials as a subset of matter,"[134] we are interested in redefining matter itself as no longer being inert – that which is given form – but active and generative, that which gives form. We want to embrace that contraction of time, that potential capacity of the material in Massumi's discussion above, and turn it into the point of departure for speculative excursions into the depths of material practices in architecture, investigations into the way that "the intelligence embedded in material itself becomes an active part of the way we generate our work".[135]

Three aspects of this positioning are particularly interesting to us. Firstly, in philosophical terms, this reassessment of the role materiality plays in architecture is an example of our determined interest in exploring the latter part of the provocative Deleuzian distinction between the realisation of the possible and the actualisation of the virtual (and our complete indifference towards the former).[136] The material and the idea are part of a continuum of potentiality and actualisation, and with our

---

[129]Brian Massumi, A User's Guide to Capitalism and Schizophrenia – Deviations from Deleuze and Guattari (Cambridge, MA: Swerve Editinos/The MIT Press, 1992), p. 10.

[130]Ibid. The original quote is "A thing has as many senses as there are forces capable of taking possession of it". Gilles Deleuze, Nietzsche and Philosophy (London, Continuum, 2006), p. 4.

[131]Aristotle, Metaphysics (London: Penguin, 1988).

[132]Katie Lloyd Thomas, Introduction: Architecture and Material Practice, in Katie Lloyd Thomas (ed.), Architecture and Material Practice (Abingdon: Routledge, 2007), p. 3.

[133]Ibid.

[134]Ibid.

[135]Jesse Reiser & Nanako Umemoto, Material Praxis, in Bernard Tschumi & Irene Cheng (eds.), The State of Architecture at the Beginning of the 21st Century (New York: The Monacelli Press, 2003), p. 34.

[136]Michael Speaks, Folding toward a New Architecture, in Bernard Cache, Earth Moves: The Furnishing of Territories (Cambridge, MA: The MIT Press, 1995), pp. xiv–xv.

increasing use of CNC milling and other material/computational technologies that close the gap between conceptualisation and manufacturing, this continuum is a more prevalent and useful model than ever.

Secondly, as should be clear from our contextualisation of the three Invincible Cities presented above, an environmental agenda is also at the forefront of our interest in materiality. To again fall back on Massumi's example, we now enjoy an unprecedented presence of signs in our materials – not only can we trace their phylogenetic past, but their entire environmental lineage, incredibly detailed knowledge of where they come from, how they have been fabricated and transported, by whom and under what economic and social conditions. We have an unparalleled responsibility to allow this increased knowledge influence the choices we make as designers of buildings and cities.

However, such actions remain in the realm of material-as-product, and it is our firm belief that more interesting results await those who dare to break with that tradition, which brings us back full circle to recent advancements in synthetic biology and our third main interest in the materiality of architecture and urbanism: its affects. While it is fairly easy to envisage a near future in which consilient design teams take control of the production of the materials that go into buildings and cities, and specify their behaviours using progress within scientific fields such as synthetic biology, it is harder to predict what forces our interest in materiality might release.

As Paola Antonelli has pointed out, in the 1960s, the plastic object "became a political symbol – serially produced, uniformly inexpensive, and available to all social classes equally".[137] The material revolution outlined above will inevitably give rise to a wide range of affects: conceptually, poetically, experientially, phenomenologically, intellectually, environmentally, socially, economically, and politically, it will make the world a different place. At the end of the day, we are interested not so much in the resulting design object (a building, a city) than in such affects, and, eventually, the sensations inside of the subjects that are exposed to objects brought about by these material potentials, manipulations, and actualisations – the "intimate perception of things".[138]

Beyond hylomorphism, we might find such a sensational, or sense-ational (as in both "exciting" and "haptic" or "pertaining to the senses") architecture, a future tradition less curious about the high-tech performance qualities so seductive to engineers of novel materials that possess exceptional properties of tensility, durability, flexibility, or lightness than about the possibilities of dramatic applications of materiomic structures to shape our everyday lives in the Invincible Cities of the future.

---

[137]Paola Antonelli, Mutant Materials in Contemporary Design. Electronic project text accompanying the 1995 Mutant Materials in Contemporary Design exhibition at the Museum of Modern Art in New York. Available online: http://www.moma.org/interactives/exhibitions/1995/mutantmaterials/MuMA1.html

[138]Luigi Alini, The Warp and Weft of Architecture: Weaving, Joining, Overlapping, Folding, in Luigi Alini, Kengo Kuma – Works and Projects (Milan: Electa Architecture, 2005), p. 8.

Acknowledgements The authors wish to thank Stamatina Th. Rassia for her angelic patience while waiting for this chapter to get finished. Also Jack Munro for his excellent assistance, Thidaa Roberts for unwavering support way beyond any call of duty, and Madelaine Levy for delivering a baby boy while her partner tried his best to return to Earth from a mental space filled with thoughts about post-sustainable urbanism in the Materiomic Age.

Qualitative Affects of Building Life Cycle: The Formation of Architectural Matter

Robert Stuart-Smith

Abstract Architectural design essentially organises matter as built form. Designers would therefore benefit from taking a more active approach to material formation; where matter is not perceived as inert, but instrumentalised for numerous designed agencies (Stuart-Smith 2011). The incorporation of generative design principles within pragmatic aspects of architectural production and consumption offers an alternative to engineered reductions in architectural expression often promoted for the sake of design efficiencies, and suggests an expansion of the domain that architectural design operates within. The Architectural Association's Design Research Laboratory (AADRL) has been exploring the design of qualitative architectural affects through the rethinking of building life cycles as a design opportunity. Beyond quantifiable methods of building life cycle analysis, new strategies are emerging that challenge our assumptions about building life cycles; redefining relationships of matter and energy within aspects of architectural production.

1 Material and Energy Flows

Godfrey Reggio's film – *Koyaanisqatsi* impacted a generation through its portrayal of the temporal cycles that operated in the flow of people, cars, cities and the weather; an environment constantly in flux. Filmed using time-lapse and high-speed photography, the world was revealed in another time scale – where flows of energy and matter became more apparent (Reggio et al. 2002). Cities and their architecture; are essentially material orders that operate at numerous physical scales of various durations. The built environment is constituted through the flow of materials and energy. As early as 1923 the architect Erich Mendelsohn suggested that

R. Stuart-Smith (✉)
Kokkugia, London, UK
e-mail: rob@kokkugia.com

S.Th. Rassia and P.M. Pardalos (eds.), *Cities for Smart Environmental and Energy Futures*, Energy Systems, DOI 10.1007/978-3-642-37661-0_5,

contemporary concepts in the physics of his time enabled energy and matter to be practically interchangeable, compelling the engineer to "abandon their mechanical theory of dead matter.... and from primal states deduce the laws determining interactions..." (Mendelsohn 1923). Mendelsohn's comments reach further than his own design agenda or time, towards the materialist philosophical discourse of Deleuze and Guattari. Deleuze and Guattari's concept of *Matter Form* discussed in *A Thousand Plateaus: Capitalism & Schizophrenia*, suggested that form should not be understood as an ideal that imposes itself on matter (*hylomorphic*), but rather arises through the interactions of its matter; as an affect of "matter-relations" (Deleuze and Guattari 1988). Architectural design is mostly undertaken within the clinical confines of the office environment, distanced far from the material concerns of architectural production or its environmental impacts, yet it essentially proposes an organisation of built material, and therefore can benefit greatly in taking a more active approach to material formation, where matter is not perceived as inert, but instrumentalised for numerous designed agencies (Stuart-Smith 2011).

2 The Organisation of Matter

Research into climate change although still debated within the scientific community, has raised global public awareness of the environmental impact of human civilization, and posed new ethical demands on the way architecture is designed and inhabited. The building industry's attitude towards energy and material resources is also changing, with increasing economic incentive to undertake more environmentally "sustainable" projects (Bergman 2012). Commonly available building products are likely to keep pace with this conceptual shift in the industry through the amelioration of their environmental and economic performance, yet much opportunity lies in rethinking the entire material order of building. Hannes Meyer's insistence that there is a one-way route for architecture to be assembled from the products available (Meyer 1928) no longer holds true, as numerous materialization alternatives are beginning to appear and expand the possible trajectories various architects may pursue their projects within. Bespoke multi-material assemblages may be specifically fabricated to designers' intentions effortlessly through means such as the polyjet 3d printers manufactured by Objet (Objet 2011). These technologies enable design decisions to be implemented at microscopic scales due to the precision of the equipment that not only affords design freedoms in form, shape and complexity; but also in material properties. The Objet machines are able to heterogeneously mix materials with a range of different properties in order to fabricate products with differentiated attributes of rigidity, elasticity and translucency.[1] It is easy to imagine similar possibilities to arrive soon at

[1]Objet's 260 Connex 3d printer can include up to 14 different materials within a single printed part with a resolution as accurate as 16 μm in dimension (Objet 2011).

nano-scales. While ethically controversial, bio-customisation technologies such as those imagined by Rayfish for the fabrication of custom designed shoes from the skins of transgenic stingrays suggest even more extreme possibilities of a bio-engineered material future (Rayfish 2012).

Adopting new technologies does not require amnesia from successful principles utilised in the past. Vernacular architecture has often managed to provide environmental performance for a minimal expense through innovative matter-form arrangements. In contrast to the contemporary reliance on appendages such as air conditioning or heating systems, the environmental performance of structures such as the igloo are intrinsic to their material organisation. The interior of the igloo maintains a constant temperature of 10–15 °C while the outdoor temperature may be as low as −20 °C (Ikaga et al. 2008). Constructed from the same frozen material as its environment, the improvement in ambient temperature within the igloo is a direct consequence of its specific organisation of matter; redistributing the found material produces a quantitative difference in temperature. Rethinking the building life cycle by designing specific relationships of matter and energy can enable negotiations to take place within the design and materialisation process that satisfies both humanist and environmental agendas, and does so efficiently and economically. This can be achieved through qualitative design affects that also contain pragmatic and performative utility.

Architecture is both a product and a consumable, however it is also capable of playing a more integrated role within a broader production process. Principles such as "Cradle-to Cradle"[2] (Braungart and McDonough 2002) have been effective in raising awareness to the materials and chemicals that constitute architecture via the utilisation of quantifiable methods of building life cycle analysis (Banfill et al. 2009). However, new strategies are emerging that develop knowledge in rethinking the building life cycle as a design opportunity via explorations into the relationships of matter and energy within aspects of architectural production. A contemporary concern for environmental performance has characteristically focused on the engineering of the construction and operation of buildings, leaving open opportunities for further exploration of more qualitative aspects of architectural design. Pragmatic processes such as the fabrication, transport, assembly, construction, demolition or the landfill of buildings can be utilised as creative participants within the conception of architecture. Designing qualitative affects of building life cycle requires an interrogation into the material formation of architecture and its temporal life cycle.

[2]Michael Braungar utilizes the term "Cradle to cradle" to describe design solutions that consider the full life cycle of their constituent products and how they these continue to be of value and harmless to people and the environment throughout their cycles.

3 Re-conceptualising Architectural Lifespan

Architectural education and practice traditionally emphasizes only a short moment of a building's materialisation and raises little concern for the on-going lifespan of buildings. Questions a student architect may receive during a review of their academic project tend to revolve around the immediate use, construction or fabrication of their proposal. Academic references to buildings also inevitably focus on seminal historical examples that have endured, and this inadvertently misrepresents the fact that the average life span of the world's buildings is comparatively short, with some cities operating under extremely brief time cycles. China's construction industry accounts for more than 40 % of the world's annual concrete and steel consumption and produces residential developments with an average life expectancy of 25–30 years (Wang 2010). While this short lifespan is largely driven by construction profits, in Japan a similar lifespan is fostered by a culture that places high monetary value on products that are new, resulting in fast economic depreciation of ageing building stock (JS International 2011). These high rates of consummation inevitably lead to much waste and loss of design opportunity due to the building industry's lack of consideration for the short lifespan of its products. Embracing and designing for short lifespans offers enormous potentials for improvements to the built environment.

Environmental design agendas are starting to gain traction in architectural practice, where some clients see immediate and long-term economic value in reducing a building's energy footprint (Building's consume 39 % of world energy, 72 % of all electricity and 40 % of the world's raw materials (Bergman 2012)). Reducing a building's energy and material needs can generate significant reductions in the initial cost of building due to reductions in the amount of required mechanical equipment, insulation, and structure (Bergman 2012). Despite this, little design effort in developed cities goes into broader issues of how buildings may design material organisations more strategically, or extend, demolish, re-use, or age building stock over time. It seems that the materialist aspect of architecture has been ignored and an emphasis on the inert properties of material are promoted through a misconception that architecture endures, and its economic value rests predominantly in the instantiation of standardised material assemblages.

Programming temporalities into the architectural life cycle could add-value to the built environment. However, it requires a different approach to material organisation that strategises time-based transformations. It necessitates the development of working methods not currently widespread in the building industry. Industry standard software or building products do not accommodate lower-level decisions of material order to be revisited so easily. Our practice; Kokkugia (Snooks et al. 2004) continuously engages in research to develop custom software, design, and fabrication methodologies to deal with these issues.

4 Designing Aspects of Building Life Cycle

At the Architectural Association School of Architecture's Design Research Laboratory (AA.DRL), in the design studio Behavioural Matter (AADRL 2012), we have been exploring qualitative aspects of building life-cycle through the development of non-linear processes of material organisation that engage with environmental ethics by design.[3] Through the writing of custom software and the making of large-scale material experiments and prototypes, we have been able to design material organisations that speculate on interesting architectural design opportunities. The research has involved designing within areas of the building process that often rest outside of the traditional domain of architecture. Each project investigates how aspects of building life cycle can contribute to a generative design process. The projects have focused on issues such as fabrication, assembly, temporality and material placement within a designed life cycle. Rather than addressing these topics as solely pragmatic problems, these have been investigated as design opportunities where they play an instrumental role in the formation of the architectural design product.

4.1 Fabrication and Assembly

Analysis of current commercial methods of constructing complex curvature concrete resulted in a research project that aimed to minimise the waste of form work within precast concrete through the implementation of a procedural computational process whereby a cast element became the formwork for the next cast element (Fig. 1). Algorithmically controlling the precise time and location of concrete pouring via a CNC machine, enabled a wide variety of complex curvature possibilities to be developed that embedded the history of their making as a design affect within the result (Fig. 2). The local fluctuation in surface character of the casts propagates to become larger spatial affects further down the fabrication line. The project *CastonCast*[4] was awarded a Holchim Prize for its inventive way of reducing fabrication waste whilst offering innovative design possibilities (Holchim Awards 2012).

---

[3]The AADRL is an abbreviation for the Design Research Laboratory. A post-professional Masters program in Architecture and Urbanism at the Architectural Association School of Architecture in London. The program runs for 1.5 academic years and the international body of students undertake their research project in teams within a design studio. Robert Stuart-Smith runs a design studio in the program titled "Behavioural Matter".

[4]AA.DRL project *CastonCast*. Students: Povilas Cepaitis, Lluis Enrique, Diego Ordonez, Carlos Piles. Supervisors: Yusuke Obuchi, Robert Stuart-Smith.

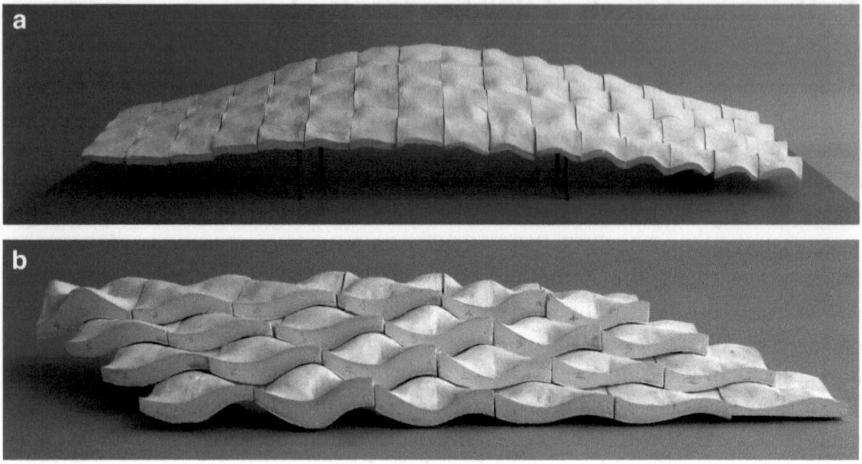

Fig. 1 (**a**) Precast elements form a complex double curvature surface with no waste of formwork (**b**) Precast elements are utilised as formwork for the next pour (AADRL research project CastonCast. Students: Povilas Cepaitis, Lluis Enrique, Diego Ordonez, Carlos Piles. Supervisors: Yusuke Obuchi, Robert Stuart-Smith)

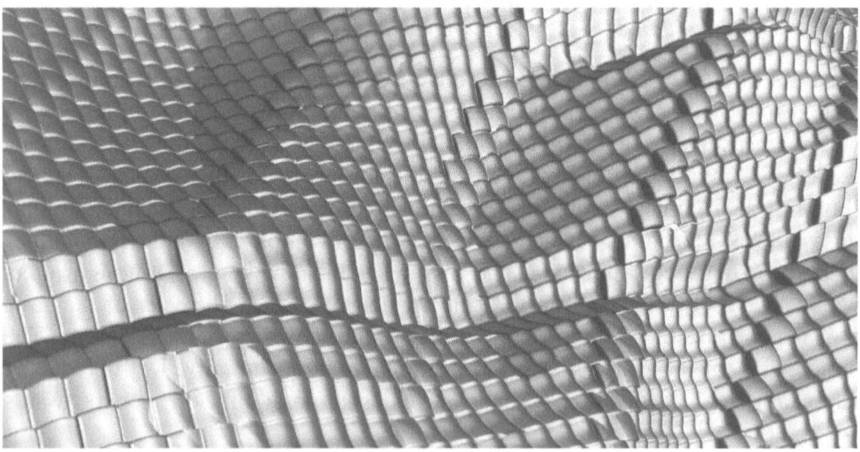

Fig. 2 The history of making is visually registered as a qualitative affect in the resultant assemblage of casts panels (AADRL research project CastonCast. Students: Povilas Cepaitis, Lluis Enrique, Diego Ordonez, Carlos Piles. Supervisors: Yusuke Obuchi, Robert Stuart-Smith)

4.2 Temporal Cycles

A series of projects have investigated the incorporation of temporal cycles within architecture that undergo qualitative changes in their designed affects whilst also offering strategic pragmatic advantages. A speculative mushroom farm was

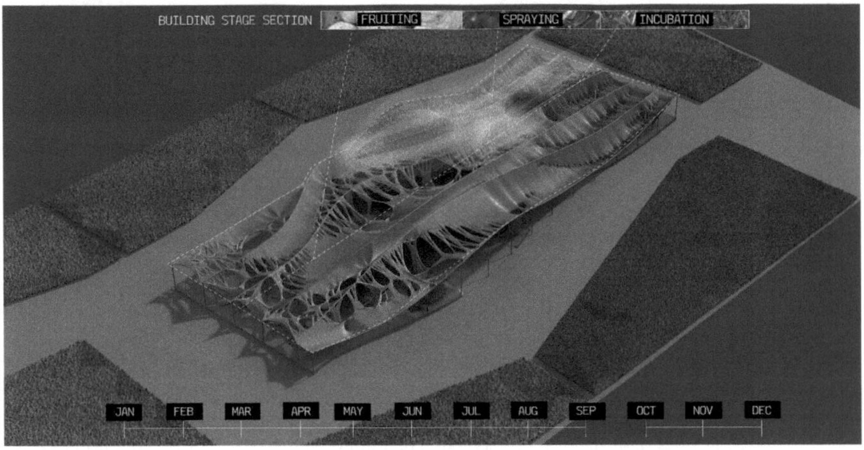

Fig. 3 The mushroom farm complex is constructed with mycelia impregnated rice-straw. The building is designed to produce mushrooms and decompose in a staged sequence annually (AA.DRL project MicroFarmX: Xin Guo, Walee Phiriyaphongsak, Bo Thammwiset, JunJie Zeng. Supervisors: Yusuke Obuchi, Robert Stuart-Smith)

designed to seasonally produce both mushrooms and its own architecture through the use of the mushroom fungus *Mycelia* as a building material. This *renewable building* project, carefully managed the annual resources of its site; creating a closed-loop cycle between rice productions, farm waste, mushroom production and architecture, which subsequently decomposed to restart the cycle (Fig. 3). It required extensive material testing and algorithmic coding in order to arise at a building material concept that would offer spatial and structural potential by using mycelia as a binding agent within a substrate rice-straw waste material. The resultant proposal was differentiated in time and space, as part of the building was planned to grow, fruit and decompose ahead of other parts in order to make full use of the time and resources of the site.[5]

Similarly a project for a public baths complex, utilised water erosion to control the collection and irrigation of water, and to transform the degree of enclosure of architectural spaces seasonally. Part of the exterior building envelope would gradually perforate to allow for natural ventilation and evaporative cooling in warmer months. The proposal incorporated both an annual and a 10-year building cycle in its strategy of cyclical building and erosion (Fig. 4). This process also created ornamental and material qualitative changes in time that benefited the project beyond its pragmatic intentions. These projects utilised algorithmic and material processes that were developed in the AA.DRL studio, to explore a detailed scale of architectural matter that was organised and harnessed for qualitative design

[5] AA.DRL project *MicroFarmX*: Xin Guo, Walee Phiriyaphongsak, Bo Thammwiset, JunJie Zeng. Supervisors: Yusuke Obuchi, Robert Stuart-Smith.

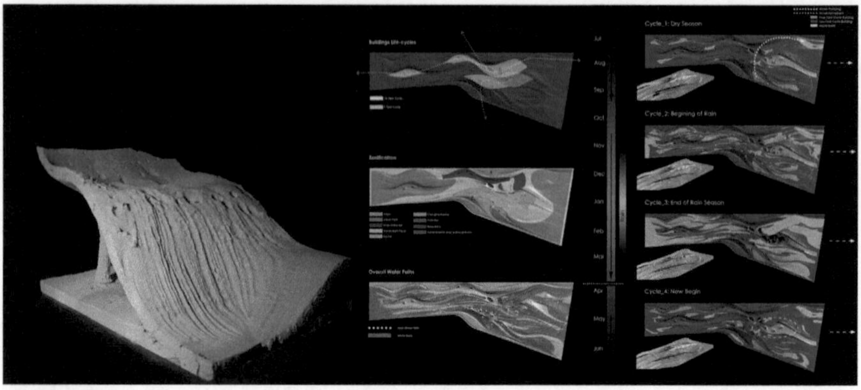

Fig. 4 (**a**) A series of material prototypes were eroded over a period of 6 months to develop control of erosion rates and material affects. (**b**) The public bath consists of two temporal cycles of reconstruction; a small part is rebuilt annually and the majority once every 10 years (AA.DRL project SWAP. Students: Wandy Mulia, Paola Salcedo, Ashwin Shah, Yue Shi. Supervisor: Robert Stuart-Smith. Studio consultants: Knut Brunier, Tyson Hosmer)

Fig. 5 Photograph of 3d printed model of digital erosion simulation. AA.DRL project SWAP. Students: Wandy Mulia, Paola Salcedo, Ashwin Shah, Yue Shi. Supervisor: Robert Stuart-Smith. Studio consultants: Knut Brunier, Tyson Hosmer

affects. Such affects were intrinsic to material formation and transformation in time (Fig. 5). The proposals did not require expensive or high-tech materials but operated through very simple and economical products readily available almost everywhere. In the public bath project, an extensive research into full-scale material prototypes and computational methodologies provided an insight into material erosion that enabled the simulation of multiple designed contingencies. The process of erosion was tested both physically and digitally on specific variations of; material

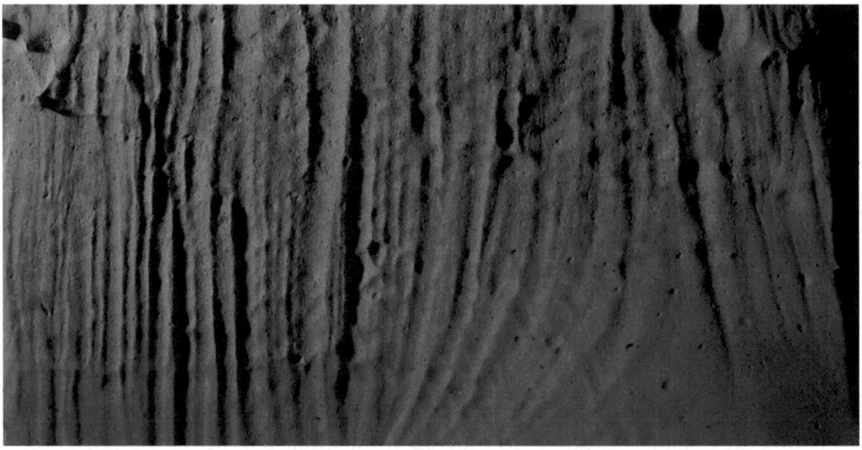

Fig. 6 A detail of a 1:1 scale material erosion test. The prototype was eroded in a custom rain machine for 6 months at the AADRL (AA.DRL project SWAP. Students: Wandy Mulia, Paola Salcedo, Ashwin Shah, Yue Shi. Supervisor: Robert Stuart-Smith. Studio consultants: Knut Brunier, Tyson Hosmer)

hardness, slope and a range of curvatures in order to generate affects that changed larger scale spaces over time (Fig. 6). Micro-scale control generated emergent human scale affects. Such architectural affects are now possible to design due to our increasing ability to simulate complex phenomena and fabricate more precise material products.[6]

4.3 *Algorithmic Material Placement*

Similar performances operate at even smaller scales in the thermal regulation of termite mounds. Curiously, these mounds are constructed without termite architects, through local, event-driven behavioural rules. Each termite has no knowledge of the overall construction they are contributing to. Highly efficient thermal regulation in the mound is the result of a time-based non-linear construction process where termites communicate with one another through local information exchanges made possible through pheromones (Bergman 2012). The bottom-up organisation of material in the mound results in the qualitative affect of air circulation. Environmental performance is intrinsically associated with the precise organisation of built material (There is no additional MEP system!). Scientists Bonabeau and Theraulaz have demonstrated how an algorithmic process can simulate the

[6]AA.DRL project *SWAP*. Students: Wandy Mulia, Paola Salcedo, Ashwin Shah, Yue Shi. Supervisor: Robert Stuart-Smith. Studio consultants: Knut Brunier, Tyson Hosmer.

Fig. 7 (**a**) Structural stress analysis. (**b**) Agent-based material placement (AA.DRL project Softkill. Students: Nicholette Chang, Gilles Retsin, Aaron Silver, Sophia Tang. Supervisor: Robert Stuart-Smith. Studio consultants: Knut Brunier, Tyson Hosmer)

termite's construction behaviour (Bonabeau et al. 2001). Kokkugia's research into self-organisational systems has involved the development of algorithmic design methodologies that utilise similar event-driven behavioural logics. Such behavioural methodologies enable solutions to emerge from often conflicting design criteria in order to strategise the placement of matter at micro-scales within architecture (Snooks and Stuart-Smith 2010). These methodologies are capable of defining specific polyvalences within architecture that allow micro and macro material concerns to be incorporated within the built product of architecture. In the AA.DRL these methodologies have been used to explore a concern for the designed organisation of matter to inform contemporary methods of fabrication such as 3d Additive Manufacturing.

The research project *Softkill*[7] advocated that the algorithmic placement of 3d-printed biodegradable plastic offered superior material efficiencies for prefabricated construction compared to more homogeneous material elements produced by heavy industries such as that of steel (Fig. 7). A sponsorship by the 3D printing company Materialise (Materialise NV 2012) enabled a series of large-scale prototypes to be developed and tested in the AADRL studio. The relatively weak structural properties of PLA were mitigated through the development of custom written Topological Structural Optimisation software[8] and agent-based methods of material placement.[9] These two algorithmic methodologies enabled a strategic porous distribution of fiber-like material to be consolidated only where material stresses were calculated to be significant (Fig. 8). Working at a tolerance of less than

[7]AA.DRL project *Softkill*. Students: Nicholette Chang, Gilles Retsin, Aaron Silver, Sophia Tang. Supervisor: Robert Stuart-Smith. Studio consultants: Knut Brunier, Tyson Hosmer.

[8]Topological structural optimisation attempts to reduce the amount of material and therefore self-weight from a structural element by iteratively undertaking structural stress analysis and removing material from regions that are not under large amounts of stress. By progressively removing material, the final material product becomes substantially lighter and perforated, increasing its strength to weight ratio.

[9]An agent here refers to an autonomous individual entity capable of making its own decisions amongst a larger population of individuals. In Kokkugia's work, this typically refers to a scripted individual who has rules relegating it's behavior relating to design intent. In agent based systems the emphasis is on the collective behavior indirectly controlled from many agents individual decisions Snooks and Stuart-Smith (2004).

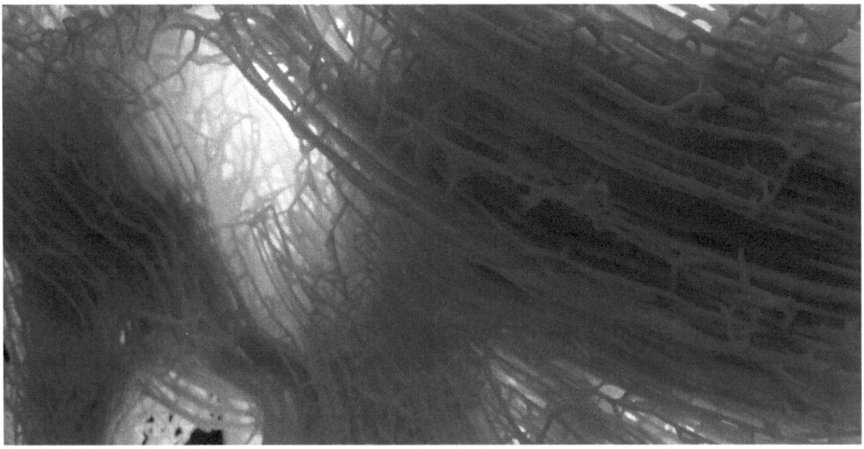

Fig. 8 Agent-based material placement consolitdates in fibers along regions of high structural stress (AA.DRL project Softkill. Students: Nicholette Chang, Gilles Retsin, Aaron Silver, Sophia Tang. Supervisor: Robert Stuart-Smith. Studio consultants: Knut Brunier, Tyson Hosmer)

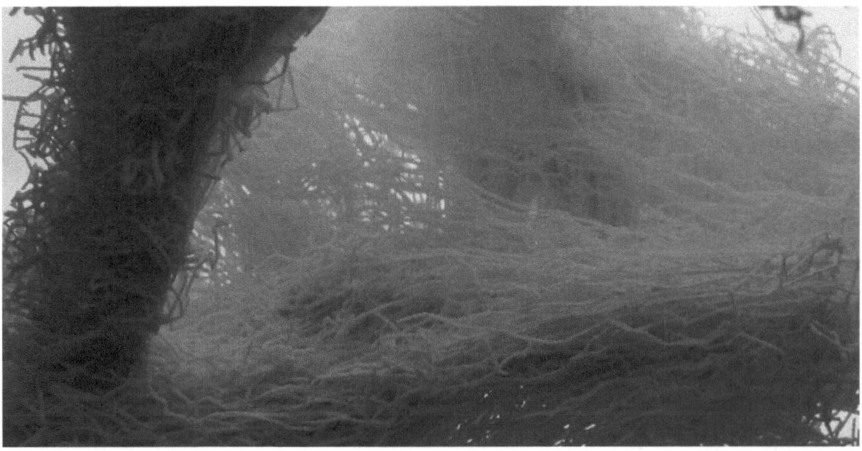

Fig. 9 Agent-based behaviour generates differentiated character and distributions as intrinsic ornamental affect (AA.DRL project Softkill. Students: Nicholette Chang, Gilles Retsin, Aaron Silver, Sophia Tang. Supervisor: Robert Stuart-Smith. Studio consultants: Knut Brunier, Tyson Hosmer)

half a millimeter, the prefabricated PLA parts achieved significant strength to weight ratios whilst also incorporating highly ornamental qualitative affects. These affects were essentially characteristics and tendencies of formation, intrinsic to the same behavioural methods that solved the structural and material concerns of the project (Fig. 9). The project demonstrated how an algorithmic design process of material placement could enable significant reductions in material consumption within the building industry through the use of mass-customised additive manufacturing whilst incorporating more detail and ornament within this same design economy.

5 Conclusion

These research projects involve an active design approach to material formation; where matter is not perceived as inert, but instrumentalised for numerous designed agencies. Architectural design is focused on the organisation of matter as built form. The incorporation of generative design principles within pragmatic aspects of architectural production and consumption offers an alternative to engineered reductions in architectural expression often promoted for the sake of design efficiencies, and suggests an expansion of the domain that architectural design operates within. Beyond quantifiable methods of building life-cycle analysis, these strategies develop qualitative design affects from the rethinking of building life cycles. By redefining relationships of matter and energy within aspects of architectural production such design processes can offer an enriched architectural product that introduces other qualitative properties into the built environment that engage productively with the flow of energy and materials that define both today's and tomorrow's cities.

References

AADRL (2012) Architectural association school of architecture design research laboratory. http://www.aaschool.ac.uk/aadrl/. Accessed 14 Aug 2012

Banfill PFG, Khasreen MM, Menzies GF (2009) Life-cycle assessment and the environmental impact of buildings: a review. Sustainability 1:674–701. doi:10.3390/su1030674

Bergman D (2012) Sustainable design: a critical guide. Architectural Press, Princeton pp 15, 24–5, 62

Bonabeau E, Camazine S, Deneubourg JL, Franks NR, Sneyd J, Theraulaz G (2001) Self-organisation in biological systems. Princeton University Press, Princeton, pp 377–403

Braungart M, McDonough W (2002) Cradle to cradle: re-making the way we make things. North Point Press, New York

Deleuze G, Guattari F (1988) A thousand plateaus: capitalism & schizophrenia. The Athlone Press, London, pp 408–409

Holchim Awards (2012) Holchim awards: next generation 1st prize Europe. http://www.holcimfoundation.org/Portals/1/docs/A11/EUR/Posters/A11EUng1UK.pdf. Accessed 14 Aug 2012

Ikaga T, Murakami S, Japan GreenBuild Council (eds) (2008) Evaluating environmental performance of vernacular architecture through CASBEE. IBEC, Japan, p 1

JS International (2011) Japan Property Characteristics Sotheby's international realty. http://www.japan-sir.jp/english/clearance/jsir.html. Accessed 1 Sept 2012

Materialise NV (2012) About materialise. http://www.materialise.com. Accessed 14 Aug 2012

Mendelsohn E (1923) Dynamics and function. In: Conrads U (ed) Programs and manifestos on 20th-century architecture. MIT Press, Cambridge, MA, pp 72–73

Meyer H (1928) Building. In: Conrads U (ed) Programs and manifestos on 20th-century architecture. MIT Press, Cambridge, MA, pp 117–118

Objet (2011) Objet 260 connex: compact multimaterial 3D printing system. http://tri-tech.net/uploads/pdf%20files/Connex%20260.pdf. Accessed 23 Sept 2011

Rayfish (2012) Rayfish http://www.rayfish.com/index.php?chapter=about. Accessed 14 Aug 2012

Reggio G (Dir), Glass P (Music), Fricke R (Cinematography) (2002) Koyaanisqatsi: life out of balance. MGM Home Entertainment, DVD

Snooks R, Stuart-Smith R (2004) www.kokkugia.com. Accessed 14 Aug 2012

Snooks R, Stuart-Smith R (2010) Non linear formation: or how to resist the parametric subversion of computational design. In: Xanthropoulos P, Zindrou I (eds) Apomechanes: nonlinear computational design strategies. Asprimera Publications, Greece, pp 140–147

Stuart-Smith R (2011) Formation and polyvalence: the self-organisation of architectural matter. In: Hallnas H, Hellstrom A, Landin H (eds) Ambience' 11 proceedings. University of Boras, Sweden, pp 20–29

Wang Q (2010) Short-lived buildings create huge waste, China-Daily. http://www.chinadaily.com.cn/china/2010-04/06/content_9687545.htm. Accessed 1 Sept 2012

Other Cities

Mark Raymond

'They taught us to look at other cities to see how they work and to look at seemingly nonarchitectural environments. For them, no issue was too humble or lowly.'

Rem Koolhaas on Maxwell Fry and Jane Drew and tropical architecture, in Advancement versus Apocalypse, Ecological Urbanism, 2010

Abstract Modern society is confronted today with three critical challenges; the depletion of fossil fuels demanded by society to function at the current level of operation; climate change, which is understood to be progressively eroding the essential requirements of biological and ecological equilibrium necessary to support human existence and population growth, which presents unprecedented social, logistical and spatial challenges; most patently expressed in the increasing demands placed on the inhabitation of cities and the impact of this phenomenon on the broader environment. It is evident that these issues must be addressed in order to maintain the sustainable advancement of society. Contemporary cities, or imagined cities or those cities outside of our conventional scope represent a critical field for investigation. Cities are fields within which innovative design practice can engage with the pressing environmental and social issues confronting modern society. We find ourselves at a significant moment, one at which design practice might reevaluate its pedagogical foundation and investigate alternative forms of practice. New forms and modes of practice are capable of cultural production that through models and strategies that embrace social and environmental challenges can be effectively deployed in the reconceptualisation of cities.

M. Raymond (✉)
Architect, Port of Spain, Trinidad and Tobago
e-mail: mark@markraymond.com

S.Th. Rassia and P.M. Pardalos (eds.), *Cities for Smart Environmental and Energy Futures*, Energy Systems, DOI 10.1007/978-3-642-37661-0_6,

1 Sustainability

The idea of sustainability has emerged as a focal concept in the strategizing that has evolved in response to the current crisis. Recently it has been proposed that the triple bottom line of economics, society and the environment that has come to triangulate our collective understanding of sustainability, be expanded to a broader conceptual basis – 'circles of sustainability'.[1] This framework emphasizes the relationship of the elements of the concept of sustainability with the idea of the social. This conceptual model of sustainability is coordinated through the interpretation of economic, ecological, political and cultural factors with a view to holistically accommodating a wider range of values. Such an expansion invites a recalibration of the environmental and developmental challenges facing society and places in focus the need to ensure that our readings and interpretation of the issues at hand are sufficiently thorough and pervasive to be able to inform appropriate and humane remedial strategies. It is a conceptual framework that offers wider possibilities with which we can view our predicament as less of a crisis and more as an opportunity to re-conceptualize and diversify the role, nature and future of the city.

2 Cities

Nowhere is the need for the active engagement of sustainable principles more critical than in the future of cities. Cities have emerged as sites of increasingly concentrated human habitation and activity where the impact of the crisis of consumption and where urban densification compounded by the deleterious impacts of climate change is most evident. It is now widely accepted that more than half of the world's estimated population reside in cities and whilst the scale and consequence of this fact is widely acknowledged it appears to have so far eluded any comprehensive remedy.

Many of those cities recognized as being subject to the urban growth and densification as a consequence of mass migration for example, continue to expand with increasing numbers of people living in spaces that are chronically inadequate in terms of physical and social infrastructure and amenity. Cities embody the crisis of the global human condition threatened by the consequences of overcrowding, inadequate infrastructure and subject to the potentially disastrous consequences of climate change such as rising water levels and dangerously unpredictable and often extreme weather patterns.

[1] http://citiesprogramme.com/aboutus/our-approach/circles-of-sustainability

Cité Soleil, Porte-au-Prince, Haiti

Informal hillside housing near Pétionville, Porte-au-Prince, Haiti

3 The Urban Age Project

In 2004, under the direction of Ricky Burdett, The Urban Age Project[2] based at the London School of Economics began an international discourse across a wide range of disciplines with a view to informing the governance of cities with a closer

[2]See The Endless City: The Urban Age Project by the London School of Economics and Deutsche Bank's Alfred Herrhausen Society, ed Ricky Burdett and Deyan Sudic, Phaidon Press; Reprint edition (London: December 6, 2010) also Living in the Endless City, ed Ricky Burdett and Deyan Sudic Phaidon Press (June 1, 2011).

understanding of the emergent and complex issues confronting them. In 2006, the Venice Biennale took as its theme the contemporary condition of the global city "Cities, Architecture and Society".[3] These initiatives together with the many other comparable initiatives that have subsequently emerged have contributed to the expansion of our concept of the city and alerted us to the urgency of addressing the issues facing its future and the challenges and meaning of sustainable development in relation to urbanism. It is a discourse that whilst nominally embracing a familiar subject has uncovered new themes and perspectives and is notable in the extent to which it has drawn from numerous and diverse disciplines not previously embraced by discourse on the city. The project has thus served to broaden appreciation beyond the conventional disciplines of for example, planning, geography and governance and engaged with a more expansive array of influences, inviting a perspective of the city not as traditionally framed – as a physical object – but as a dynamic, continually evolving organism accommodating and reflecting the challenges, values and attendant complexities of modern society.

This interest and approach has drawn from a variety of disciplines and thinkers allowing us to focus attention through a wide range of inquiry and research. We have come closer to a more effective appreciation of the contemporary city however perplexing and bewildering that reading remains in the face of the enormity of many of the pressing challenges. Such inquiry has also served to illustrate the problem of not only the complexity and sheer breadth and scale of these challenges but perhaps more importantly invited us to investigate how this understanding might meaningfully inform restorative or remedial strategies. Until recently we have tended to collectively imagine the city as a complete and resolved object embodying and symbolizing cultural, political and social values whilst accommodating a harmonized and equitable concentration of human activity. The complexity of the contemporary condition and the threats to its sustainability now appear to reveal a more diversified and complex conception.

4 The Global City: Myths and Realities

In an effort to identify the pertinent contemporary themes and issues in the study of contemporary cities, interpretations and readings have been ultimately selective in their focus. Our understanding of cities has tended to implicitly characterize the phenomenon of megacities as typical of universal conditions.

Whilst the themes and influences impacting on cities might in more generalised terms be seen as common, it is evident that the manner in which they influence or impact upon the diverse range of social, political, economic and environmental circumstances in which cities are located varies considerably. It would seem

[3]Cities: People, Society, Architecture: 10th International Architecture Exhibition - Venice Biennale, Dir. Ricky Burdett, Rizzoli (November 21, 2006).

Shenzhen, China

appropriate that in addressing the current issues facing cities and in our attempts to conceptualise appropriate and responsive models, that the apparent diversity and appreciation of a range of conditions be included in any progressive formulation. This would allow for a more comprehensive understanding of the complexity of how cities might evolve across a range of sites and how deleterious influences might be mitigated. Such an approach may also serve to sharpen our appreciation and identification of the points of intersection between the different manifestations of the city.

A predominant reading of cities is currently predicated on a model consistent with Saskia Sassen's identification of the concept of the Global city.[4]

Whilst ostensibly the idea of the 'global' may be understood to be inclusive, it is ultimately a contested term excluding a range of critical sites, issues and themes. Sassen's global cities embody the characteristics of an emergent global culture that has proliferated concurrent with the demise of nation-states and in response to movements in labour and capital. Sassen investigates the nature and characteristics of and the relationship between cities that exist in the context of transnational culture. This is a view which whilst it pertinently addresses the major and significant operational forces accommodated and symbolized by instances of the contemporary urban condition exemplified by the idea of specific and prominent cities is not a an analysis which can be applied universally. Many of the narratives unfolding around cities respond to the global city concept by tacitly applying it

[4]Saskia Sassen, The Global City, Princeton University Press; 2nd edition (September 1, 2001).

to all cities. The current concept of the global city may therefore still be seen to privilege certain conditions in its scope and focus. Sassen's focus has a particular and specific orientation and such a grouping or clustering of cities clearly serves an appropriate structure for the nature and basis for her inquiry. However we must question the relevance of the often universal transposition of this thinking. Sassen's concerns may present a universal resonance and value that we must consider but the notion of the global is not exclusive to her frame of reference in relation to cities. It is an understanding of the scope of cities which whilst apparently expansive is limited as numerous 'other' sites, or cities which do not readily or ostensibly conform to what is proposed as global are excluded. The focus endeavours to define global as globally connected and technologically sophisticated or advanced but is significantly and ultimately measured economically. I refer here to excluded or omitted sites that although they are subject to the vagaries of globalisation are not centrally engaged or are perceived as peripheral in relation to the influence of transnational frameworks. One would hope that the scale, nature and far-reaching implications of the challenges facing contemporary society demand that any set of concepts or ideas falling under the rubric of global must by definition be inclusive.

We must learn to interpret the presence of other entities, systems or models which may be obscured by any given socio cultural perspective. In accepting other realities we open ourselves to the entertainment of other perspectives or readings of the operations of globalization and transnational culture in our negotiation and definition of the spatial and cultural parameters which impact on and inform design practice.

Hong Kong

An example of an alternative reading relevant to our reconceptualization of cities and spatial constructs is advanced in the work of the architect Keller Easterling, in this case one which is skeptical about the forces and impact of flows of capital and global culture. Easterling[5] charts the emergence and proliferation of the tourist resort, the retail chain, call centres and ports – examples of what she terms 'spatial products'. These products are denuded of conventional design values and derive from political and economic expediency. Spatial products are seen as part of a newly configured flow and confluence of impulses, ideas and initiatives materialised in certain typologies of built form and which exist in the world often specifically equipped to resist the peculiarities of place and which serve particular economic, political and social interests as required by their economically dictated programmes. This occurs invariably at the exclusion of any concession to the public domain. This practice has the capacity to defy conventional criteria in terms of the public dimension of design production. Emerging from globalised culture and the operation of transnationalism it draws upon optimised industrialised technique for its materialization and can be deployed globally in ways which resist - often deliberately - localised or domain specific engagement. Spatial products also serve to constitute another dimension of the manifestation of aesthetic regimes. It is a phenomenon that challenges conventional notions of design practice and production. It operates in ways which challenge the conventional ethical assumptions surrounding design practice and those spatial parameters which we habitually rely on in our identification of cities.

Metropolitan centres such as New York, London and Tokyo have so far retained their primacy within the emergent global order it appears largely as a consequence of their predominance in the management of financial markets. This status results in an hierarchical ranking of other cities in the context of global networks. It is a system that creates a range of status in which many other cities occupy a conceptually interstitial space somewhere between the concept of the global city and cities defined by previously defined notions of underdevelopment.

If the challenges facing society are to be effectively addressed, truly sustainable and incorporated into comprehensive and socially equitable strategies, the current discourse on the future of cities must be more expansive and inclusive. The process of globalization in whatever manner this might be understood, whether as positive or negative force or indeed both, has resulted in a radical transformation of global movement both physical as well as in terms of the flow of information and the deployment of political, economic and social power.

We are thus confronted with a range of major questions and challenges. How might other cities be incorporated into the collective effort to address the broader pressing issues particularly if the focus lies with cities with often radically different cultural criteria which have governed their respective development? What are the conceptual models and modes of design practice which might effectively be pursued

[5]Keller Easterling, Enduring Innocence Global Architecture and its Political Masquerades The MIT Press 2005.

to ensure their consistent and sustainable evolution and development? How might these practices be, structured, directed and deployed to foster inclusive models and strategies in locations other than the conventional sites which traditionally dominate this discourse? How might design practice be conceptualized to practically address the environmental challenges which face the contemporary condition and future of the city?

5 Engaging Design Practices

The realm and field of design practice is where the greatest transformative scope, possibility and potential may be realised in addressing the current condition. The complex and rapidly transforming confluence of, on the one hand the urgent demands of environmental crisis and on the other our reevaluation of the cultural value and significance of the city, has stimulated wide ranging responses from the field of architecture, urban and landscape design. Within the functions of society and beyond the confines of fine art, design practice possesses more than any other field the projective and imaginative capacity to assimilate pertinent characteristics of specific environmental and cultural conditions, human needs and desires and to creatively negotiate the projection of sustainable futures.

The current configuration of design practice is however polarized and is directed to, and by, economic factors that divert attention from the broader social benefits that it has the capacity to provide. This is a phenomenon which in its most advanced and extreme manifestation can be observed in the emergence of an architectural and design culture more closely aligned to the world of haute couture than the liberal professional mode engaged with critical social issues under which the modern movement and its respective practices previously operated; as Peter Davey the former editor of the Architecture Review wrote; 'We live in a world permeated by the cult of celebrity and dominated by the electronic media, which demands constant novelty. The more unusual the gesture, the more enhanced an architects brand. The cult of celebrity has been so successful that most of the limited international competitions are open only to a small group – perhaps no more than 100 – who are almost forced to become increasingly demonstrative and *outré* to ensure that they retain their place in the hierarchy of the celebrated.'[6] The downstream consequence of this culture is the relegation of architectural and urban design production to one of commodity. Design has commonly evolved into a process which is driven by the idea of the branded product rather than a responsive activity fuelled *inter alia* by the socially liberative capacity of design.

[6]Bling, Blobs, Burgeoning: Problems of Figure, Peter Davey, p74 The Architectural Review, EMAP Construct March 2005.

In his introduction to a series of essays published in the Harvard Design Magazine on Commodification and Spectacle,[7] Kenneth Frampton expands upon this critique of contemporary practice and identifies the tendency as the consequence of an economic and political order which has governed the trajectory of architectural culture and which threatens fundamental social values and ideals; 'international monopoly capital has increasingly challenged the authority of the nation-state, which still ostensibly embodies the democratic precepts of the free world.' Frampton accuses contemporary architectural and design critique and practice of evading the 'the psycho-political substrate underlying the compulsions of our commodified society. It is as though they would prefer to avoid a critical confrontation with socioeconomic causes that are directly responsible for the environmental degradation of the late modern world.'

Frampton suggests that what exists is an oppressive system and that 'architecture has become a brand in itself, particularly for the "signature" architects, whose mediatic overvaluation finds a direct correspondence in the systematic undervaluation of other equally if not more talented architects whose work has yet to be confirmed by the mediatic consensus as a discernable and desirable brand.' He is critical of this tendency asking 'Where is the anachronisitic culture of architecture to situate itself in the face of all this digitally dematerialized misrepresentation?' Frampton invokes architects to address this 'psycho-political substrate' which determines so much of architectural production stating 'At this juncture one can hardly emphasize enough how the substance of political process needs to be articulated within the field ......political consciousness, in the broadest sense, ought to be as much part of design education as any other component in an architectural curriculum....'

This commentary on the commodification and disenfranchisement of architecture represents part of a narrative articulated by Barry Bergdoll in his introduction to Small Scale Big Change: New Architecture of Social Engagement.[8] Bergdoll charts the perceived demise of the promise of modern architecture, contrasting the elevated and celebrated role of the architect and the promise of architecture at the advent of the modern movement with its current status '..the credo of the modern movement, despite what historians now recognise as its great variety of positions and practices, was that a new architecture could ultimately serve for the large scale transformation of the inherited order – whether the physical order of cities and suburbs or for the most committed, the transformation of inherited social, political, and even economic structures.'

This demise and increasingly indeterminate role is also linked to the process of globalisation and capital 'Even more significant than direct challenges to the

---

[7]Kenneth Frampton, The Work of Architecture in the Age of Commodification, Introduction to Commodification and Spectacle in Architecture, WIlliam S Saunders Ed, Harvard Design Magazine Reader, Harvard University Graduate School of Design University of Minnesota Press, 2005.

[8]Barry Bergdoll Introduction to Small Scale Big Change: New Architectures of Social Engagement. Ed Andres Lepnik Museum of Modern Art, New York. 2010 The Museum of Modern Art, New York.

profession, however, has been the steady erosion of the real power of architects as shapers of the environment. As the ebbs and flows of globalisation have brought both development and its discontents to nearly every corner of the planet, the role of the architect in the spatial arrangements of global capital and the concomitant commodification of space has been largely left undetermined.'[9]

As a form of holding pattern, Frampton proposes a monastic or resistant response, a critical and strategic withdrawal from ethically unacceptable or socially detrimental modes of practice. Bergdoll more optimistically observes that practitioners and architects alike are 'rediscovering the critiques of orthodox modernism of the 1970s and rethinking them afresh in relationship to challenges, both environmental and social that seem exponentially more pressing 40 years later'. He also identifies potential lessons to be learnt from architects working outside of the conventional capitalised frameworks of architectural production; 'One important lesson to draw from these architects as they work against both the forces and the assumptions of globalisation, is that the flow of knowledge can move in multiple directions, that new perceptions about the needs of a severely challenged developed world can be found in practices developed in the under developed world, particularly as the issue of appropriate technology becomes the most urgent mantra for architectural practice everywhere'.

From this it would appear that notwithstanding the disjunction between capital flows, the limited opportunity for architects and designers to project ameliorative alternatives in an attempted remediation of the crisis facing society and the current dire need for such issues to be addressed that operational space nevertheless exists for such practice. Furthermore there appears to be a role for design that is currently overlooked or overshadowed with the concentrated focus on the impact of the contemporary crisis on global or megacities. Whilst current circumstances may not permit the embrace or even accommodate more critically aware forms of design practice there is significant potential in what are commonly perceived as peripheral locations in other cities or sites.

6 Contemporary Models and Strategies for Design Practice: Education

Schools of architecture, urban design and landscape architecture have been increasingly engaged with the development of strategies for the remediation of the environment. This is a trajectory that has resulted in a welcome move away from the proliferation of the obsessive and digitally-driven experimentation with surface and form supported by spurious and often facile rationalisations which appeared to have characterized much academic production over the past 20 years.

---

[9]Barry Bergdoll Introduction to Small Scale Big Change: New Architectures of Social Engagement.

Current preoccupation reflects a range of investigation and inquiry that addresses the programmatic aspects of architectural, urban and landscape design. It is an approach that has expanded to incorporate fields such as environmental and civil engineering in a search for strategies which are no longer object-based but founded on the exploration of the responsive and programmatic application of overlapping disciplines. Work has begun to appear that is grounded in a more ethical conception of architectural production in contrast with the singular and virtuoso culture promoted by mainstream architectural culture. The prevalent culture identified by Peter Davey in the description quoted earlier in this essay depicting the image of an elitist stable of architects perpetually competing for attention in the mediatic circus of the global architectural world. Notwithstanding the improved content of current academic and research-driven work, much of it is still visibly inscribed by a culture which persists in its privileging of virtuosity, a tendency which often unfortunately often distracts from the essential value and credibility of many of the propositions being advanced. The relationship between theory and practice calls for clearer focus on pedagogic objectives and a reform of the nineteenth century model on which much architectural education is essentially founded. Pedagogic frameworks might more effectively address themselves to structures which focus on the process of design and the value inherent in research, experimentation and craft – not in its associative sense of handicraft but in the sense of developing skill through close attention to dedicated application over time.

Schools nevertheless remain fertile fields for the cultivation of new theories and strategies and would also perhaps benefit from more progressive approaches to education, encouraging those that foster not only the development of creative and imaginative responses and design skills but which direct focus and attention to the application of ideas to the specific functional requirements of social life and encourage interdisciplinary exchange and interaction. It is pedagogic structure which largely still underpins the formation of architects and designers and in a world where the practice and production of architecture, urban design and the architecture of landscape play an increasingly significant role in determining the future of society pedagogic principles must evolve to ensure that design education meets this challenge.

7 Architectural Strategies

Kenneth Frampton's seminal essay espousing Critical Regionalism[10] was originally proposed at a time when the principles underlying modern architectural culture and production were under threat by a preoccupation with scenographic representation

[10]Kenneth Frampton, "Towards a Critical Regionalism: Six Points for an Architecture of Resistance" in Hal Foster ed., *The Anti-Aesthetic: Essays on Postmodern Culture* (San Francisco: Bay Press, 1983).

in the form of post modernism. Frampton asserts that Critical regionalism is not a style but a resistant form of practice that critically interprets its socio-political sphere of operation and looks to the peculiarities of place for the generation of architectural form. It is a strategy that encourages architects to evade the mainstream imperatives that drive the commercial trajectory of most practices and to identify modes of operation which are environmentally and socially more responsive. Such 'critical practice' in its challenge to the prevailing modes of practice and production reasserts both the idea of design practice as craft and as ethically motivated activity directed by progressive principles and geared towards social advancement.

Such approaches might be usefully employed in addressing the challenges faced today in endeavoring to embrace more humane considerations in the design of cites and the propagation of approaches which while sensitive to the demands of global culture and its attendant amenities and benefits might accommodate a heightened sense of place in the articulation and formulation of built or urban form. It is an approach that might contribute more profoundly to a more diverse and engaged interpretation and evolution of the idea of the global.

Another valid but seemingly 'archived' architectural model or formulation which might usefully inform responsive urban strategies is that of what is known as tropical modernism. The British architects Maxwell Fry and Jane Drew who assisted Le Corbusier with the design of Chandigarh worked in West Africa and developed an approach to the production of modern architecture in tropical climates that has come to be known as tropical modernism. Their work and approach was documented in their publication Tropical Architecture in the Humid Zone.[11] Hannah Le Roux has described the formation and significance of this tendency in the 1950s and 1960s and its proliferation and resonance in the British Commonwealth in the post war period.[12]

Tropical modernism whilst culturally loaded as a consequence of its relationship and location within the context of colonial and post-colonial culture offers specific and tested strategies for the development of architectural design and planning principles as applied to the particular circumstances of tropical climates which reveal a clearly articulated and functional approach to issues of sustainability. Rem Koolhaas in his contribution to Ecological Urbanism[13] refers to the teaching of these principles that constituted a specialization offered at the Architectural Association School of Architecture in London where Koolhaas studied in the 1960s. He acknowledges its contemporary relevance and potential in addressing current conditions; 'I was fascinated by its teachers, who taught us an incredible respect for the landscape..... They were also interested in the tropics as a special domain,

[11]Maxwell Fry and Jane Drew, Tropical Architecture in the Humid Zone, Reinhold, 1956.

[12]Hannah le Roux, The Networks of Tropical Architecture, The Journal of Architecture, Volume 8 Autumn 2003.

[13]Rem Koolhaas, Advancement versus Apocalypse, Ecological Urbanism, eds Mohsen Mostafavi with Gareth Doherty Harvard University Graduate School of Design, Lars Müller Publishers 2010 pp 60–61.

which is now the front line of the tensions and impossibilities that we are confronted with. The studies also examined how an architecture could emerge that would persist in this climate without the degree of artificiality that we now take for granted.'

It is an approach born out of a rigour and adherence to functionalist principles that offers instructive principles notably absent from current architectural discourse with the current focus on product rather than process and reliance on proprietary systems over craft based technique. Tropical modernism accommodated a combination of both scientific and craft based understanding of the construction and formation of the architectural and urban environment that could be usefully incorporated into a contemporary application. It represented an ethical approach to the design of cities both environmentally sensitive and programmatically and functionally appropriate.

8 Landscape Strategies

The concept of Landscape Urbanism and in particular the theoretical positions posited by one of its leading proponents James Corner provides arguably the most pertinent strategy for the engagement of the notion of sustainability as a primary focus in the formulation of design strategies appropriate to current conditions. It is a strategy that also contains the potential for translation and implementation over a wide range of sites. In his essay Terra Fluxus,[14] Corner references the current conceptualizations of the discrete practices of architecture, urbanism and landscape and proposes that they are 'moving toward a shared form of practice

Botanical Gardens, Port of Spain, Trinidad

[14]James Corner, "Terra Fluxus" in Charles Waldheim, ed. *The Landscape Urbanism Reader* (New York; Princeton Architectural Press, 2006) 21–33.

for which the term landscape holds central significance'. He cites the scope of landscape as offering the 'capacity to theorize sites, territories, ecosystems, networks, and infrastructures, and to organize large urban fields.' Whilst optimistic in scope, Corner presents a compelling case for landscape as a potent agency of design influencing the built environment and the application of the practice and representation in sites outside of the conventional domains of the constructed form of architecture.

Corner charts what he describes as 'provisional themes' for the practice of landscape urbanism as a strategy for addressing 'the complexity of the rapidly urbanizing metropolis'. These are in summary, process, surface staging, operational or working method and the imaginary. The first two themes might be viewed as issues related to productive technique relevant to Corner's own specific practice, the last two themes perhaps offer broader potential in recognizing relevant issues for the remediation of the current environmental condition through design practice.

In addressing the issue of operational practice or working method Corner proposes that the breadth and range of activities embraced by landscape practice, when effectively configured, provide unique and privileged access to and creative potential within critical fields. His position recognizes – similar to Frampton's critical practice – that although immersed in a world in which design practice whether architectural, urban or landscape is dependent upon its function as a service profession nevertheless has capacity through its potential to assimilate, process and consolidate data, observation and to creatively process and then project. It is thus a unique societal operation allowing us to effectively see into the future through the breadth of such scope and interaction. Whilst Corner claims this scope for landscape practice it is not the exclusive domain for such a strategy and is a position that could be adapted and embraced in the conceptualisation of all domains of design in addressing the current condition.

The imaginary is the last of his provisional themes but resonates equally with the idea of the operational or working method. This theme challenges practice to draw upon the material and imaginary context of the human condition and project and invest this capacity into cultural production. Again an invocation not restricted to its application in landscape practice nor domain specific. Corner invokes us to deploy the imagination in locations other than the conventional and prescriptive sites to which it is otherwise condemned. The imaginary can be harnessed to challenge the commodification of architectural production and representation by positing alternative readings to the homogenised imagination which has come to govern the representation of design.

An example of this imaginary and its provocative capacity can be seen in the work of Didier Madoc-Jones and Robert Graves (GMJ) who in their project Postcards from the Future,[15] in a series of compelling images, depict a future London.

[15]http://www.postcardsfromthefuture.co.uk

Buckingham Palace Shanty, Postcards from the Future, Didier Madoc-Jones and Robert Graves, (GMJ) (Photography by Jason Hawkes)

Parliament Square Paddy Fields, Postcards from the Future, Didier Madoc-Jones and Robert Graves, (GMJ)

It is a London where conventional pre-conceptions are overturned and replaced with juxtapositions of familiar iconography; conflating a range of contemporary issues and integrating them into a vision of a provocatively constructed imaginary future. London is alarmingly confronted with the consequences of environmental, social and cultural transformation.

Counter to the often doom-laden and apocalyptic prophesizing of many commentators of environmental and social degradation these *mises en scène* refreshingly and vicariously articulate a form of sustainability but one which does not conform to our current notions of social, spatial and environmental order. The images remind us of the capacity of the imagination to project alternative circumstances – to imagine other cities.

Trafalgar Square Shanty, Postcards from the Future, Didier Madoc-Jones and Robert Graves, (GMJ)

9 Coda

Contemporary cities, or imagined cities or cities outside of our conventional scope represent a critical field for investigation. Cities are fields within which innovative design practice can engage with the pressing environmental and social issues confronting modern society. Social polarization, and the ostensibly limited scope for critical practice do not preclude its engagement or pursuit. We find ourselves at a significant moment, one at which design practice might reevaluate its pedagogical foundation and investigate alternative forms of practice. New forms and modes of practice are capable of cultural production that through models and strategies that embrace social and environmental challenges can be effectively deployed in the reconceptualisation of cities.

Urban Parangolé

The Syncretic City

Lindsey Sherman, Michael Contento, Alfredo Brillembourg, and Hubert Klumpner

Abstract In the last century, rapid urbanization throughout São Paulo's greater metropolitan region has resulted in a population of nearly 20 million. While this makes São Paulo one of the most vibrant cities in South America, it also creates a condition in which established mobility infrastructure is no longer effective. Asymmetrical urbanization has also perpetuated marginalization and the development of informal settlements. As a result, São Paulo is experiencing both physical and social immobility. In order to address these complex conditions, it is necessary to expand the definition of sustainability beyond environmental concerns to include both economic and social sustainability. Urban Parangolé generates a framework for this comprehensive approach – a framework that allows conflicting urban forces to engage in a productive coexistence without a homogenizing effect. Integrated mobility infrastructure can serve as the interface for this renewed interaction. Using São Paulo as a testing ground, Urban Parangolé fills a dual role. It is a set of innovative mobility strategies that also serves as a framework in which to insert ecological, social, and economic viability into the city. Urban challenges are no longer seen as inevitable symptoms of growing cities, but as a formative basis for alternative and sustainable urban models.

> Before anything it is necessary to clarify my interest for dance, for rhythm, in my particular case it came from a vital necessity for a dis-intellectualization... it was therefore, an experience of greater vitality, indispensable particularly in the demolition of preconceived ideas and stereotypification... there was a convergence of this experience with the form which my art took in the Parangolé.
>
> - Hélio Oiticica

L. Sherman (✉) • M. Contento • A. Brillembourg • H. Klumpner
Brillembourg & Klumpner Chair of Architecture and Urban Design, D–Arch, ETH Zurich
Neunbrunnenstrasse 50, ONA /J25 CH-8050 Zürich, Switzerland
e-mail: sherman@arch.ethz.ch; contento@arch.ethz.ch; brillembourg@arch.ethz.ch; klumpner@arch.ethz.ch

S.Th. Rassia and P.M. Pardalos (eds.), *Cities for Smart Environmental and Energy Futures*, Energy Systems, DOI 10.1007/978-3-642-37661-0_7,

Urban Parangolé[1] is a framework that responds to the human need and desire for movement – it engages us in a dance with the city. This notion pays homage to the work of Brazilian artist Hélio Oiticica and extends the central tenet that 'life is movement' from the body into the city. Here, urban mobility serves as a set of strategies to engage the city.

A smart city is a syncretic city – one that integrates its conflicting forces to generate a productive coexistence without giving in to a homogenizing tendency. Urban mobility, as a network, provides an important interface where this effective vitality can flourish. However, in the city of São Paulo, one condition that arises from ongoing, asymmetrical urbanization is a debilitating immobility – both physical and social. While the city experiences rapid expansion, inadequate mobility infrastructure can no longer handle the increased demands of its growing population.[2] Continued growth of informal settlements also questions the notion of urban progress. Sustainable futures for São Paulo are tied to systems of physical mobility and opportunities for social mobility:

> Over the last 200 years, too many transport and urban planners have ignored most of these issues while technicians, engineers, and architects worked in isolation, without taking into account society's wider goals. They have optimized single elements such as vehicles, roadbeds and alignments, public transport and buildings while treating them as separate, isolated areas. This fragmented approach ignored larger developments and has placed the future of cities seriously at risk.[3]

Beginning in the 1930s, government investment focused on the growth of extensive automobile infrastructure, a trend that has diminished investment in alternative modes of mass transit and resulted in current infrastructural limitations and congestion. Between 1970 and 2012, the number of vehicles on São Paulo's roads increased sevenfold, while the area of road surface less than doubled. On average, citizens lose a cumulative total of 27 full days per year stuck in traffic.[4] Even as urban transport networks slow, recent investment in public transportation has fallen by one third.

At the same time, the most immediate and visible evidence of asymmetrical urbanization has been the growth of informal settlements. They are the result and the eventual casualty of the same urbanization processes. In a city with nearly

---

[1]This essay expands on the exhibition of the same name developed by Urban-Think Tank for the Audi Urban Futures Initiative, 2012.

[2]While the center of the city is losing significant population, the share of the population in the peripheral areas is still growing very fast, representing 19 % of the population in 1991 and 30 % in 2000. As a result, the region still demands strong public investment in transportation and other urban infrastructure, with considerable environmental impact for the city. See: Haroldo Torres, Humberto Alves and Maria Aparecida De Oliveira, "São Paulo peri-urban dynamics: some social causes and environmental consequences." Environment and Urbanization 2007 19: 209, 210, accessed January 17, 2013, http://eau.sagepub.com/content/19/1/207

[3]Hermann Knowflacher, Philipp Rode and Geetam Tiwari, "How Roads Kill Cities," *Living the Endless City*, edited by Ricky Burdett and Deyan Sudjic. Phaidon Press Ltd, 2011: 343.

[4]Raquel Rolnik and Danielle Klintowitz, "Mobilidade cidade de Sao Paulo," *Estudios Avancados* 25(71), 2011.

1,500 favelas holding a population of over three million people,[5] these areas are acutely familiar yet remain out of bounds – marginal in their physical locations and marginal in their socio-economic exclusion. Despite a dynamic cultural and social environment within their limits, they are fundamentally disconnected from the city physically, socially, and economically. Informal settlements account for the fastest growth rates in the city, yet they lack even basic social equipment and provisions for education, health, employment, and leisure. This sets the groundwork to expand the definition of sustainability beyond environmental responsibility to include both social and economic imperatives.

These conditions of informality and immobility are quickly becoming the norm, not the exception. As such, it is necessary that we shift our understanding of these challenges as inevitable symptoms of outmoded urban models to foundational potentials for new trajectories. This is the territory of Urban Parangolé and the Syncretic City.

Urban Parangolé is a new mobility system that actively fills a dual role – as a set of innovative mobility strategies that address inadequate infrastructure and as a robust framework to insert ecological, economic, and social viability into the city. It uses São Paulo as its testing ground to read and act in an existing context. Using the challenges of the real city as a formative basis, Urban Parangolé unfolds a latent territory within the urban fabric. It is in this space where the project reads and affects the varied and conflicting forces of urbanization toward new alternatives. It is not a rejection of the city; it is a resistance through syncretism. In this way the city is a catalytic field of potential new trajectories that counter current models of unsustainable growth. It does not create a temporary smart city, but, through an ongoing resistance from within, it encourages a sustainable "learning" city (Fig. 1).

However, one cannot begin to discuss sustainable cities unless formal and informal areas are seen as equally vital. In other words, one is not simply the negative result of the other; they are two sides of the same coin. Otherwise, any notion of sustainability risks being confined to the formal city, leaving vast urban areas out of the discussion. This is not to say, then, that Urban Parangolé privileges one over the other. It does not romanticize the informal and fully resist the forces of urbanization, nor does it seek to erase informal settlements in a total accommodation of urbanization. It creates a unique platform for the integration of formal and informal, of top-down planning and bottom-up initiatives. It is precisely this elusive critical territory between the formal and informal that serves as a new point of contact for architecture and urban design in the city. Critical territory here is not meant only as a physical space, but also as a territory of critical thinking. Each can learn from the other. It is not an evolution from exclusion to inclusion, but from exclusion to collaboration. To simply include marginalized populations while implementing existing ideas of sustainability is not enough. These areas are instrumental in understanding possible new solutions. One must not only learn from and act in the formal city but also learn from the informal city.

[5] As presented by the Secretaria Municipal de Habitação (SEHAB) of São Paulo, August 2010.

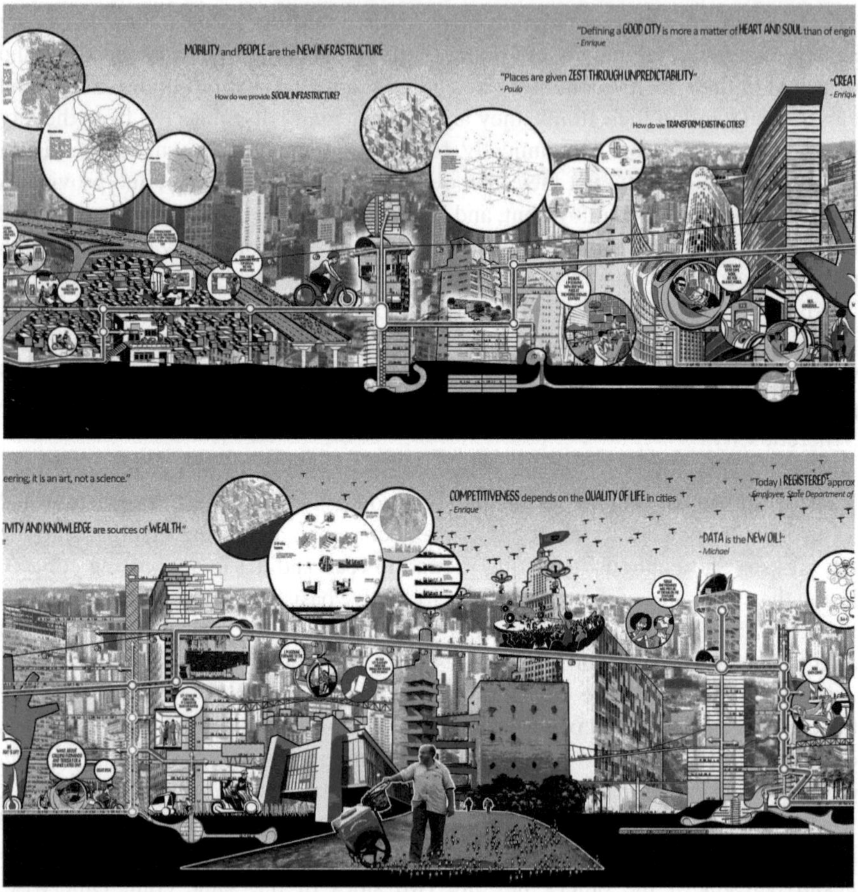

Fig. 1 The Syncretic City combines the strengths of existing practices to produce completely new forms of urban lifestyle, a city that is productive, healthy, and vibrant (Urban-Think Tank Archives/Audi Urban Futures Initiative 2012)

Inherent to Urban Parangolé is a fundamental repositioning of the role of the architect into an animator of change – an agent provocateur. Designers must operate as an interface between top-down and bottom-up processes, forces that seldom meet without new forms of moderation. A new collaborative sensibility provides the medium for trans-disciplinary practice and purpose-oriented design strategies.

Using the territory between the formal and informal as a new point of contact further redefines sustainability as a more inclusive endeavor. It is important to shift our notion of sustainability as a technical validation of fundamentally unsustainable urban models to the fundamental imperative of the initial design process. In this sense, sustainability does not rely on technology that simply dampens the negative effects of conventional models, but rethinks their basic demands. It is more than an attempt to sustain what is not working – it is a shift from sustainability to

Fig. 2 The Syncratic City builds the new city on and with the old (Urban-Think Tank Archives/Audi Urban Futures Initiative 2012)

productivity. Urban Parangolé provides the framework within which this new definition can be applied at multiple scales – infrastructure, landscape, building, and the individual.

The congested ground plane of São Paulo has become increasingly difficult to navigate, locking inhabitants into rigid routines based on necessity, on oscillating between limited, isolated points. As a set of mobility strategies, Urban Parangolé responds to the human need and desire for movement by generating seamless loops of connectivity that address the limitations of existing infrastructure. It envisions a set of multi-modal mobility networks to ensure access to all parts of the city – from urban cable cars to automated traffic systems, from high speed to slow speed, and from individual transport to high capacity public transport. Movement is not confined to the necessary travel between point A and B, but is opened up into new urban strata – activating underutilized space in the city fabric including rooftops, air space, areas below ground vacant buildings and the ground level. It transforms basic movement into a dynamic activity. Movement is more than physical; it is not simply a unit of time and distance. It is rather a dynamic, productive journey. A set of digital technologies accompanies each system and new space. Each user becomes a conduit of relevant information. People and infrastructure exist in innovative ways; people and mobility replace conventional infrastructure (Fig. 2).

In response to both physical and social congestion, Urban Parangolé liberates the spaces of the city, engaging all possibilities of the three-dimensional urban field. It unfolds a latent territory in the urban fabric and introduces mobility and accessibility to these areas. The urban fabric and voids are re-organized as a porous volume allowing for strategic placement of public space and mobility conduits. It takes as its starting point a thorough understanding of the city that occupies this zone and injects it with purpose-oriented design imperatives. The syncretic city builds the new on and with the old.

Urban Parangolé does not limit this new territory to mobility infrastructure, but prepares it as a new framework for additional programs and social infrastructure. Systems of mobility and their relationship to the city can be "improved not only by investment in rapid and efficient transport systems, but by a better distribution of economic and social activities in urban space."[6] It is a flexible field that enables the city to support a spectrum of innovative programs and typologies that are accessible for self-determination. Mobility is an interface in which top-down and bottom-up forces can interact. It is a space to consider "integrated transport networks as an opportunity to rebalance social imbalances of our urban fabric."[7] The understanding of the city now shifts from a fixed entity to a permeable and resilient field of new possibilities that emerge from the existing fabric. It does not simply resist the forces of the city, but catalyzes new operations from within by re-directing them. The goal is not a complete, terminal resistance but a perpetual, critical engagement with the existing layers of the city. New possible trajectories emerge, signaled by new prototypes and alternative urban models (Fig. 3).

Urban Parangolé utilizes the strengths of existing practices in a productive comingling of the existing and the new. Central to the success of the future city is a playfulness that presents inhabitants with choices for multiple pathways of motion and relates state of the art transformative technologies to local culture. Chance encounters are encouraged, allowing the individual and the collective to engage with each other and the urban environment in new and productive ways.

> If we want to understand transport in cities, we have to understand human behavior in the urban environment. No society can exist without the movement of people, goods and information. This is generally regarded as a requirement for evolution, be it through the facilitation of trade or, most importantly, for human interaction.[8]

This productive coexistence is the Syncretic City, one that can accommodate the motivations of diverse groups and users. It is comprised of several key elements – new urban strata, multi-scalar mobility networks, and networked hubs. These elements are placed in strategic locations within the city as catalysts. This generates a system of activated nodes that transform the city from within (Fig. 4).

The new urban strata appropriate underutilized spaces of the city: airspace – used for mobility in an increasingly densified city; rooftops – maximized for use and diversity of spatial resources; ground plane – liberated from its former use of transit and parking; underground – reprogrammed for extended use. This stratified

---

[6]Alexandre Gomide, "Mobility and the Urban Poor," *South America Newspaper: South American Cities: Securing an Urban Future*, Urban Age, 2008, 53.

[7]Fabio Casiroli, "Getting to Work," *Living the Endless City*, edited by Ricky Burdett and Deyan Sudjic. Phaidon Press Ltd, 2011: 387.

[8]Hermann Knowflacher, Philipp Rode and Geetam Tiwari, "How Roads Kill Cities," *Living the Endless City*, edited by Ricky Burdett and Deyan Sudjic. Phaidon Press Ltd, 2011: 340.

Fig. 3 Urban Parangolé is a set of strategies that form a transferable framework, a new urban model that provides social, ecological, and economic sustainability through mobility (Urban-Think Tank Archives/Audi Urban Futures Initiative 2012)

territory promotes a fluidity of movement, integration, and expanded limits. It is here that new movement is possible and that new urban models take shape.

In order to utilize these areas, Urban Parangolé creates additional building rights in the form of up-zoning rights. An example of this method is the use of Certificates of Additional Construction Potential (CEPAC). The municipality sells up-zoning rights to developers as a way to finance costs of the infrastructure necessary to support development. This invigorates investment in downtown areas and facilitates public/private partnerships. Bearers of these rights may use them to increase density or alter the use of a plot. Urban Parangolé allows payments to be offset by donating public space and thoroughfares to support the mobility program. Connecting the newly formed public spaces completes the mobility loop and allows fluid movement

Fig. 4 Urban Parangolé sees the city as a catalytic realm of potentials (Urban-Think Tank Archives/Audi Urban Futures Initiative 2012)

throughout the city. This newly unfolded territory is connected through a network of high-speed arteries comprised of public and automated traffic.

Within this new system are also multi-scalar, multi-functional hubs that anchor the network, ensure the fluid movement of people and resources, and provide space for transformative processes. As the peaks of connectivity, these hold potential for increased interaction and exchange between diverse social groups. Each hub develops organically and implements a set of programs relevant to the community needs. As such, the hubs are generated through a careful reading of the city and provide productive opportunities. This encourages strong social and economic engagement. Public, private, and shared mobility systems interface with each hub.

There are a variety of hub typologies classified by scope and function within the overall mobility system: Macro Hub – regional transit node; Mezzo Hub – local transit node with flexible mobility configurations and programs; Micro Hub – mobile system that responds to fluctuating demands.

While still a vision, the concepts of Urban Parangolé and the Syncretic City are evident in our work and are transferable and adaptable to other urban areas. Three specific projects – the MetroCable in Caracas, Venezuela, the Centro de Acçao Social por Música in São Paulo, Brazil, and Torre David in Caracas, Venezuela – exhibit specific fundamental concepts and begin to illustrate the transformative potential of Urban Parangolé. Furthermore, these projects represent various stages of the research and design process – realized projects, unrealized visions, and the potentials of the existing city.

1 MetroCable: Multi-modal mobility system and framework for social infrastructure[9]

The MetroCable, located in the San Agustín area of Caracas, Venezuela, is an urban cable car system that ties into existing public transportation. In addition to providing a transport network, it is a framework in which to insert ecological, economic, and social viability into the community. It is a new urban model that introduces an alternative mobility system into a topographically challenging informal neighborhood of approximately 45,000 people. The MetroCable liberates the ground plane by utilizing the airspace of the city. Ideally suited to the steep terrain, it is a minimally invasive and flexible response that supports inclusivity (Fig. 5).

At the heart of the San Agustín community exists a culture that primarily lives off the grid, without connection to municipal water, sewage, or electricity. Self-sustainability is a way of life in San Agustín and is therefore woven into the foundation of the project. The MetroCable provides social, physical, and cultural integration, utilizes a flexible kit-of-parts, and embraces ecological sustainability.

San Agustín is situated on a steep slope that rises 100 m to its ridge. Despite its central location, extreme physical and social barriers disconnect it from the formal city. To address this inaccessibility, the city proposed an access road following the contours of the hill. This would require the removal of many homes and fracture the existing social fabric. In protest to the government's proposal, Urban-Think Tank organized an informal public symposium in July 2003. During this meeting a new mobility system was proposed, one that would minimize demolition, preserve the pedestrian-oriented community, and introduce historically absent social infrastructures. Through active community participation the government eventually abandoned their plan and adopted the MetroCable proposal, successfully merging top-down planning with bottom-up initiatives.

This new mobility network preserves the dynamic quality of the existing neighborhood while simultaneously plugging it into the existing public transport system. It integrates necessary mobility infrastructure with necessary social infrastructure. Each station is equipped to accept various plug-in programs, such as replacement housing, educational centers, medical facilities, shops, and sport areas. These plug-in programs are embedded into the transport system, becoming hubs of intensity that provide social services and other community needs. In this way, the MetroCable transcends the physical and social borders separating San Agustín from the city (Fig. 6).

The MetroCable embraces the potentials for growth, adaptation, and transferability. Using pre-fabricated components, it is designed on modular principles to form a kit-of-parts, or toolbox. Not only are these elements flexible and able to

[9]Designed by Urban-Think Tank, Caracas, Venezuela, 2007–2010, Client: CNO Construtora Norberto Odebrecht, C.A. Metro de Caracas.

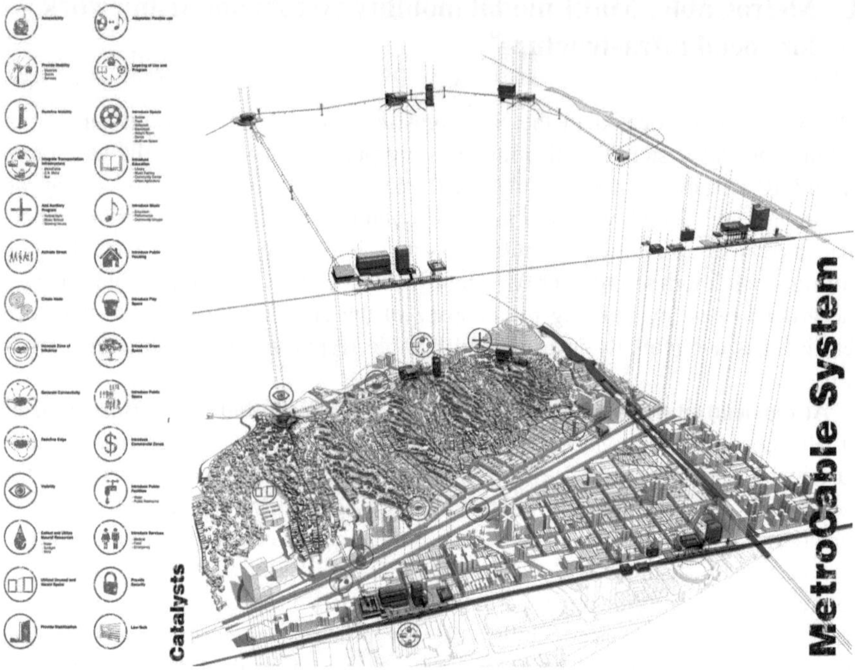

Fig. 5 The system is comprised of five stations, moving 1,200 people per hour. Two stations connect into the existing public transit system, providing a direct link to the formal city, while the three stations along the mountain ridge meet demands of community access and provide missing social infrastructure (Urban-Think Tank Archives)

adapt to changing needs, they also form a prototype that can be (and is currently being) introduced to other locations. Designed as an inexpensive shed building, the MetroCable has a standard set of design components, such as structural elements and access ramps, that are easy to manufacture, erect, and produce. In addition to these standard components, the toolbox consists of a series of elements that directly respond to the community's specific needs.

While the cable system itself will draw power from the municipal source, each station is designed to be energy-independent, with a single wind-driven turbine to provide sufficient power for station needs. The shell of each station has been designed to accommodate solar panels, which will eventually provide sufficient power for San Agustín. The stations and adjacent structures require no mechanical HVAC. The facades have been designed as "wind-catchers," – louvered walls that direct prevailing breezes into the interior, enabling fresh air to enter and exhaust the stations. During daylight hours, the station interiors will rely exclusively on natural light, admitted from the sides and through translucent portions of the shell. At night, LED lights will light the stations and the public streetscape. The impact of the MetroCable extends beyond its physical limits to address the needs of the community.

Fig. 6 El Mangito station sits along the ridge of the San Agustín neighborhood (Iwan Bann)

2 Centro de Acçao Social por Música: Multi-functional Hub[10]

The Centro de Acçao Social por Música in Grotão, Paraisópolis, is an example of a multifunctional hub – a node in the larger framework of Urban Parangolé. Its introduction into the site provides immediate forms of social and cultural exchange, while simultaneously serving as a catalyst that encourages new uses and direct connections to the surrounding areas. The project consists of a terraced landscape, music school, and community center, all conceived as one integrated building forming the new Hub (Fig. 7).

Paraisópolis is located in southwest São Paulo, adjacent to one of the wealthiest neighborhoods in the city, Morumbi. However, this marginalized community is effectively separated from the formal city. Adding to an already challenging geography of hills and rivers, a new zoning law enacted in the 1970s discouraged legal development in this area (due primarily to prohibitive lot sizes). As a result, there was a boom of invasion that marked the beginning of what is now São Paulo's

---

[10]Designed by Urban-Think Tank, Paraisópolis, São Paulo, Brazil, 2009-, Client: Secretaria Municipal de Habitação (SEHAB).

Fig. 7 The terraces provide a framework to integrate previously fragmented areas with new social infrastructure and diverse programs to strengthen collective identity and ensure positive growth for the future (Urban-Think Tank Archives)

second largest informal settlement. Today there are approximately 80,000 people living in Paraisópolis – an area covering less than one square kilometer – without adequate social infrastructure, public space, economic opportunities, or mobility.

Within this isolated community, increased erosion and dangerous mudslides have designated several areas, including the project site, as high-risk zones.[11] This designation came after a heavy rainfall and severe mudslide destroyed a large area of homes, leaving the site as a primarily inaccessible void in the otherwise dense fabric. The Centro de Acçao Social por Música fundamentally transforms this void into a productive hub and dynamic public space through social design – a process of analyzing the local effects of rapid growth and improving marginalized settlements through social infrastructure. In order to handle the challenging topography, new section profiles were necessary to retain the hill and stop further erosion. In addition to stabilization, the new terraced landscape transforms Grotão into a 'natural arena' that encourages diverse community participation. The intervention opens the edges of the void to re-establish connections within the fragmented urban fabric. This connection is strengthened by a public elevator system that facilitates circulation

[11] High risk zones are defined primarily by physical dangers, such as erosion, structural failure, and physical instability, and by social vulnerability, a metric used by the city to describe the social and economic health of an area.

Fig. 8 Fábrica de Música is a catalyst for positive change. A hub of innovative and cost-efficient technologies, it serves as a prototype (as well as a set of prototypes) that can be adapted to other sites (Urban-Think Tank Archives)

through the site and topography. New programs are introduced within this system where they were once categorically neglected. Localized moments of this program, which include sports facilities, urban agriculture, public space, transportation infrastructure, and the music school, are connected to all boundaries of the area by the landscape of activated terraces. Urban agriculture encourages a new micro-economy, replacing the previously unusable space with an active and productive surface that feeds directly back into the neighborhood. The project is a dynamic hub with both fixed and flexible programs.

Despite the strong cultural dynamic that fuels daily life in this favela, rapid urbanization and challenging topography have resulted in a community that has developed without the necessary social infrastructure and physical equipment. There are no major hubs or networks within the neighborhood to accommodate the residents' needs. Therefore, the priority of the intervention becomes more than equipping this peripheral neighborhood with water, sewage networks, lighting, and services. These physical infrastructures must be coupled with additional social infrastructure in the areas of education, safety, culture, public space, and sports. This coupling ensures stronger social sustainability as a vital catalyst for the area, expanding music and cultural programs into the favela while forming a new network that serves the youth from all levels of society. This proposed urban model aims to translate a society's need for equal access to employment, technology, services, education, and resources – fundamental rights for all city dwellers – into spatial solutions (Fig. 8).

These new infrastructures are designed to meet highly effective standards of ecological sustainability through simple, low-tech means. The building and landscape work as one comprehensive unit that utilizes the varying conditions of the wet/dry season cycle. Prevailing winds in combination with an in-floor cooling system provide efficient tempering. The waste heat of the floor system is stored in the terraces (effectively a large heat sink) during the day and is emitted through hybrid-PV panels on the roof of the music school at night. Excess heat is vented through a wind-supported solar chimney.

Water is also reused on site; what was once seen as an inescapable danger is now a valuable resource. Constructed wetlands filter the water through the site for irrigation, gray-water applications, or further rapid sand filtration. An on-site water tank distributes excess water to the sewage system during the rainy season and stores it for later use during the dry season. Additionally, building materials and operability maximize light and minimize solar gain. Through the combination of active and passive systems, the Centro de Acçao Social por Música exceeds the expectations set forth in recent guidelines for environmental sustainability proposed by the city of São Paulo (Fig. 9).

This comprehensive system of social infrastructure, public space, active and passive building systems, mobility, and productivity is a prototype that can be utilized to address other high risk zones and similarly challenging spaces, both within São Paulo and globally. It is an adaptable framework that encourages flexibility, critically re-thinking the design process for these conditions. This project proposes that architects eschew their conventional role in traditional hierarchies to serve as an enabling connection between the opposing forces of top-down planning and bottom-up initiatives. By creating common ground for these two forces, we eliminate divisiveness and generate productive interactions.

3 Torre David: Latent territory of Urban Parangolé and the Syncretic City[12]

Torre David is representative of the territory of Urban Parangolé found within the existing city. The building is an abandoned and subsequently invaded 45-story office tower in Caracas, Venezuela. It is now home to over 3,000 people. In our trajectory of urban research on informality, Torre David presents a shift from the marginalized fringes of the city to the urban core. The tower also presents a profound shift in type – from the nearly complete separation of informal settlements from the

[12]Research Project conducted by the Urban-Think Tank Chair of Architecture and Urban Design, ETH Zürich, with Schindler and the Chair of Architecture and Sustainable Building Technologies (SuAT), ETH Zürich, 2012.

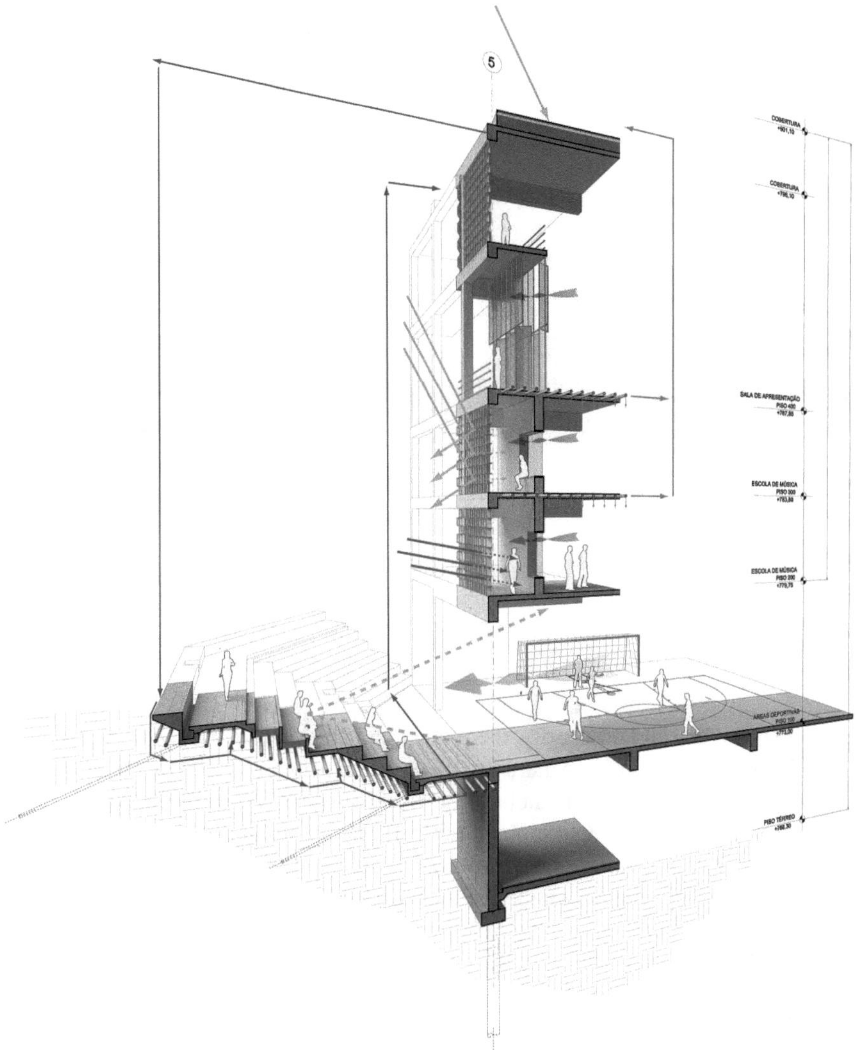

Fig. 9 The buildings and the landscape work together to create a comprehensive combination of active and passive systems, ensuring an integrated, low-tech sustainable solution (Urban-Think Tank Archives)

formal city to an informal settlement within a nearly complete formal structure. It is a physical combination of the formal and informal, a territory where new thinking is possible. This occupied tower represents how a neglected urban space has transformative potential (Fig. 10).

In January 1990 construction began on Centro Financiero Confinanzas (known as Torre David after the developer, David Brillembourg) in the heart of the city's political and financial center. As a set of mixed-use towers intended to represent the

Fig. 10 (*Left*) Torre David today, a vertical slum without elevators, air conditioning, and any formal infrastructures. (*Right*) The tower transformed through micro-scale interventions that the have improve living conditions and foster urban communities (Urban-Think Tank Chair/ETH, Zurich)

epitome of luxury and prosperity, the project had an expected cost of approximately US$82 million. However, in January 1994, Venezuela was hit by a series of bank closures that brought the financial sector to its knees. As a result, Torre David was seized before its completion and the building sat vacant for over 12 years. On September 17, 2007, during heavy rainfall, a series of events resulted in hundreds of families seeking shelter outside the locked gates of the tower. The two guards on duty allowed the families to enter, marking the beginning of the current occupation of Torre David. Today, this vertical village *is* a mixed-use tower with residences, micro-economies, and various programs, formed now by incremental, bottom-up processes (Fig. 11).

The tower is a doppelgänger. It maintains the image of the original tower but haunts the city with an alternative set of desires. It sets out a new agenda not wholly outside the lines of current urbanization but with alternative trajectories born out of its unique occupation. The tower already contains new modes of thinking and ways to organize space and building systems (water, electricity, circulation, public space). The tower is the interface between the formal and informal – the territory of Urban Parangolé.

Torre David is not an anomaly; it is evidence of a global condition of informality. Our research on the conditions of the tower has resulted in a new understanding of urban territories and an affirmation that they can foster more comprehensive urban sustainability – ecological, social, and economic (Fig. 12). As such, the tower represents an opportunity to examine and test new urban models within the

Fig. 11 Interior view of an apartment in Torre David (Iwan Bann)

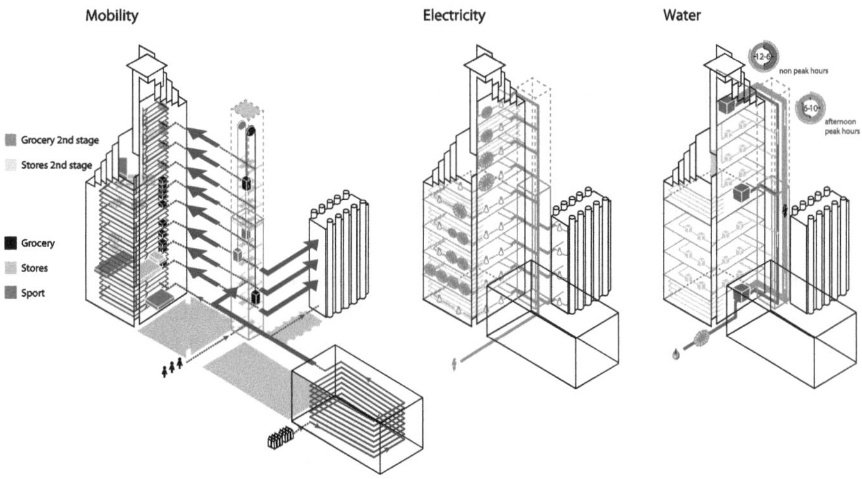

Fig. 12 Micro-scale interventions rethink tradition methods of vertical mobility, electricity generation/infrastructure, water supply, and energy storage (SuAT, Urban-Think Tank Chair/ETH, Zurich)

existing urban fabric. The research engages the possibility of alternative energy and mobility infrastructures by developing several decentralized system prototypes for the tower – a wind energy system, a hydro storage system, and a counterweight elevator. These micro-scale interventions have the potential to improve lives and foster urban communities (Fig. 12).

Small-scale wind energy systems can utilize the height of the tower to produce a supplementary energy supply and significantly shave peak demand. In conjunction with the wind energy system, a pumped pico hydro storage system takes advantage of strong community organization to coordinate water storage, distribution, and electricity production. The potential energy of the water that is pumped to and stored on the upper levels of the tower is utilized during regularly scheduled distribution times. The wind energy system provides the initial energy to pump the water to the storage tank.

A new elevator system also re-envisions traditional methods of circulation in high-rise buildings. It utilizes community organization by operating similar to a public transportation system – working on a fixed schedule as opposed to individual demand. The system uses a counterweight principle to balance people, goods, materials, and waste. It stops every five floors and is connected to an exterior ramp system that provides access to the other floors. The combination of these decentralized systems allows the tower levels to be understood as packages, or bundles, each with its own unique energy and mobility demands.

These small-scale interventions point the way to alternative models that address the challenges of informality, energy, and mobility. Their significance lies in their role as catalysts that re-think the city by building the new onto and within the old. Their positive effect is a consequence of a multidisciplinary approach that reorganizes existing conditions and knowledge into useful tools and new modes of operation. It is the beginning of the Syncretic City.

4 Urban Parangolé: A Field Guide

Design the Syncretic City Seek to foster effective interaction between the formal and informal, between the conflicting forces of the city, to create flexible spaces and inclusive environments. It is not a desire for homogeneity; it seeks productive coexistence.

Redefine Mobility Mobility is more than transportation lines. It has the potential to become a productive activity, one that also serves as a framework for necessary program and social infrastructure.

Redefine Sustainability It is essential to expand the definition of sustainability to encompass social and economic imperatives as formative potentials.

Engage the Existing City Urban Parangolé perceives the city as a catalytic realm of potential new trajectories. The new city will be built on and with the old.

Expand The architect is an animator of change – an interface between top-down planning and bottom-up initiatives. Designers provide the necessary platform for an interdisciplinary practice with purpose-oriented social design strategies.

Unfold Find and occupy the latent territory of the city by shifting the understanding of urban challenges from inevitable symptoms of outmoded models to new opportunities.

Integrate Integrate social infrastructure with physical infrastructure; integrate mobility and sustainability; integrate environmental concerns with social and economic imperatives.

Challenge the Status Quo Conventional practices must be questioned. Alternative urban models must be explored in order to address the asymmetrical urbanization process.

High Performance Buildings: Measures, Complexity, and Current Trends

Bryan Eisenhower

Abstract Most people are intimately involved with the built environment, while unfamiliar with detailed aspects of its design and operation. Buildings are everywhere and are designed and equipped using an agglomeration of many different design elements or puzzle pieces. With the recent trends towards a more energy efficient world, there has been an attempt to make buildings more efficient by using highly efficient pieces of the design puzzle. Not always does the integration of these subsystems result in an efficient building as a whole. The goal of this chapter is to highlight some of these boundaries and current engineering trends to surpass these obstacles. The discussion will be focused on large commercial buildings in the United States, while similar concerns are prevalent in other building types and in other global locations. We start by highlighting different ways that performance is measured and review the different design elements and equipment choices that are available to construct a building. The large number of interacting components creates complexity and a challenge to obtain a high performance structure. Specifically, technology barriers to realizing high performance buildings through this integration process lie in the ability to create useful models, data analysis tools, and effective control strategies. The chapter concludes with some current applied research in building systems that address the complexities in building systems and methods being developed to overcome the barriers that lie in the way.

B. Eisenhower (✉)
University of California, Department of Mechanical Engineering, Center for Energy Efficient Design, Santa Barbara, CA, USA
e-mail: bryane@engineering.ucsb.edu

S.Th. Rassia and P.M. Pardalos (eds.), *Cities for Smart Environmental and Energy Futures*, Energy Systems, DOI 10.1007/978-3-642-37661-0_8,

1 Measuring Performance

In order to gain insight into how a building is performing, building performance measures are typically normalized so that it can be compared with other buildings of similar type. The normalization includes a numerator and denominator, while it is up to the analyst as to what these two quantities contain. One of the most common metric pairs is dividing the annual energy consumption by the floor area of the building resulting in the Energy Use Intensity (EUI). The EUI is insightful in terms of energy use, but the purpose of the built environment is manifold and often energy use is not the primary concern. For instance, the purpose of a storefront property may be to get customers to visit, browse, and purchase merchandise – therefore an intensity of visits or transactions per floor area may be a more insightful metric. In an academic setting, the purpose may be to provide comfortable environments for lecturing and studying and therefore an intensity based on degrees earned may serve the analysis better. Typically, a combination of many metrics is best and depends on the building type or purpose.

To get an understanding of how relevant different types of buildings are, the United States Energy Information Administration has performed the Commercial Buildings Energy Consumption Survey (CBECS) to capture the cross section of building types in the United States. In 2003 the top four types of buildings (by floor area) were: office (14 %), mercantile (13 %), education (11 %), and enclosed malls (8 %) (U.S. Energy Information Administration 2003). Since almost half of the building stock falls into either retail or office-like environments, it should be clear that a sole measurement of energy may not be the best indicator for a high performance building. One measure that is often used for buildings with a high density of human occupants is indoor air (or environment) quality and thermal comfort.

Quantifying Thermal Comfort Occupant satisfaction takes many different factors into account including lighting or noise levels, indoor air quality, and thermal comfort. Some of these factors can be directly measured using various equipment or implied by taking surveys. Lighting levels traditionally stay fairly stagnant as the building ages, while noise levels may change as equipment wears or space is re-purposed. On the other hand, both indoor air quality and thermal comfort are quantities that can vary drastically from the first day the building is occupied, based on how the building equipment is operated.

Thermal comfort analysis has been a topic of study since the seminal work of Povl Fanger in the 1970s. In his work and research to follow, it was found that factors influencing human thermal comfort include metabolic rate, humidity, air speed, radiant temperature, air temperature, and clothing. Using these quantities, a satisfaction number can be calculated based on measured or modeled conditions in an environment. In most climates, conditioning an environment to have optimal thermal comfort is costly from an energy standpoint and therefore it is important to make the environment as comfortable as needed without excessive cooling or heating.

Energy Consumption in Buildings To better understand the role of building design and operation in terms of energy consumption, it is important to understand just how much energy buildings consume in the context of energy expenditures. Over the past 30 years, commercial and residential buildings have consumed approximately 40 % of United States Energy (U.S. Department of Energy 2012), which has not changed considerably (the mean over those years is 37 %, with a variance of 0.032 % with no discernable trend). In 2010, the source split for this energy was about 25 % fossil fuels, 1.5 % renewable, and 72.9 % electricity (of which the majority is generated by fossil fuels in the U.S.).

Looking at the end use split, which identifies which components in a building consume this energy, it is evident that about 50 % of the energy in buildings is consumed by heating, cooling, ventilation, and lighting. Advances in the design of this equipment is making them more efficient at a component level. However, regardless of how efficient any of this equipment is, energy consumption is still driven by the way that this equipment is used. For instance, even the most efficient cooling technology will waste large amounts of energy if it is left on at times when cooling is not required due to lack of occupant presence. In light of this, advanced control methodologies are being developed to optimize such operational strategies. The rest of this chapter begins to address this approach, while first highlighting challenges due to the fragmented and puzzle-like nature of designing and equipping a building.

2 Elements of Building Design and Operation

Generally speaking, as there are unique purposes and design constraints for every building, almost every building is a unique design in its own right. Buildings are one of the only commodities that are still predominantly designed and built with significant human involvement. They are designed and equipped by aggregating different architectural concepts and approaches with different types of mechanical equipment – assembled to accomplish an environment suited for the building owner. The integration is not always effective – a high performance architectural design may not cooperate with the function of high performance equipment, which presents a challenge when designing and operating a building. With this in mind, we briefly introduce some common elements of architectural design and equipment selection that contribute to this puzzle. This short review will only give an overview of some of the elements, the focus is to highlight that indeed there are many different elements of a building, all of which must be effectively integrated at the building design, construction, and commissioning stages.

Architectural Elements Siting is one of the first design considerations as it defines the boundary for all the finer details of a building. Other than aesthetic concerns, there are many design parameters in siting, one of the most obvious design parameters is the orientation of the building. The elongation of a design can be leveraged, (e.g. along East-West orientation to capture free sunlight during certain times in the year), or the aspect ratio can be adjusted – a relatively short building may be beneficial in a hot and arid climate while in a colder climate, a taller version may capture and contain rising heat through the floors (Friedman 2012). Another important concern is local wind currents as wind can amplify natural ventilation, and create cooler ambient temperatures around the building – it is important to assess adjacent buildings, tree breaks or other barriers because of this. Natural shading is an important design consideration as it will influence lighting needs as well as cooling and heating loads (from solar heat). Similarly, water in the vicinity of a building will cause solar reflection and evaporative cooling near the building.

The second aspect of building design is the structure or envelope which constitutes constructions, shading, and fenestration (e.g. windows, doors, skylights). For exterior constructions the designer must consider the amount of isolation that is desired from the ambient environment. The interplay between typical outdoor conditions and heat generated inside the building define this balance. Typically, a significant degree of isolation is needed, which is achieved using constructions with highly resistive insulation. There are many types of building materials that can be used for this insulation that have different characteristics ranging from their surface attributes (influencing convection and radiation) to its thermal mass.

The thermal mass of a building influences its dynamic response to changing conditions. Constructions with more mass tend to absorb and release heat at slower rates than constructions that are lighter. Because of this, cooling needs in the summer can be delayed in time, past the mid day peak of solar radiation, to later in the day when the building may not be occupied. The design and control of buildings using thermal mass has been discussed in Braun et al. (2001) where the operation of the cooling system takes into account thermal transients to optimally schedule the building cooling system. Recent advances in phase change materials are adding a new design element to capture the same dynamic effect in smaller packages.

Shading is a design element that trades freely available visible light with heat from solar radiation which include natural obstructions (e.g. foliage), or internal or external window shades. Consideration must be taken as to sun angles at different times of the year as well as foliage cover when plants are used. Man-made shading devices can be automated to change their angles or position based on different times of the year or day.

The main goal of windows is to add a visual aesthetic to the indoor environment and free light, while limiting conductive and radiant heat transfer through the glass. To impede conduction, multi-layered glass is designed into a window frame that establishes stagnant gas (air, argon) which resists thermal conduction. Beyond this, spectrally selective glass coatings are used to permit visible light while rejecting radiation and switchable glazing designs are available that change the properties of the glass as needed by command. A significant concern is air leakage and conduction through window frames which is a factor in the design of high performance windows.

Lighting and Mechanical Equipment In addition to the design elements that make up the structure of the building, the lighting and equipment play a key role in its functionality and comfort. How a building is lit depends on its usage, different lighting levels are needed for an office environment, data center, or agricultural terrarium. Harnessing visible light from the sun is useful as it saves lighting costs, while it may also introduce unnecessary cooling requirements. Different design approaches are used to amplify freely available light without too much heat gain including light ducts, internal atria, light shelves, etc. (Baker and Steemers 2000). The idea in these designs is to bounce light off of reflective surfaces to distribute it throughout the core of the building. There are numerous evolving technologies for electrical lighting including compact fluorescent and Light Emitting Diode (LED) bulbs. All of the lighting technologies can be coupled to electronic management systems that control light levels during occupied times or the balance between electrical and natural light when available (through dimming).

Other mechanical equipment addresses the need for fresh and conditioned air. Fresh air is needed for occupant satisfaction and is typically measured by the number of Air Changes per Hour (ACH). An ACH is the number of times within an hour that the volume of air a room is refreshed. Typical values for ACH are 4–6 for an office building and up to 600 for a Class 1 clean room and are regulated by the American Society of Heating, Refrigerating and Air-Conditioning Engineers standard 62.2 (ASHRAE 2010). Historically fresh air is brought into an environment through non-mechanical natural ventilation, which can be enhanced by designs that amplify buoyancy effects or local wind currents. Electronically controlled louvers or fans can also enhance this approach. The issue is that natural ventilation is only effective when the climate is mild and conducive to thermal comfort inside the building. It should be noted that a building also is never air tight due to construction tolerances and doorways, a leakage rate of 0.5–10 ACH is not uncommon depending on pressure differential due to mechanical equipment, local wind profiles, and occupant activity.

Mechanically conditioned air is needed to remove excess humidity or cool or heat outdoor air to meet indoor comfort requirements. This conditioning is achieved in ducted systems by distributing air that is conditioned in a central location (e.g. on the roof) throughout the building. In extreme climates, conditioning this air is energetically intensive and hence recirculation is used that blends fresh outdoor air with air that is exhausted from occupied areas. This is performed in an economizer mixing box which uses control logic to actuate dampers in a way that only the optimal amount of fresh air that is needed is supplied to the occupants, which limits energy consumption. Newer technologies are being developed that include energy capture within exhausted air to precondition the incoming fresh air through various types of air-to-air heat exchangers (e.g. a heat recovery desiccant wheel Smith 2007).

Two main types of ducted systems exist; single, and dual-duct arrangements. In single duct systems, in the summer months, air is cooled to a specified setpoint in a central location. The air is then distributed to different locations within a building reheated as necessary at the termination of the duct. This is needed as some areas of the building will require more cooling than others. In a dual-duct arrangement, a cold and hot air duct pair traverse the building and a mixing box is located at the terminal location that mixes these two streams as needed to obtain the required cooling (or heating) that is needed. A third type of air distribution is underfloor air distribution in which the entire area underneath an occupied space becomes the plenum (because of this, it is often called ductless) and conditioned air is introduced into the occupied space through vents in the floor.

The control of ducted systems includes scheduling and regulating dampers, louvers, and fans to maintain air flows at desired levels. Fan power is a significant contributor to energy costs and therefore smart controls are desired to minimize energy costs. Beyond control of fresh air mixing in the economizer, other strategies include flushing the building with extra air during unoccupied times in order to leverage beneficial conditions during nighttime hours, or optimized setpoint scheduling for the delivery temperatures of either single, or dual-duct air supply temperatures.

Ventilation is important for indoor air quality, while heating and cooling equipment are used to address other aspects of thermal comfort. Heating can be achieved by passive means, by harnessing solar energy. As mentioned, orientation, shading, and window size and selection all play roles in capturing or preventing solar heat. A Trombe wall is a specific design element, which uses solar radiation to heat a thermal mass that is typically just inside the envelope, with physical dimensions and layout that amplify convective transport between this wall and the rest of the environment. Energy intensive heating can be achieved through the use of various boiler devices, which typically use fossil fuels or electricity to heat water or create steam. Heat pumps, move energy between ambient conditions (air or earth for example) and may be used for heating in milder climates. Electric heaters which use resistive elements to produce heat are used at smaller scales because they are typically less efficient than other heating approaches.

In terms of cooling, similar energy intensive equipment is operated to generate cold air or refrigerant which is distributed either through the building or is used to cool air at a central location (that is then distributed through the building). One such system is a chiller plant which utilizes a refrigerant that is compressed and expanded in a heat exchanger that is then exposed to water that is subsequently chilled. A ground source heat pump is another similar technology that uses such a cycle to move heat from a building to the earth – leveraging the relatively stable temperatures just under the surface (or within bodies of water).

Design and Equipment Integration Even with this brief introduction to the elements of building design, it is clear that many different technologies play a role in designing and equipping a building. Integrating all of these design elements into a system of systems that functions optimally is a significant challenge. A schematic of heating and cooling loops for a typical equipment design is presented in Fig. 1, which shows multiple loops that transport conditioned air into a occupied zone, as well as water loops for chilled and hot water distribution and a refrigerant loop that is used to create this chilled water. Even for one zone, the schematic

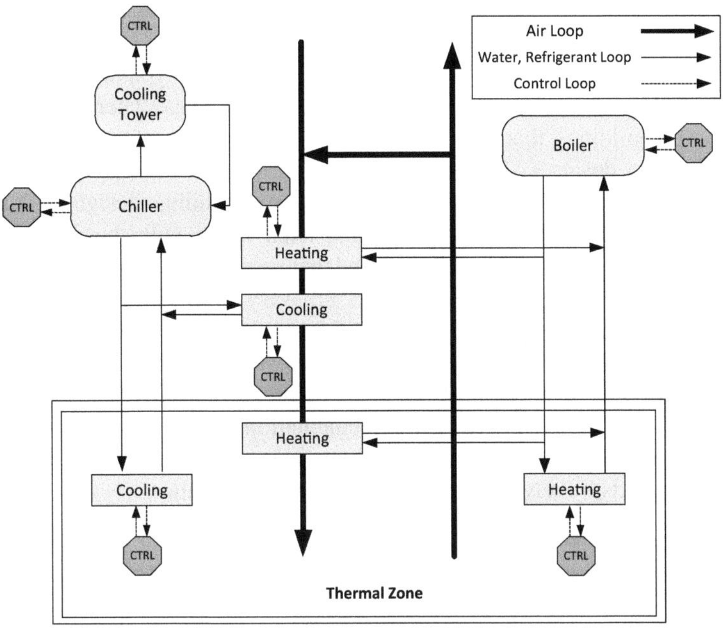

Fig. 1 Schematic of one approach to heating, ventilation, and cooling a building. These systems are spread throughout the entire building, operate on different time scales with different media (e.g. air, water, refrigerant), and are manufactured by different companies. The control and management systems are also heterogeneous in design and implementation

is complex – especially when considering the wired control loops that actuate fans, dampers, valves, pumps, compressors, and other mechanical devices. The complexity is further compounded when considering that there are often many zones on a floor of a building, and the loops are separated by large spatial scales. In addition to this, they respond at different time scales and are exposed to different disturbances from weather, occupants, and other un-predictable sources. Most of these systems are acquired from different component manufacturers, which have different design intents – that do not consider the building as a whole – only their specific component. All of these reasons make the integration of these equipment in a productive way a challenge.

Despite these challenges, with the recent rise in environmental awareness there have been initiatives established, in both private and governmental arenas, that seek to address some of the concerns of designing high performance buildings. The Architecture 2030 Challenge (Arhitecture 2012) is one such approach with the intent to promote collaboration between government, component manufactures, and building designers and operators to achieve carbon neutrality in buildings by 2030. Similarly, the United States government put forth over $25 Billion in 2009 for energy efficiency measures through the American Recovery and Reinvestment Act of 2009 (American Recovery and Reinvestment Act 2009), much of which was for making the building stock more efficient.

High performance buildings are certainly achievable but typically require longer design and commissioning cycles than standard buildings. There are plenty of examples of buildings that have EUI's that are a fraction of a similar building with a standard design (ASHRAE 2012). In Turner and Frankel (2008) a study was performed that investigated high performance building designs – buildings that were intended to be highly efficient and rated with Leadership in Energy and Environmental Design (LEED). It was found that surprisingly, many of the buildings that were intended to be high performers fell very short of their target. In fact, some buildings that were predicted to use 25 % of the energy that a standard design consumed 50 % *more* energy than a standard design. Among many performance obstacles that were identified, effective modeling, monitoring, and control were three technical barriers to successful implementation of high performance building designs. These three topics will be discussed below in the context of current engineering trends for advanced building design and operation.

Building Energy Modeling Building energy modeling is a process that takes the design elements of a building (as listed above) and creates a virtual physics-based model capable of calculating energy consumption and comfort throughout an annual climate cycle. Reasons for energy modeling vary from performing compliance checks (e.g. for LEED rating), to performing design alternative trade studies in architectural firms, to research where model-based design optimization, performance monitoring, energy diagnostics or optimal control is performed. One of the issues is that although there are many available energy modeling softwares, that accurately capture the relevant physical equations, enormous amounts of parametric information must be defined, all of which is significantly uncertain (Trcka and Hensen 2010). Current trends in the buildings community is to capture this uncertainty and systematically address the output from energy models as uncertain or probabilistic rather than a well defined answer (Eisenhower et al. 2012a).

Energy modeling is also evolving to capture more information than annual energy performance. Current efforts at the Lawrence Berkeley National Laboratory are underway to develop a model library that is capable of capturing fast timescale transients with sufficient fidelity to perform control system analysis at the building level (as opposed to studying control dynamics at the component level as in the past) (Wetter 2009). The use of energy models for optimization has been performed in the past on limited sets of parameters (Kämpf et al. 2010) while methods have been recently developed that handle large numbers of optimization parameters in rapid fashion (Eisenhower et al. 2012b).

Data Analysis The management systems illustrated in Fig. 1 create so much data that often only a fraction, if any, is investigated with any regularity. Data aggregation methods are one approach to resolve the issue of the enormity of data that is generated in building control and sensor systems. Approaches to this aggregation include principle component analysis (PCA) which highlights components in the data that contribute the most signal energy (see Du et al. 2007 or Ruch et al. 1993). While these methods have been used for some time with a degree of success, this approach masks important information from expected temporal variation in the dynamic response of a building. That is, buildings are exposed to periodic disturbance from climate, occupancy, and control system scheduling and regulation. These disturbances are evident in data and current research is being performed to use spectral decompositions on building system data as a means to aggregate large amounts of data (Eisenhower et al. 2010).

Beyond this, a second approach to utilize data in a more meaningful way is to implement automated fault detection routines into the computers that control the equipment in the building (Katipamula and Brambley 2005). The objective of these methods is to use algorithms to identify when and how equipment is performing sub-optimally. The fault is then prioritized and a list of possible actions are performed to account for the degradation (e.g. tolerate the problem, shut down the system, or reconfigure its operational mode).

Advanced Building Control Control in buildings is composed of hierarchical loops, each with different purposes, and advances are being made at all of these levels. At the finest level (the *inner* loops), the objective of the control system is to actuate valves, fans, pumps, compressors and other mechanical equipment. Advances are being made in terms of exploiting the interaction between different control loops through the synthesis of Multi Input, Multi Output (MIMO) controllers (see Anderson et al. 2008; Jain et al. 2010).

At a more coarse level, the *outer* loop objectives may be defined to identify what the *right* settings are for the inner loops. For instance in the dual or single duct ventilation systems described above, the optimal selection of the supply temperatures plays a key role in being able to maintain a thermally comfortable environment without wasting too much energy. Recent methods are being developed to integrate models into the process of identifying these optimal targets using model based control (Corbin et al. 2012; Ma et al. 2012; Oldewurtel et al. 2012).

3 Summary

While buildings are ubiquitous, each is unique in the way they are designed and equipped. The vast number of design elements and choices available to create an occupiable structure lead to many efficiency pitfalls when integrating the different puzzle pieces of the design. Recent advances in modeling, data analysis, and control are leading to better building designs and operation strategies. At this point many of these approaches are isolated cases, while with time for sufficient maturity, these methods will be scalable for larger stocks of the built environment. These advancements are helping to contribute to a world focused on efficiency that depends on more renewable resources ensuring a energy secure future and stable environmental outlook.

Acknowledgements The author would like to thank Professor Igor Mezić (University of California, Santa Barbara) and Professor Zheng O'Neill (University of Alabama) for their comments on this manuscript.

References

Anderson M, Buehner M, Young P, Hittle D, Anderson C, Tu J, Hodgson D (2008) MIMO robust control for HVAC systems. IEEE Trans Control Syst Technol 16(3):475–483

Arhitecture 2030 (2012) http://architecture2030.org/. Accessed Aug 2012

ASHRAE (2010) ANSI/ASHRAE 62.2-2010,ventilation and acceptable indoor air quality in low-rise residential buildings (ANSI/ASHRAE approved). American Society of Heating, Refrigerating and Air Conditioning Engineers, Atlanta

ASHRAE (2012) High performance buildings. American Society of Heating, Refrigerating and Air Conditioning Engineers, Atlanta

Baker N, Steemers K (2000) Energy and environment in architecture. Taylor and Francis, London/New York

Braun J, Montgomery K, Chaturvedi N (2001) Evaluating the performance of building thermal mass control strategies. HVAC&R Res 7(4):403–428

Corbin CD, Henze GP, May-Ostendorp P (2012) A model predictive control optimization environment for real-time commercial building application. J Build Perform Simul 6(3):159–174

Du Z, Jin X, Wu L (2007) PCA-FDA-Based fault diagnosis for sensors in VAV systems. HVAC&R Res 13(2):349

Eisenhower B, Maile T, Fisher M, Mezić I (2010) Decomposing building system data for model validation and analysis using the koopman operator ibpsa national conference. In: Proceedings of the third national conference of IBPSA-USA, New York

Eisenhower B, O'Neill Z, Fonoberov V, Mezic I (2012a) Uncertainty and sensitivity decomposition of building energy models. J Build Perform Simul 5(3):171–184

Eisenhower B, O'Neill Z, Narayanan S, Fonoberov V, Mezic I (2012b) Methodology for meta-model based optimization in building energy models. Energy Build 47:292–301

Fanger PO (1970) Thermal comfort. Danish Technical Press. Copenhagen

Friedman (2012) A fundamentals of sustainable dwellings. Island Press, Washington, DC

H.R. 1–111th Congress: American Recovery and Reinvestment Act of 2009: In GovTrack.us (database of federal legislation) (2009). Retrieved 30 Aug 2012 from http://www.govtrack.us/congress/bills/111/

Jain N, Li B, Keir M, Hencey B, Alleyne A (2010) Decentralized feedback structures of a vapor compression cycle system. IEEE Trans Control Syst Technol 18(1):185–193

Kämpf J, Wetter M, Robinson D (2010) A comparison of global optimization algorithms with standard benchmark functions and real-world applications using energyplus. J Build Perform Simul 3(2):103–120

Katipamula S, Brambley MR (2005) Methods for fault detection, diagnostics, and prognostics for building systems a review, part I and II. HVAC&R Res 11(1):3–25

Ma Y, Kelman A, Daly A, Borrelli F (2012) Predictive control for energy efficient buildings with thermal storage. IEEE Control Syst Mag 32(1):44–64

Oldewurtel F, Parisio, A, Jones C, Gyalistras D, Gwerder M, Stauch V, Lehmann B, Morari M (2012) Use of model predictive control and weather forecasts for energy efficient building climate control. Energy Build 45:15–27

Ruch D, Chen L, Haberl J, Claridge D (1993) A change-point principle analysis (CP/PCA) method for predicting energy usage in commercial buildings: the PCA model. J Sol Energy Eng 115:77–84

Smith PF (2007) Sustainability at the cutting edge. Elsevier, Amsterdam

Trcka M, Hensen JL (2010) Overview of HVAC system simulation. Autom Constr 19(2):93–99
Turner C, Frankel M (2008) Energy performance of leed for new construction buildings. Tech. rep., U.S. Green Building Council
U.S. Department of Energy (2012) 2011 buildings energy data book. Energy efficiency and renewable energy. http://buildingsdatabook.eren.doe.gov/
U.S. Energy Information Administration (2003) Commercial buildings energy consumption survey (CBECS). http://www.eia.gov/consumption/commercial/
Wetter M (2009) Modelica-based modeling and simulation to support research and development in building energy and control systems. J Build Perform Simul 2(2):143–161

Eco-cities: The Role of Networks of Green and Blue Spaces

Fabiano Lemes de Oliveira

Abstract The chapter analyzes how networks of green and blue spaces are being used in the design of new eco-cities. These projects aim to define a new city typology, create models of sustainable development that would contribute to the resolution of the global need for more urban spaces, increase well-being and decrease ecological footprint. In this regard, green and blue infrastructures are playing fundamental roles in current eco-city design. Initially, the text presents our contemporary urban challenges, the emergence of the idea of the eco-city and the new meanings that networks of green and blue spaces are acquiring. Four recent case studies are analyzed and compared in order to draw conclusions on the significance of these systems of open space in the construction of a theory and practice of eco-city design.

1 Introduction

The challenges that cities need to face are huge. The world's population is growing at unprecedented rates, impacting significantly on the nature of our urban and natural environments. By 2050 we will be nine billion people on the planet, of which 70 % will be living in urban areas. That is the reason why Peirce et al. (2008) have rightly called the twenty-first century, the 'Century of the City'. Recognizing this trend is crucial to taking action, and hopefully by the time we have gone halfway through the 2000s we will be able to say that we are living in the 'Century of the *Sustainable and Equal* Cities'. The instruments, paradigms and methodologies that would lead us to this outcome are currently in the process of definition, development, testing and evaluation. In this regard, new standalone cities with the

F. Lemes de Oliveira (✉)
University of Portsmouth, Portland Building, Portland Street, Portsmouth PO1 3AH, UK
e-mail: fabiano.lemes@port.ac.uk

S.Th. Rassia and P.M. Pardalos (eds.), *Cities for Smart Environmental and Energy Futures*, Energy Systems, DOI 10.1007/978-3-642-37661-0_9,

prospect of becoming new flagship models of urban sustainability – the eco-cities – have been proposed around the world, particularly in the Far East. We are also seeing the return of attention to the creation of networks of open space, not only now for leisure, but also and foremost to support integrative social life, healthy lifestyles and sustainable cities. This chapter concerns both contemporary preoccupations. It looks into the nature of networks of green and blue spaces in current eco-city proposals, and is particularly interested in the roles that these systems play in helping these initiatives achieve more sustainable urban environments. In essence the work investigates how networks of open spaces are being used in the theoretical and practical construction of the eco-city as a new city typology claiming to become models of sustainable urbanism for the future.

In its first part, the chapter presents our contemporary challenges and the paradox of the need for more cities at the same time we need to lessen our ecological footprint. It then discusses the emergence of the eco-city ideal and why and how green and blue infrastructures have returned to the minds of the built-environment professionals. The work then analyzes how some of the newly created eco-cities organize their green and blue infrastructures, and how they envisage their role in contemporary life and in the future urban environment. Finally, the chapter establishes some comparisons among the case studies and discusses its findings on how networks of green and blue spaces are essential elements of planning more sustainable and equal cities.

For the purpose of this chapter, we focus only on proposals for the creation of new standalone cities, as opposed to the building of neighborhoods, districts or retrofit projects. To find the case studies, we undertook an analysis of the relevant literature, searching for the key words 'eco-city', 'eco-town', 'eco development', 'sustainable development'. It was also part of the criteria that proposals had a serious attempt at planning networks of green and blue spaces, were contemporary projects and that there were sufficient available information. The cases selected to be analyzed here have a significant size, population, mix of uses and a clear boundary. Moreover, they can be considered to some extent self-sufficient entities. With regard to the networks of open spaces, we considered the designer's description of their approach when possible, combined with our visual analysis of the proposals. During the collection of data, it became apparent that many terms were used in the different design proposals to mean the same type of space. Based on the above and to allow for comparison and facilitate comprehension, a reduced set of terms for the typologies was defined in this study.

2 Times of Paradox: Cities, HDIs and Ecological Footprint

Why is it relevant to discuss the development of eco-cities today? What are the conditions of possibility that allowed them to emerge as one of the most ambitious attempts at promoting sustainable development on a large scale for our contemporary and future cities? For a start, recently the world's population reached seven billion people and the prediction is that we will be more than nine billion by 2050. To put this in historical perspective, just 50 years ago only 3 out of 10 people lived in urban areas. Today, the number is 5 out of 10 and by the middle of the century this will grow to three quarters of the globe's population (UN-HABITAT 2008, p. IX). To accommodate this phenomenon, and in order to avoid the proliferation of slums and other forms of deregulated urban sprawl, coordination in national planning systems and a proactive approach to densify our cities, promote urban expansions within clearly defined frameworks and create new cities is essential.

However, as well as the need to cater for more urban dwellers, we also need to reduce our ecological footprint in the face of climate change. It is widely held that cities are responsible for 80 % of our total CO_2 emissions (UN-HABITAT 2010; WWF 2010). Urbanization processes and climate change are therefore directly linked, which brings cities to the forefront of any solutions aimed at creating a sustainable future. Ecological footprint, carbon dioxide emissions and the need for resources are not equal across the planet (Ewing et al. 2010). Although only 20 % of human beings live in developed countries, they are responsible for 80 % of the global CO_2 emissions (UNITED NATIONS 2007). As the UNDP *Human Development Report 2010–11* shows, countries with very high HDI tend to account for larger carbon dioxide emissions (UNDP 2011). It has been found that as a country's HDI grows beyond a certain threshold, its impact on the environment also grows substantially. In essence, we need more urbanized areas, which in turn can increase our CO_2 emissions and ecological footprint. We also want more equal and better living conditions for all – that the HDI, particularly for those in development, rise – which in turn is bound to worsen our environmental problems. The questions then is, how can we promote more urban areas to accommodate the predicted population growth, improve wealth and quality of life across the globe without increasing our ecological footprint and carbon dioxide emissions, and indeed, even by bringing them down? As we know, it is no longer enough just to avoid extra strain over the environment, we need to drastically reduce the pressure we are already putting on the globe. It is in this context that eco-cities are being presented as possible resolutions to the paradox *urbanization–well-being–reduced ecological footprint*.

3 The Emergence of the Eco-City

Since 1970s, when our ecological overshoot (WWF 2010, p. 34) started to become increasingly noticeable, there has been a worldwide debate around the need to think of more holistic approaches towards urban development and the protection of the environment. The Bruntland Report, *Our Common Future* (1987), set forth the concept of 'Sustainable Development' as meeting the needs of the present without compromising the ability of future generations to meet theirs. 'Sustainability' would be comprised of three pillars: environmental, economic and social. Placing sustainability as a core preoccupation in the international political agenda was fundamental to seeing the discussions advance and attempts to implement change emerge. From Rio 92 Agenda 21, the Kyoto Protocol 1998, Johannesburg 2002 and Rio + 20, among others, different ideas on ecological development emerged; and more recently, discussions around how to define indicators and parameters to construct sustainable cities and urban spaces have manifested themselves in planning and urban design.

It is widely accepted that urban areas in all their manifestations need to be addressed, both by retrofitting and redeveloping brownfield and vacant intra-city sites, as well as by creating structured expansion and new cities from scratch. A large amount of research has been done on the relation between urban form and sustainable indicators (e.g. Beatley 2000; Jenks and Burgess 2000; Jabareen 2006; Roaf et al. 2005; Jenks and Dempsey 2005, etc.). But what would be the difference between a sustainable development and an eco-city? If the very same definition of "city" is a complex matter, that of "eco-city" is even more problematic and imprecise. Since Register's book Eco-city Berkeley (1987), the term has been brought to attention. Today, the word "eco-city" – because of the very lack of a clear definition and the current market value of green credentials– is being used loosely as a label in all kinds of circumstances. It has been appropriated by existing cities as a complementation of their names, such as the case of the *ÉcoCité* programme in France; by new neighborhoods and districts in existing cities, like BO01 in Malmö, Sweden, Logroño in Spain, Changxingdian in Beijin, Vauban and Hafen City in Germany or Whitehill-Bordon in the UK. Today there is a boom of new cities designed from scratch that have also adopted the label. These proclaim to aim at becoming new role models of sustainable development for the future, going beyond what has already been achieved in the last decades. The majority of eco-city projects are taking place in China and other non-European countries, which have started a movement to build cities from scratch that has not been seen in the world since the British New Towns of the post-war period. In reality, the impact of these projects could be extremely significant for achieving the levels of sustainability expected for the next decades, taking into consideration that the population in these countries is expected to reach eight billion by 2050 and that levels of urbanized land will tend to grow at similar rates.

There have been some attempts to define "eco-city" (e.g. Kenworthy 2006; Roseland 1997; Wittig 2007; White 2002),[1] but far from being clear-cut statements, they tend to set out general principles related to the Bruntland Report's concepts of environmental, economic and social aspects of sustainability. Wittig (2007), for instance, defines eco-cities as those in which ecological requirements are combined with socio-economic conditions. According to Rogers (2011), they must be compact, clearly defined and contained – the greenbelt being a good instrument to achieve the latter. Rogers also highlights the importance of good public transport and cycling and walking networks. In this chapter, we define an eco-city as a new urban

[1] Since Register's book there has been a series of International Ecocity Conferences. The last one was held in Montreal in 2011.

settlement of a significant size and population, which includes diverse uses, has a clear boundary and in which there is a fundamental attempt to promote sociable and equitable spaces and economic vitality, using minimal natural resources and generating minimal ecological footprint. Eco-cities should aim to:

- Promote healthy and less-consumerist behaviours,
- Reduce ecological footprint to levels below the world average biocapacity per person,
- Increase resource efficiency,
- Reduce waste,
- Be low carbon,
- Aim for relative self-sufficiency,
- Be socially inclusive and integrated,
- Provide good public transport and quality facilities,
- Promote walking and cycling,
- Work both at regional and local levels,
- Be compact and aim for medium-high density,
- Achieve medium-high density,
- Adopt mixed-uses and diverse social mix,
- Facilitate economic efficiency,
- Promote biodiversity and greater access to green spaces, and
- Engage in sustainable food production.

Eco-cities are today bubbling up, particularly in the Far East. They are experimental undertakings that have encountered fertile testing grounds, which are being promoted both by local governments and by private developers. Lian et al. (2010) draw attention to the fact that they not only serve as pilot projects where innovative ideas can be put to test, but also as investment attractors, which has led to an emerging consensus in the region regarding the eco-cities' potential.

4 Green and Blue Infrastructure in Eco-City Design

The history of integrated and cohesive systems of greenery is large and dates back to Frederick Law Olmsted in the nineteenth century. Evidently, the concept has acquired different facets since, but it has persisted in the minds of city planners since its original formulation. 'Park Systems' have historically been put forward as necessary elements of the modern city, in respect that they would sew the urban fabric together; fill in urban voids; control or direct urban expansions; and bring nature, sunlight and fresh air back into the cities. 'Recreational Systems' became more prominent as integrated systems of open spaces in the 1930s, with a focus on active recreation, health and well-being. Nowadays, networks of open space are being presented, particularly in the UK and US, under the term "green infrastructures" (Benedict et al. 2006; CABE 2004) and have become part of a set of necessary strategies in planning policy and practice aimed at achieving urban

Table 1 Summary of the development of the idea of networks of integrated open spaces

	Park system	Recreational system	GI
Origin	Olmsted	Several	Several
	Mid-nineteenth century	1930s	1970s
Aims and objectives	To connect parts of the city	As before +:	As before +:
	Green spaces as elements of planning	To provide space for active recreation	Environmental approach (biodiversity, water management, etc.)
	To control/direct urban sprawl	Open spaces should serve to promote physical as well as mental well-being	Combat climate change
	To bring nature back into the city		To be elements of social integration

sustainability, and promoting and enhancing ecological habitats. In most, if not all, eco-city projects observed there is a clear intention to create networks of green and blue spaces from the outset, as intrinsic elements of planning and covering a vast number of functions (Table 1).

Green infrastructures are usually conceived of as a network of interlinked open green spaces, public or private, including a variety of uses and typologies. Blue infrastructures have not yet being sufficiently studied. Spaces where the water features prominently are often included in green infrastructure projects, but tend to have less prominence than their green counterparts. Recognizing the need for integrated networks of blue spaces, coordinated with the other urban systems, is a fundamental step towards tackling climate change and social inequality (ARUP 2011; WWF 2010). Sustainable water resource planning and management are, as well as SuDS, fundamental aspects of eco-cities not only with regard to water supply and treatment, food production, recharging of water tables, drainage and flood control, but also as environmental amenities that have a positive psychological effect on citizens. The benefits of green and blue infrastructures are manifold, but can be summarized as in Table 2.

The typological elements vary substantially, but most of them fall within the categories of "hubs" or "links", although some elements perform both roles. Hubs are larger or more hierarchically significant elements within the structure. They help anchor it and act as significant reference points like, for instance, parks, squares and lakes. The linear elements promote connection and indicate movement – as green or blue ways – but should be quality places in themselves as well. Some are more directly linked to human activity, like neighborhood squares, and others perform distinct functions within the system, in which human presence and activities are more strictly controlled. From the case studies observations discussed below, it can

Table 2 Benefits of green and blue infrastructures

Climate and energy	Ecology	Place and society	Economy
Reduce urban heat island	Improve biodiversity and ecology	City form and sprawling control	Add value to place
Promote air-flow	Allow for wildlife corridors	Urban connectivity, integration and encourage pedestrian and cycle mobility	Generate renewable energy and resources
Implement sustainable urban drainage systems	Promote the preservation or enhancement of ecological habitats and natural settings	Connect urban and rural settings at regional scale	Accommodate waste treatment
Minimize acid rains	Recharge the water tables	"Knit" or "fill in" the urban fabric	Sustainable food production
Reduce surface water runoff and flood risk	Improve air quality	Psychological benefit	Rainwater harvesting
CO_2 storage	Allow for wildlife corridors	Provision of high quality public spaces	Promotes healthy living

be seen that there is a wide range of elements in use. These can be summarized and classified as below, but by no means are we suggesting that these are the only elements currently in use. Simplifying the terminology and grouping blue and green elements together facilitates comparisons among the case studies and reinforces the intrinsic link between the two systems. We can identify:

- Green/bluebelt,
- Green/blueways (including rivers, streams, brooks and canals),
- Central park/lake/river,
- Major park/lake,
- Neighbourhood park/lake,
- Green squares/courtyard,
- Pond/fountains and other localized water features,
- Sea,
- Green roof/walls,
- Street planting,
- Gardens,
- Urban agriculture,
- Agricultural land,
- Ecological habitats, and
- Others.

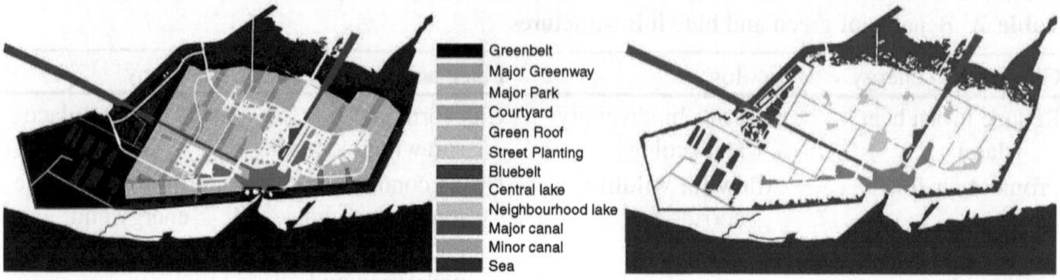

Fig. 1 Map of *green* and *blue* infrastructures in Dongtan

5 Eco-cities and Green and Blue Infrastructures: Case Studies

How are green and blue infrastructures being used in current eco-city projects in the Far East? In what follows, the chapter looks at four recent eco-city proposals that are representative of a systemic approach to embed the aforementioned networks of open spaces into the inception stage of planning and during the design process. The case studies are Dongtan, Changchun JingYue and Tianjin, all in China; and Songdo, in South Korea.

China's rate of population and urbanization is unprecedented. There is an expectation that nearly 300 million Chinese will be moving to urban areas in the next decade. The immense rural flight and the need to keep the country's economic growth is driving the construction of a vast number of new cities. As there is strong evidence that there is a direct correlation between urbanization levels and the wealth of a country (UN-HABITAT 2008, p. 6), urbanization is a key word in China's 12th five-year plan. On the other hand, China has also committed to reduce its carbon dioxide emissions by 40 % from 2005 levels. To this end, supposedly eco-city developments have been announced in over 100 locations across China (Stover 2011).[2] Nowhere else is the paradox of the need for more urbanized area, better living conditions and reduced ecological footprint is so strongly manifested.

Dongtan was one of the first examples in China of an attempt to develop a new city in a sustainable manner. It was designed by Arup in 2006, but after successive revisions the plan has yet to be implemented. It was intended to be a zero carbon city for up to 500,000 people (in 2050) and serve as an example to the world to inform sustainable masterplanning. As such, Dongtan was from the beginning intended to be a model for the future. The eco-city would occupy an area of 86 km^2 and the developed land would eventually cover just 40 %.

In the regional context, the whole proposal stands as a "linear city" composed of six urban nodes connected together by an arterial line and surrounded mainly by eco-farms. There would also be an "eco-park" and a "wetland park" (Fig. 1).

[2]However, very little information can be found about the majority of these projects.

As Lord Rogers stated, one of the common principles of eco-cities is their compactness and defined shape (Rogers 2011). In the first development phase of Dongtan, this is not only achieved by the creation of a greenbelt, but also and foremost by the definition of a bluebelt. In fact, the blue infrastructure project defines the city's urban structure at large. The designers intended to break down the city into three villages and several neighborhoods. To achieve this, two perpendicular canals were laid out, from the city center towards the wetlands. The core of the city itself is clearly marked by water, with the central lake its main feature. Minor canals were then introduced to help structuring the neighborhoods. A series of neighborhood lakes were placed across the city, reinforcing communities' sense of identity and creating moments of interest, anchor points within the mass of buildings.

With regard to the green infrastructure elements used in Dongtan, the greenbelt is its predominant feature. This works together with the bluebelt to define the confines of the city, providing a large park where renewable energy and waste management facilities can be integrated and a buffer zone between the urban area and the wetland created. Cutting through the urban fabric from one side of the greenbelt to the other, staggered north–south greenways were introduced. They bring sea breeze into the city, and serve as ecological corridors and as public spaces for the residents. Two major parks were placed on either side of one of the major canals, as the main gateway to the city. Courtyards, tree-lined streets and green roofs complement the network.

The green and blue infrastructures for Dongtan work together as a system aiming at creating sustainable and likeable communities. They shape the form of the city, control its sprawl, tie it back to the territory, define its spatial arrangement at different scales and play crucial roles in the creation of an environmentally sustainable city. The characterization of the city center with water highlights the importance it plays in the whole vision for the site. The canals tie the city back to the territory and define the districts, while the minor canals work at a smaller scale, visually delineating neighborhoods. The lakes are main hubs in the system, placed along the main route, in conjunction with the greenways. They function for the neighborhoods in the same way that the central lake does for the whole city. The greenways, minor canals and lakes work together to construct a sense of place for the different parts of each village. In addition, the greenways bring cooling breezes into the city. The connection between the city and the countryside is provided through the greenways and the canals. Food production is a main function of some of the open spaces, as well as energy production, water management and recycling (Fig. 2).

Several other eco-cities have emerged in the last decade alongside the high-profile flagship example of Dongtan. Green and blue infrastructures are also crucial in Changchun JingYue, in China, planned by AS&P from 2003 to 2007 to accommodate 500,000 people over an area of 60 km^2. They have been integrated into the general vision for the city, particularly in the application of AS&P's decentralized concentration approach. The creation of a multi-centered strategy is a shared approach between different eco-cities – such as in Dongtan – potentially creating compact and dense urban settings, and yet allowing for more direct

Fig. 2 Map of *green* and *blue* infrastructures of Changchun JingYue

integration with extra-urban green spaces. In Changchun, the network of open spaces created the framework for the articulation of six distinct districts, each of them with their own center of activities. They were laid out along a central park and linked together with a light rail tramline. The districts were clearly demarked by perpendicular green and blueways. These elements cut through the urban fabric from a green/bluebelt on Changchun's western boundary to the city's eastern edge. A series of squares, courtyards, green roofs, tree-lined streets, lakes and ponds complements the system. Environment is regional, as Ian McHarg (1969) showed. Both in Dongtan and Changchun there is a clear attempt to make use of the network of open spaces at different scales: to articulate the city to the territory and to define the nature of the districts.

Waterways are also key to the design of another eco-city: Tianjin. This is a Sino-Singapore undertaking, aiming like the previous discussed examples, at becoming a low-carbon/low-energy urban development. This would be for 350,000 people, over 30 km^2, by 2020 and a model that could be replicable (Singapore Government 2010). The Tianjin eco-city is one of few formally to state "blue" network planning as a key principle. The network of blue spaces includes the presence of a river, a central lake, minor blueways and lakes. Similarly to Changchun, a green axis with a tramline connecting the city's multiple centers was laid out. Tianjin will have three centers and four districts. The key feature of its network of green spaces is the central park – or Eco-core, as it is officially described – from which greenways along streams cut through the land dividing the districts up, in a manner similar to Dongtan and Changchun. They would also work as eco-corridors. The minor greenways split the urban fabric up even more to create neighbourhoods and to offer green squares. The latter are embedded into the 'eco-cell' concept of the masterplan. It consists of a 400 m × 400 m cell with a green square at its core. Four of them would make an 'eco-community', and several of those an 'eco-district' (Fig. 3).

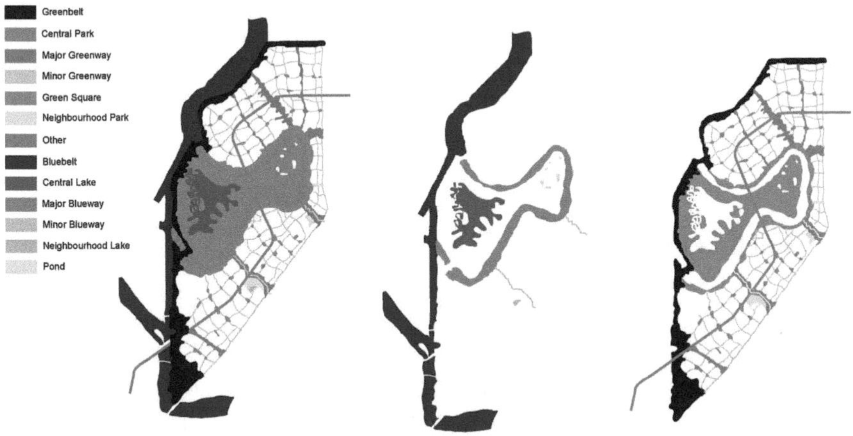

Fig. 3 Map of *green* and *blue* infrastructures of Tianjin Eco-city

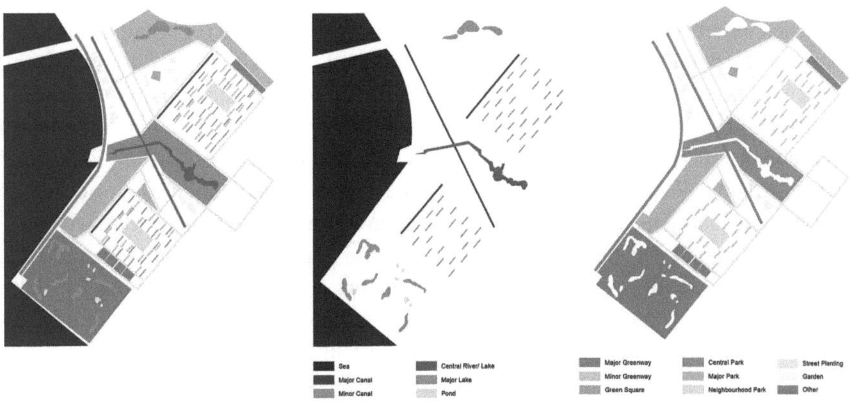

Fig. 4 Map of *green* and *blue* infrastructures of Songdo

Designed in 2004 by KPF in South Korea, Songdo is another example of a city that wants to be a model of sustainability for the future. It is intended to be inhabited by 65,000 people, over 6 km^2, of which 34 % will be open spaces. The designers call it a 'new garden city' (KPF 2008, p. 40), and with regards to its green spaces, make direct reference to precedents like New York's central park and Hudson River Park, and to Savannah's green squares. The water edge and the central park, with its lake, feature here, as in Tianjin, as the main elements of the system at city scale. The central park divides the city into two districts. There is a major open space in each of them, and a network of greenways and canals forming the neighbourhoods and blocks. According to KPF, the 'green areas form the fabric of the neighborhoods' (KPF 2008, p. 38) (Fig. 4).

Extending on that, the research showed that not only do the green spaces form the fabric of these eco-cities, but also their blue counterparts. The same would apply to other ongoing eco-cities in the planning phase or under construction, like Masdar city (AECOM 2011), in the UAE, and Caofeidian Eco-city, Wanzhuang and many others in China.

6 Conclusions

There have been a growing number of designs for eco-cities around the world, most of them hoping to become role models for the world in the creation of a new typology of city, the Eco-City. This involves more than meeting some indicators and principles of sustainability. It is about a new city concept that is able to present solutions to our current challenges. This chapter has analyzed how some of the most current eco-city examples have embedded the creation of green and blue infrastructures into the design process from the outset. They have become intrinsic elements of the eco-city design, both at city scale and regionally. A holistic approach to design is been sought and developed, as there is no current general framework to test which would enable us to finally resolve the paradox of growth–well-being–reduced ecological footprint. It is clear that there is a general attempt to establish integrated approaches between different disciplines, professionals and local residents. We are now constructing the theory and practice of eco-cities, in which coordinated networks of open spaces have a fundamental role to play.

In this chapter we have discussed how these spaces can be at the same time shape makers, quality places and resource centers. The urban framework for these eco-cities, as well as their image, are highly defined by green and blue spaces, which undoubtedly support many principles set out for eco-cities, such as: healthy living; reduced ecological footprint; promotion of walking and cycling; preservation of wetland and fauna; conservation and promotion of biodiversity; the need for compactness, access to fresh air and quality green spaces; and the production of food – as well as places for the generation of renewable energy, local biomass and for recycling.

Despite the fact that new approaches are being tested, there seems to be little evidence of new typologies of green and blue spaces emerging. Instead, both form and function of well-know elements are being re-semanthicized. However, it is noticeable that blue spaces are becoming ever more structural to eco-city design and are being cohesively planned as systems not subjugated, but at the same level of importance in relation to the networks of green spaces. Green and blue belts, parks, lakes, green and blue ways, squares, green roofs, street planting, wetlands, woodlands, allotments and other elements have emerged as key typologies of open spaces in use in these first significant eco-city projects. Green and blue belts clearly defined the urban fringes, playing an important role in the production of energy and management of waste. Green and blue ways tended to cut through the urban fabric defining districts or neighborhoods, connecting different parts of the city, making links to the countryside, offering a vast array of possible uses for people as well as working as ecological corridors. Parks, lakes, squares, courtyards and other hub elements complemented the systems, at both city and local community scale.

From conception to completion and management, there is still a long way to go. There are clear threats inherent to this process, which are still to be tackled effectively if the eco-cities are indeed to be part of the solution to our challenges. Criticisms over the lack of consultation with the local community and localization of imported ideas, as well as affordability and the sustainable credentials of the deliverance raise questions. If there are examples of some attempt to accommodate references to local architectural history in the new masterplans, very little in this regard is visible when it comes to landscape and urban design.

A systemic, holistic approach to eco-city design needs to have at its core a strategy for green and blue infrastructures, without which any attempt to tackle effectively the preoccupations of our ever increasing *urban* world will be incomplete. Despite the difficulties inherent in assessing such pilot cases, an interesting chapter of our century's urbanization processes and methodologies is being written: One in which a significant contribution to resolving – or mitigating – the sustainability question, and catering for all its pillars, is expected.

References

AECOM (2011) Masdar city public realm. http://www.aecom.com/What+We+Do/Design+and+Planning/stories/Masdar+City. Accessed 3 Mar 2011

ARUP (2011) Water resilience for cities: helping cities build water resilience today, to mitigate the risk of climate change tomorrow. ARUP

Beatley T (2000) Green urbanism learning from European cities. Island Press, Washington, DC

Benedict MA, McMahon E, Conservation Fund (Arlington VA) (2006) Green infrastructure linking landscapes and communities. Island Press, Washington, DC

CABE (2004) Green space strategies: a good practice guide. CABE, London

Ewing BGS, Oursler A, Reed A, Wackernagel M (2010) Ecological footprint atlas 2010. Global Footprint Network, Oakland

Jabareen YR (2006) Sustainable urban forms: their typologies models and concepts. J Plan Educ Res 26:38–52

Jenks M, Burgess R (2000) Compact cities: sustainable urban forms for developing countries. E. & F.N Spon, London

Jenks M, Dempsey N (2005) Future forms and design for sustainable cities. Architectural Press, Oxford

Kenworthy JR (2006) The eco-city: ten key transport and planning dimensions for sustainable city development. Environ Urban 18:67–85

KPF (2008) New city Songdo: green city. Kohn Pedersen Fox Associates PC, New York

Lian KK, Gunawansa S, Bhullar L (2010) "Eco-Cities" and "Sustainable Cities" – whiter? Soc Space, Lian Centre 3:84–92

McHarg I (1969) Design with nature. Natural History Press, New York

Peirce NR, Curtis W Johnson, Peters FM (2008) Century of the city: no time to lose. The Rockefeller Foundation, New York

Register R (1987) Ecocity Berkeley: building cities for a healthy future. North Atlantic Books, Berkeley

Roaf S, Crichton D, Nicol F (2005) Adapting buildings and cities for climate change: a 21st century survival guide. Elsevier Architectural, Oxford
Rogers R (2011) In conversation with Lord Rogers. In: Conversation with. http://www.ecocit.org/main.php. Acessed 1 Sept 2011
Roseland M (1997) Dimensions of the eco-city. Cities 14:197–202
Singapore Government (2010) Sino-Singapore Tianjin Eco-city. http://www.tianjinecocity.gov.sg/. Accessed 3 Mar 2011
Stover PL (2011) China's ecocities initiative. http://thechinabusinessnetwork.com/China-Green-Business/Chinas-Ecocities-Initiative.aspx. Accessed 3 Mar 2011
UNDP (2011) Human development report 2011. Sustainability and equity: a better future for all. Palgrave Macmillan, New York
UN-HABITAT (2008) State of the world's cities report 2010/11: bridging the urban divide. Earthscan, Malta
UN-HABITAT (2010) Climate change strategy 2010–2013. UNHABITAT, Nairobi
UNITED NATIONS (2007) Urban population, development and the environment. United Nations, Department of Economic and Social Affairs, Population Division, New York
White RR (2002) Building the ecological city. Woodhead Publishing, Cambridge
Wittig R (2007) Principles for guiding eco-city development. In: Carreiro MM (ed) Ecology, planning, and management of urban forests: international perspectives. Springer, New York
WWF (2010) Living planet report 2010: biodiversity, biocapacity and development. WWF, Gland

Decarbonising City Precincts: An Australian Perspective

Jessica Bunning, Colin Beattie, Vanessa Rauland, and Peter Newman

Abstract The need to decarbonise the economy can be greatly assisted if precinct scale city development can be a focus. The new low carbon technology for energy, water and waste are favoured by the precinct scale, especially the use of trigeneration, renewables, recycling wastewater, collecting rain water, waste to energy plants and automated solid waste collection. However, to make this work will require much more attention to the proper frameworks for carbon accounting and acknowledgement of best practice at precinct level. Governance will need to refocus on this smaller scale of delivery as it is not available at the moment. New models are developing that enable low carbon precincts to operate with a degree of independence within a broader centralised utility structure. Australian illustrations, especially from the City of Sydney, are showing that it is a feasible transition though some kind of new carbon credits for urban development need to be created.

1 Introduction

The shift in the economy to low carbon will require many different scales of activity and innovation. There has been much attention at the building scale and even more at the national scale but much less at the precinct scale. Neighbourhoods are based on precincts, and new developments within brownfields and greenfields areas are often occurring at the precinct level. Yet, the tools to help deliver low carbon outcomes are not readily available at this scale. This chapter will seek to provide some perspective on the way precincts can be decarbonised, the measurements required, the technologies that are appropriate at this scale, and the governance systems needed to make precinct decarbonisation feasible and the potential for some

J. Bunning (✉) • C. Beattie • V. Rauland • P. Newman
Curtin University, Curtin University Sustainability Policy Institute (CUSP), Bentley, Australia
e-mail: jessbunning@hotmail.com; colbeattie@gmail.com; V.Rauland@curtin.edu.au; P.Newman@curtin.edu.au

S.Th. Rassia and P.M. Pardalos (eds.), *Cities for Smart Environmental and Energy Futures*, Energy Systems, DOI 10.1007/978-3-642-37661-0_10,

kind of precinct scale carbon credits to become available. Australian examples will be used to illustrate.

2 Precinct Technology Scenarios

Decarbonising city precincts requires the appropriate technologies that can best be applied at the scale of a precinct. Fortunately many of the recent technologies in decarbonising the built environment appear to work best at a precinct or distributed scale (Newman et al. 2009). These technologies involve trigeneration, renewables, recycling wastewater, collecting rainwater, waste to energy plants and automated solid waste collection. These have been discussed in detail in many sources (Beattie and Newman 2011; BioRegional 2009; CSIRO 2009; Kenway et al. 2008; Rauland and Newman 2011).

A series of scenario planning exercises have been done in Australian precincts in recent years using the Kinesis modeling tool (Beattie 2011; Kinesis 2012). The results are summarized in Table 1 to show the size and density of the precinct, the main technologies and practices adopted for the scenarios and the results in terms of reduced carbon and reduced car use. The modeling covers costs as well though not on a comparative basis. The Cockburn Coast costing came out at just $5,600 per dwelling extra or around 1 % of a typical mortgage.

As can be seen from this data there are very significant and low cost reductions in carbon, which are feasible when the precinct scale of action is the focus for decarbonising the economy.

Precinct scale approaches are outlined in recent literature from global and local sources (BioRegional 2009; Jones 2004, 2010a; Kenway et al. 2008; Kinesis Consortium 2012; Taper et al. 2010; VicUrban 2010; Woking Borough Council 2011). The most significant Australian local government leading this work in recent years has been the City of Sydney.

The City of Sydney set a bold target of reducing greenhouse gas emissions across the entire local government area by 70 % from levels in 2006, by 2030. Across Australia this kind of leadership from local government is rare. Although all levels of government need to show leadership it is difficult to demonstrate anything innovative at a precinct level unless a local government is heavily involved. It is at local government level that planning powers exist which can set targets that developers will need to meet. Combined with control over the public realm it is possible to coordinate and deploy the green infrastructure needed to provide carbon reducing opportunities to new (re)developments and existing precincts alike. The City of Sydney is developing six plans around energy water and waste, with the following four directly impacting the city's carbon emissions.

Table 1 Modeled emissions reductions in selected existing and proposed developments or precincts in Australia

				Energy/GHG reductions					Transport		
Technologies and practices adopted	Area (ha)	Density (hh/ha)	Comm. area	Low carbon (Trigeneration)	Renewables	Building energy efficiency	Renewable fuel	**Sub-total**	Reduced private vehicle use	Cap. cost	Comment
City of Sydney	2600	36 (avg.)	32 mill. m^2 (including residential)	**20 %**	**19 %**	**20 %**	**6 %**	**65 %**	**n/a**	**n/a**	The City of Sydney is based on a number of existing precincts – the results shown are combined
Cockburn Coast, WA	332	60	148,149 m^2	**55 %**			**0 %**	**55 %**	**35 % (GHGe)**	$5,600/Dwelling or $44/$m^2$ of fl. area	Based on the District Structure Plan developed by Landcorp
Stirling City centre, WA	215	40	886,723 m^2	**21 %**	**5 %**	**17 %**	**0 %**	**43 %**	**34 % (VKT)**	$76/$m^2$ of fl. area	Based on the Structure Plan developed by the Stirling Alliance
North Port Quay, WA	245	42	257,460 m^2	**0**	**111 %**	**19 %**	**47 %**	**177 %**	**13 % (GHGe)**	$200/$m^2$ of fl. area	Hypothetical model pushing boundaries of RE to achieve a zero carbon outcome

Fig. 1 Mapping the GHG emissions profile of the City of Sydney (Reproduced from Kinesis Consortium (2012))

2.1 Trigeneration Master Plan

The key component in meeting the challenging target that the City of Sydney has set itself is through the deployment of a decentralised energy network that can service both new and existing buildings with low carbon electricity and zero carbon heating and cooling. The technology that makes this all happen is trigeneration. In this instance the trigeneration system is made up of reciprocating engines the smallest of which is capable of driving a 4 MWe generator. The waste heat that comes from the cooling jacket and the exhaust of the engines is captured and used to heat water, which flows around the precinct providing an opportunity for any building that connects to it to harness the available thermal energy through a simple heat exchanger. In order to meet the cooling demand the heat can be converted through an absorption chiller to 'coolth' effectively displacing electricity normally used for heating and cooling and reducing electricity demand.

The obvious starting point was to establish which parts of the city were responsible for the highest intensity of Greenhouse Gas emissions. Figure 1 demonstrates the amount of emissions by the height of each stack on a city block while the colour indicates the intensity of emissions per square metre of lot area – the darker the colour the higher the intensity. A total of 15 Plants across four Low Carbon Infrastructure Zones will be constructed by 2030 with a total capacity of 360 MWe of low carbon electricity, now expanded to 447 MWe including additional stand-alone projects.

This in itself is hardly innovative in a global context. Decentralised energy networks including trigeneration are relatively commonplace around the world, in particular in Northern Europe. The innovative thinking that is happening in Sydney is around the deployment of a system in the face of fierce criticism not just from developers but from advocates of alternative solutions to reduce CO_2 in the power sector (Dodson 2011; Wright 2012).

2.2 Renewable Energy and Waste

Precinct scale trigeneration under-pins a number of other strategies that the city needs to implement in order to achieve their ambitious target. Renewables in the form of solar PV play an important role and the city has shown leadership in this area by installing a 48 kW system on the roof of the Town Hall. This is one of many new solar PV projects making up a total of 1.25 MW that will contribute towards the renewable energy target the city has set itself. All of the city's energy requirements are to be met by renewables (30 %) or low carbon energy (trigeneration providing the balance of 70 %). The Renewable Energy Masterplan will also include an analysis of wind (onshore and offshore), hydro, geothermal, Concentrating Solar Thermal (CST) and marine, all of which can contribute to the greater city's target, when the plan is released for public consultation in 2013. Renewable electricity will make up approximately half of the target set by the master plan, sourced fairly evenly between urban and rural areas.

2.3 Renewable Gas

The key, and possibly the most innovative part of the renewable energy plan will focus on renewable gas. Decentralised energy networks may be conventional in many parts of Europe but the idea of producing the thermal energy with a renewable fuel is a relatively new concept.

There are many potential sources of fuel including municipal waste, commercial and industrial waste, sewage and landfill gas. Additional sources become increasingly available should opportunities beyond the city boundary be explored and the City of Sydney is looking at what potential lies within a 125 km radius of the city. In doing so, officials cite the *proximity principle* (Willis 2008) which is intended to reinforce links with the nearby rural community and to minimize any inefficiencies that may occur in the delivery of low carbon energy across long distances. This allows additional feed stocks like agricultural and forestry residues, and livestock manure to be considered from the bioregion of the city.

Gas, as a carrier of energy, has significant advantages over electricity. Its flexibility allows it to be converted to LNG and transported to where it is needed should it be necessary, as was done in the UK at Newlands Corner, Surrey where the feedstock was not located near any grid (gas or electricity) (Johansson 2008). In this instance, after being converted to LNG, the gas is transported into the city where it could provide a source of fuel for delivery trucks and public transport vehicles, contributing to reducing emissions from the second largest source of emissions behind primary energy – emissions from motor vehicles.

A number of alternative strategies for deploying the renewable gas could be considered. Finland, Denmark and Germany are all developing long term strategies injecting renewable gas into the grid with the long term goal of cleaning up the gas grid – particularly important in Europe where gas generally has a greater penetration into the energy supply chain than in Australia. For Sydney, natural gas is seen as a transitional fuel for the city's trigeneration network, and renewable gas provides the link to drive emissions down in line with the city's long-term goals. Thus as each new precinct is developed within the trigeneration network, renewable gas will help to contribute to an ultimate goal of 'greening of the gas grid' (Prendergast 2012).

2.4 *Automated Solid Waste*

Sydney also has an automated solid waste plan that is as symbolic as it is practical, as no other part of Australia has yet adopted this technology. The technology is described by Allan Jones, the CDO for Energy and Climate Change at the City of Sydney, as "moving waste collection into the twenty-first century" (Jones 2010b). For many centuries sewerage has been collected through a serious of underground pipes (arguably, the *first* green infrastructure) in ancient cities like Rome and Istanbul. It's not difficult to see that the collection of solid waste is simply the next step in the process of designing clean, modern cities now, and into the future. Solid waste will become an energy commodity and as soon as it is viewed from that perspective the way in which it is managed will undoubtedly change for the benefit of all. Automated waste collection will become standard practice but for the City of Sydney it will demonstrate (very publicly) how 95 % of the emissions from garbage trucks can be saved (Jones 2010b).

The example of the City of Sydney shows how they are demonstrating a way to decarbonise a city – precinct by precinct. There is considerable potential being demonstrated by cost effective precinct scale technology, but it does not always mean change will happen.

3 Carbon Claims and the Need for a Standardised Framework

New terminology has emerged in recent years to help describe the carbon reductions achieved within urban land developments around the world. Terms such as Carbon Neutral, Carbon Zero (or Zero Carbon), Zero Emission, Carbon-Free, Low Carbon,

Carbon Negative, Carbon Positive, Climate Positive and Climate Neutral are now being used by developers looking for new ways to differentiate themselves in an increasingly ambitious world. However, the myriad of names and their varying definitions, or lack thereof, have led not only to confusion, but also scepticism and mistrust, represented by the term 'greenwash' or the emerging term 'carbonwash'. The lack of consistency in definition, including the methodology for calculating emissions or how the carbon reduction is ultimately achieved, largely remains unclear.

3.1 Defining Carbon Reductions

Despite the widespread use and prevalence of the new carbon terms within the public domain, there remains no universal definition, nor any widely accepted international certification system for recognizing the achievements. Murray and Dey (2009, p. 238) note that the term carbon neutral – one of the more commonly used terms – currently appears to be 'defined by popular usage'. While the broad intention of the terms is to create a reduction of carbon in the atmosphere in relation to the inputs and outputs of a product or service, or in this case a precinct of a city, an increasingly number carbon claims (i.e. those including 'zero', 'negative', 'positive', 'free' and 'neutral' etc.) go beyond merely reducing carbon. They aim to reach a point where the development has no net carbon associated with it. This is generally achieved through a three-step process as illustrated in Fig. 2.

The first step involves measuring the carbon footprint of the development, as it is commonly accepted by those working in the field, 'you can't manage what you don't measure'. The process of identifying emissions' sources is essential in order to understand where reductions can be made. For land development, this involves identifying a 'business as usual' (BAU) scenario, which can be undertaken using various new modeling tools such as the Kinesis CCAP Precinct modeling tool (Beattie et al. 2012) which is further described below.

The second step involves identifying and implementing initiatives and measures that aim to reduce emissions to the greatest extent possible. For the built environment, this could be through better building design, low carbon transport infrastructure or switching to renewable sources of energy.

The last step involves offsetting as many of the remaining unavoidable emissions associated with the precinct development as possible. Such 'unavoidable' emissions include the embodied emissions in materials of buildings and infrastructure. Several iconic precinct developments have demonstrated how offsets can be generated using a city's surrounding bioregion (Newman et al. 2009).

The overall reduction process for urban development should be cyclical and iterative, requiring constant monitoring, reducing and offsetting of emissions in order to ensure carbon claims remain accurate.

Despite broad consensus over the general three-stage process, significant variation still exists around how the task is undertaken. For example, clarity is required around: how the carbon footprint is calculated (i.e. what sources are included); to what extent emissions are reduced before being offset; and which offsets

Fig. 2 The carbon reduction process

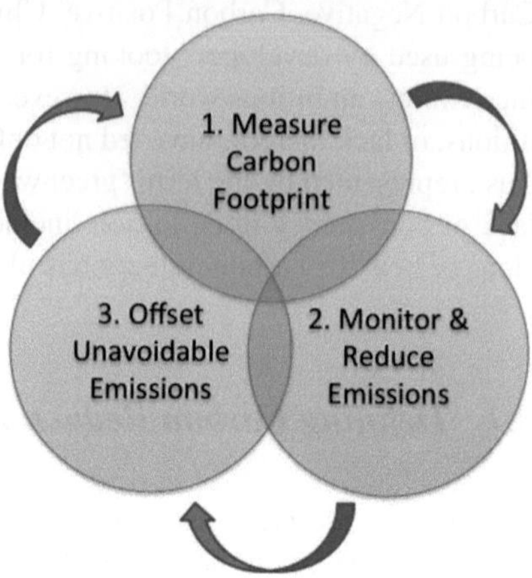

should be considered eligible. Without a consistent framework for determining the emissions associated with urban development, as well as the specific process for achieving the carbon reductions, claims will continue to vary, making them incomparable and susceptible to allegations of greenwash. Internationally accepted standards, guidelines and protocols are gradually being developed to measure carbon reductions in a variety of areas such as businesses, projects and even local governments (ICLEI 2009; WRI and WBCSD 2004; Arikan and Desai 2012; WRI and WBCSD 2003). However, there remains no universal standard for assessing emissions attributed to precinct level urban development. Kennedy and Sgouridis (2011, p. 5259) also identify this, stating:

> Given the complexity of material and social interactions on an urban scale, we find that currently there are no concrete definitions upon which these claims can be measured and compared. Therefore, in order to make the ambitious targets of low and zero carbon emissions meaningful concepts in the context of urban planning, a carbon accounting framework needs to be rigorously defined and adapted to the urban scale.

3.2 *Developing a Framework for Assessing Precinct-Scale Emissions*

Carbon footprints and inventories for precinct scale emissions can vary considerably depending on the inputs and decisions made on various issues. Therefore, when determining these emissions associated, several issues need to be considered and clearly articulated to ensure transparency and comparability of carbon claims. Some of the main issues identified are outlined in Table 2. These clearly demonstrate the need for clarification on several levels.

Despite the numerous city-wide and local government GHG frameworks that currently exist, relatively few take into account land development, specifically in terms of the carbon emissions associated with urban form and the built environment.

Table 2 Key considerations for carbon footprint and definitions

Issue	Description
Gases	Which gases are included in the analysis? I.e. just carbon-related or multiple greenhouse gases?
Measurement units	What unit is the carbon footprint measured in? Weight (i.e. tonnes) or area (i.e. global hectares)? A footprint has traditionally been measured by land area
Accounting methodology	Is an inventory approach or a Life Cycle Analysis (LCA) used? If LCA is chosen, which methodology, i.e. Process Analysis (PA), Environmental Input–output (IEO) Analysis, Hybrid EIO-LCA?
Boundaries	What are the overall system boundaries? I.e. Are they purely operational emissions or are supply chain emissions included such as the embodied emissions in materials?
Scopes	Which scopes of emissions are included? Direct and indirect? Scope 1, 2 and 3?
Timeframe	If LCA is used, what is the lifetime given to the embodied emissions in buildings and infrastructure?
Emission reduction	Are there targets set for emission reduction? What are the allowable emission reduction options?
Offsets	Are offset included in the claim? If so, is there a limit? What offsets are eligible? Are they certified, credible offsets?

This is largely due to the boundaries identified and the accounting methodology adopted. City frameworks generally use an emissions inventory approach, which focuses primarily on operational emissions and assesses the *activities* occurring within the legislative or jurisdictional boundaries of the city. This approach does not measure or consider the emission abatement opportunities possible from different types of urban design or variations in the building and resource management technologies. For example, significant emission reductions can occur: when alternative (low carbon) materials are used in construction; if more energy efficient building designs are adopted; if integrated public transport infrastructure is provided instead of relying on private transportation; and if new approaches for managing resources e.g. small-scale renewable (or low carbon) energy generation or alternative water and waste management systems are adopted (Rauland and Newman 2011; Newman et al. 2011). Thus, it is argued that a life cycle approach is more useful and appropriate in assessing the emissions associated with land development.

3.3 Proposed Framework for Precinct-Scale Land Development

The following framework is proposed as a way to account for the emissions associated with precinct scale urban development. It involves six key areas of development that can be influenced by a developer. Traditionally, developers have

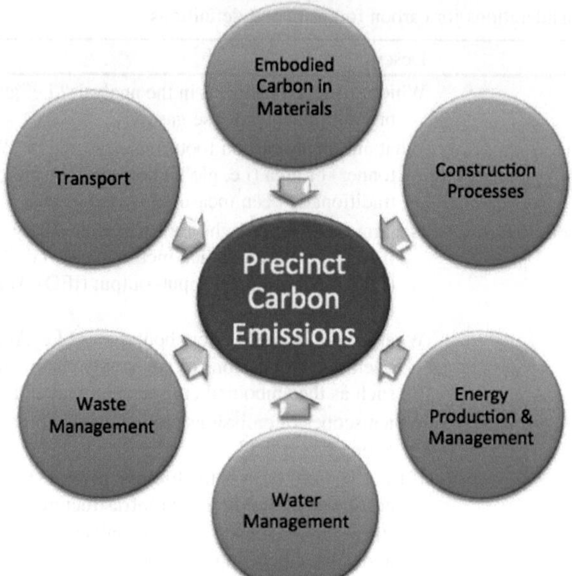

Fig. 3 Key emission sources within precinct development

neglected some of the areas identified, as they have commonly fallen outside their operational control, and therefore, their responsibility (i.e. energy production and transport infrastructure). However, integrated, small-scale, precinct infrastructure based around transit, is beginning to be identified as a priority by developers and increasingly being targeted in the design of developments (Newman et al. 2009; Newman and Matan 2013).

The six key emission sources, illustrated in Fig. 3, are proposed as integral factors to be included in any carbon analysis of a precinct development, regardless of the final carbon claim. The sources have been chosen based on the significant amount of abatement possible within each if alternative practices are adopted.

3.4 The Role of Offsets in Achieving Carbon Reductions

Another important issue to consider when making carbon claims within the built environment relates to offsets. It is common within the development industry to classify surplus renewable energy generated onsite and fed into the Grid as offsetting, which can then be used against other emissions sources within the development, such as the embodied emissions in materials (Williams 2012). The term 'netting' (Williams 2012), is often used to describe the process of balancing the carbon consumed onsite with sufficient carbon free renewable electricity produced onsite and exported to the grid over a 1 year time period. In this situation, a development becomes 'net zero' carbon on an annual basis.

However, this concept and method of balancing emissions is generally not accepted under official carbon certification schemes such as the Australian Federal Government's National Carbon Offset Standard (NCOS) Carbon Neutral Program, which uses internationally recognised protocols for offsetting (Commonwealth of Australia 2011). The reason is that energy-related offsets are not considered to be additional, voluntary forms of abatement in Australia, as the energy sector is a covered sector under the Federal Government's national Carbon Pricing legislation. Thus, any action to reduce emissions from this sector is argued to have occurred regardless because of the legislation already in place (Commonwealth of Australia 2012a). The only eligible domestic offsets under this scheme come predominantly from the Carbon Farming Initiative (CFI) and involve regional land management initiatives such as tree planting and camel culling (Commonwealth of Australia 2012b).

There are currently no specified limits regarding the amount of offset credits that can be purchased or used to offset a precinct's emissions; only guidelines that stipulate the type of offsets eligible under the standards exist. Thus, the only fully articulated guideline and certification scheme in Australia that could potentially recognise a precinct scale development's carbon reduction contains a rule that makes it particularly difficult for a developer to achieve their claim, by not recognizing their onsite carbon reduction.

4 Governance for Low Carbon Precincts

If precincts are the appropriate scale to achieve major reductions in carbon, then it is essential that a parallel set of innovations can enable precincts to become the appropriate scale for governance. A shift needs to happen from a reliance on centralised, top-down, decision-making from governments to mitigate climate change to an increasing openness for broader community participation and decentralised, precinct-oriented governance to instigate rapid and decisive change. The first signs that this is happening are beginning to be recorded. Councils are being given greater autonomy to respond and flexibility to collaborate and cooperate with organisations and enterprises to build capacity for low carbon development of their precincts (Tanner et al. 2009; Duit et al. 2010).

However the process of change is very slow and hence this final section will try and show how barriers to change in governance can be overcome.

Local authorities are typically responsible for many urban planning decisions that directly impact a cities' carbon footprint. In particular they are much closer to the changes at a precinct level than the older institutions of utilities that operate at a much larger scale. Therefore, it makes sense that greater attention be given to facilitating the way councils implement and manage their infrastructure networks and mechanisms for service delivery. Developing better local governance that supports cheaper, more efficient, integrated low carbon, energy, water, waste and transport services (like those illustrated above) will help cities to reduce their overall impact on carbon emissions and become more resilient to the impacts of climate change.

Four key principles for assisting this process are outlined.

4.1 Strengthening Decentralised Inter-Municipal Collaboration

Collaboration between neighbouring municipalities will be important as the concept of precinct scale governance of infrastructure spreads. There are likely to be important advantages for providing a diversity of load for 'green' technology options. For instance, trigeneration could perform much better if it is able to serve a much more diversified load of multiple sites across municipality boundaries, benefiting from a mix of residential, commercial and retail end-users. The financial returns generated from such projects, through savings on network charges and electricity bills can be in turn re-injected into more low carbon urban development projects for the precincts (Jones 2012).

Strengthening local-level inter-municipal ties and linkages at finer scales between neighbouring municipalities, rather than through isolated, hierarchical and generalized systems will help to build capacity at the local level to detect, quickly respond, and deal with cross-boundary urban challenges (Leck and Simon 2012). Infrastructure networks often provide shared resources across municipalities and are interconnected and mutually interdependent in nature therefore, restructuring less carbon intensive infrastructure systems lends itself to fostering greater municipal collaboration between precincts.

This multi-layered and cooperative governance style mirrors the emerging networked and localized 'distributed, systems' approach for water, waste and energy services. It involves specialised 'niches' of employees working with a diverse network of organizations to solve complex local issues, which require a high degree of information and resource exchange that fosters mutual learning, cooperative management and innovation (Biggs et al. 2009).

4.2 Public-Private Infrastructure Partnerships

Pooling together financial resources from multiple municipalities will also help to build capital and attract private sector funds for investment in precinct scale green infrastructure projects. Many organisations, such as pension funds and insurance companies are seeking to demonstrate a low carbon profile either through their shareholders or their boards of directors through their investments and operations

(Jones 2012). This is being done as a combination of altruism and innovation for the emerging new green economy. So already a strong incentive exists for private sector engagement with the public sector, and local government and precinct governance can benefit from this interest.

4.3 Building New Roles for Precinct Communities and Synergies with Other Sectors

Increasing concerns in Australia about energy and water security, waste management, rising carbon emissions and increasing utility bills has prompted communities to become more actively engaged in the management of their local resources. This trend is leading to new roles and responsibilities emerging for local businesses, councils and communities.

Traditional consumers within precincts are becoming producers and providers of local power, water and waste services (Biggs et al. 2009). As an example, Western Australia's Independent Power Producers Association promotes local enterprises becoming utility providers. This holds great potential for land developers wishing to develop independent, integrated power, water, waste utility services; and for attracting property owners interested in buying into low carbon precincts using recycled water or renewable power.

There is increasing evidence that this precinct scale market for innovative infrastructure provision does exist with citizens putting pressure on governments to have the flexibility to choose which sources they buy their energy and water from, for example brown coal or trigeneration; and communities are seeking to have more influence and greater ownership of their resource assets by becoming owners of water systems (Biggs et al. 2009) or community shareholders of wind farms.

There is also potential for these small precinct producers to trade across local networks, which along with other municipalities could involve building synergies with other sectors, such as agriculture (Biggs et al. 2009), industry, farming and fishery. For instance, precinct wastewater could be used to service nearby industries, or surplus precinct renewable energy could power nearby fishing or leisure vessels at marine harbours. Similarly, resources used by certain sectors situated on the periphery of precincts could be re-used to meet the needs of those nearby communities, for instance agriculture or farming waste could be used as renewable gas to power precinct trigeneration schemes.

Further consideration needs to be given by communities to their resource assets in terms of the inflow and exchange of resources with their precinct and the surrounding area, as well as the processing of those resources within the local biosphere. Finding these synergies and trade-offs at a fine grain level will help to strengthen the entire resilience and continuous supply of resources and fundamentally drive the economic viability and overall sustainability of low carbon precincts.

4.4 Removing Barriers for Low Carbon Precincts

Australian communities, like most around the world, are ready to move forward from simply discussing ambitious carbon mitigating strategies to actually implementing resilient, low carbon infrastructure networks and creating low carbon precincts. However, Australia's present planning and regulatory barriers actually limit and constrain innovative action and progress towards decarbonising precincts. Regulations that are often costly, inefficient and favour centralised business-as-usual power, water and waste services need to be updated to accommodate low carbon, distributed options for production, distribution and consumption of resources (ClimateWorks Australia 2011).

Unjustifiably high costs for obtaining a precinct electricity or water licence need to be adjusted and complicated, arduous processes that differ on a case by case basis such as for grid connection or establishing a grey water scheme, need to be standardized and streamlined. This will build confidence in the market and stimulate more investment from the private sector. The CEO of the Property Council of Australia, Peter Verwer (Edis 2012) claims that property owners are "frustrated by approval processes and regulatory regimes that block them at every turn".

More guidance is required for municipalities and businesses in choosing business or funding models that are appropriate for implementing and managing low carbon solutions to meet local challenges; this includes better defined pathways for obtaining approvals and complying with regulations. At present, there is not enough experience from people and tested case studies of established low carbon precincts in Australia therefore, so much uncertainty exists around: the risks and benefits involved (Edis 2012); as well as who should get involved and what their roles should be (Yu et al. 2011). This effects the overall viability and widespread adoption of low or zero technology options (Yu et al. 2011). Furthermore, these uncertainties in the market drag out decision-making and implementation processes, creating additional costs and dis-incentives (Edis 2012).

If Australia is to succeed in the transition towards a low carbon economy and society, it will need to reconcile these barriers and move from centralised water, waste and brown grid systems to the creation of 'green grids' that are integrated with low carbon water and waste solutions that complement traditional centralised systems (ClimateWorks Australia 2011). These should coexist to gradually evolve into localised interdependent 'eco-resource networks' for 'Eco-Precincts'.

4.5 Some Emerging Governance Models

Increasing interest from municipalities in developing low carbon precincts that provide low or zero carbon power, water and waste services to residents has meant new business models and financing mechanisms are emerging to support the operation and management of these decentralised networks for precincts. Two examples from Australia will be provided.

4.5.1 The City of Sydney, Low Carbon Precinct

The business model for the City of Sydney's Trigeneration Masterplan involved procuring the private distributed energy company, Cogent Energy to design, finance, build, operate and maintain the city's trigeneration network. This enabled the city to remove 230 buildings and 21,000 street lighting columns off the grid and create a Low Carbon Precinct (Jones 2012). The council acts as an 'anchor' customer for the trigeneration network and owns the thermal reticulation network (TRN), which connects up the energy centres and distributes thermal energy across the city (Kinesis 2012). Cogent Energy is responsible for retailing the electricity and thermal energy and for operation and maintenance of the TRN (Kinesis 2010).

Through a fund reimbursement financial mechanism the council forward funds for the construction of the thermal network and gets increasing returns on its investment as each customer connects to the network. Once the network load reaches 20 MW the council receives all its money back and starts to benefit from returns on investment (Jones 2012). This mechanism has allowed the city to overcome regulatory and economic hurdles for private sector, by using funds in a targeted way to ensure a larger more economically viable system is connected (Jones 2012).

Using Cogent Energy with its specialised software technology and expertise in managing and operating the export of electricity from trigeneration, the council can apply the same mechanism set up by Cogent Energy for the trigneration system (i.e. market compliant metering, data collection and aggregation, settlements and billing) to export solar electricity. The energy sales agreement for supplying trigeneration to council buildings means generating at a few sites will supply low carbon electricity to all of the council's buildings and street lights. The council owns the solar PV and the generated solar electricity so this comes out of the council's collective bill first before the supply of trigneration electricity owned by Cogent. By 'piggy backing' onto the same trading system Cogent has set up for trigeneration the cost of carbon abatement is reduced from $65 a tonne to $30 a tonne, which significantly improves the economics of renewable energy (Jones 2013).

4.5.2 Fremantle Community Wind Farm

A Community Wind Farm is in the process of being launched to supply renewable energy to Fremantle. A community owned model was employed, to set up a standard propriety limited company with sponsorship from Fremantle citizens and equity/debt financing from corporate sponsors (Ally 2012). This model allows people in the Fremantle community to raise money to build the wind farm and also to become shareholders of the company. The wind farm will generate electricity for the Fremantle community and sell energy to others around them, creating a return, which will be then invested back into community projects and provide a dividend back to community shareholders and investors (McCulloch 2012).

The company already has a generation licence to generate the renewable electricity and is in negotiations with retailers to be able to sell the green power to an off-taker (Ally 2012). As yet, there are no policies to incentivise retailers, so the project relies on either internal advocates within large energy providers who are personally invested and want to see the project get off the ground or community support to prompt policy change. As the community owns the bulk of the company, there is external pressure on the government to change whatever policy is needed. It also builds other revenue sources for the community and community groups just through the revenue generated. So there are multiple benefits (McCulloch 2012).

To overcome policy complications the company has used a very coordinated approach, involving different people tackling the policy issues, different people tackling the business model and how it would work and who the partners are with the project, and different people again working with the community and entering the community buy in (McCulloch 2012).

Both case studies from Sydney and Perth demonstrate that although barriers to distributed energy still exist in Australia, new governance models are emerging to overcome these obstacles, which can also be applied elsewhere. However, these are temporary measures to solve very complex issues and ultimately in the long-run, Australia will have to change and update its centralised governmental rules and regulations to accommodate distributed low carbon infrastructure systems if it is to achieve its carbon emissions reduction targets, for precinct scale city development will inevitably be a major driver in this change.

5 Certifying Carbon Reductions in Land Development

Precinct scale land development is going to play an important role in helping to decarbonise cities and nations in the future. While measures are predominantly undertaken at the local level, it requires carbon reductions in various different sectors of the economy (i.e. the transport, resource management, building, industrial and manufacturing sectors). However, despite the potential for considerable carbon reductions, there are few incentives to assist developers to make low carbon choices. This is in contrast to what is occurring in the energy markets, where carbon credits are available to help drive the transition to low carbon fuels and technologies at the front end of the economy.

At the end-user part of the economy there are only incentives provided by the voluntary carbon market. The introduction of the Australian Federal Government's NCOS Carbon Neutral certification program has brought many benefits to the voluntary carbon market. It has increased the credibility and integrity of carbon claims, thereby creating additional marketing opportunities for businesses and organisations to demonstrate their commitment to the environment. It has also demonstrated to participants of the program that, reducing your carbon footprint can also produce significant financial savings, particularly from energy efficiency improvements. As a result, countless businesses, organisations, local governments and even schools[1] have embraced the opportunity to be recognised for their carbon reduction achievement (Low Carbon Australia 2013; SimplyCarbon 2012).

However, despite the rigor and stringent guidelines currently underpinning the new NCOS Carbon Neutral program, there remains no specific methodology for calculating the emissions associated with city land development, nor any developer who has sought certification. Furthermore, the issues surrounding eligible offsets as discussed above, remain a key challenge for developers wishing to pursue this status. The current requirements for offsets under the NCOS Carbon Neutral program guidelines make it difficult and expensive for developers to achieve this status using conventional offsetting methods (i.e. purchasing certified offsets).

It is hoped that a new and unique framework for calculating the emissions associated with land development at the precinct scale can be created based around the six areas identified above, as well as a new system for claiming carbon reductions within the built environment, which offers an alternative method for producing offsets based on emission abatement opportunities delivered onsite. With this as its basis, the possibility should now be pursued for developing a carbon trading market for urban development in a parallel scheme to that provided for energy utilities and large industry.

6 Conclusions

The precinct scale has been demonstrated to have high potential as the focus for delivering low carbon outcomes in the built environment. It can be measured if given a well-developed framework as suggested above. The technologies are available and the scenarios developed so far in Australia show mass reductions are feasible at low cost. The barriers are generally in governance, though the first examples are appearing to demonstrate how these can be overcome. The City of Sydney has proven leadership in all of these areas. To accelerate the transition using precinct scale urban development it is suggested that a new kind of carbon credit scheme for precinct scale urban development needs to be developed in parallel to the mainstream carbon credits market.

Acknowledgements The authors would like to acknowledge the Australian Research Council for their financial support in the research behind this chapter.

[1]South Fremantle Senior High School was the first school in Australia to be certified as Carbon neutral and did so on a cost neutral basis

References

Ally J (2012) Interview with Jamie Ally (General Manager, EMC Engineering) by Jess Bunning. Tape recording, Perth, 29 June 2012

Arikan Y, Desai R (2012) Global protocol for community-scale GHG emissions C40 cities climate leadership group and ICLEI Local Governments for Sustainability

Beattie C (2011) A carbon assessment and design tool to assist in planning low carbon development. Paper read at MODSIM 2011 international congress on modelling and simulation, MSSANZ, Perth Australia

Beattie C, Newman P (2011) The density trade-off. In: State Of Australian Cities. Melbourne

Beattie C, Bunning J, Stewart J, Newman P, Anda M (2012) Measuring carbon for urban development planning. Int J Clim Chang Impacts Res 3(4):35–52

Biggs C, Ryan C, Wiseman J, Larsen K (2009) Distributed water systems: a networked and localised approach for sustainable water services. Victorian Eco-Innovation Lab (VEIL) and University of Melbourne. http://www.ecoinnovationlab.com/uploads/attachments/234_Distributed%20Water%20Systems.VEIL.pdf. Accessed 1 June 2011

BioRegional (2009) BedZED seven years on, the impact of the UK's best known eco-village and its residents, BioRegional, Surrey, England

ClimateWorks Australia (2011) Unlocking barriers to cogeneration: project outcomes report, a project by ClimateWorks Australia with Seed Advisory. Property Council of Australia, Melbourne

Commonwealth of Australia (2011) Carbon Neutral Program Guidelines: National Carbon Offset Standard. Edited by Department of Climate Change and Energy Efficiency. Version 2.0: available online. http://www.lowcarbonaustralia.com.au/page/program-publications-and-related-reference-material. Accessed 15 Sep 12

Commonwealth of Australia (2012a) National Carbon Offset Standard. Edited by Department of Climate Change and Energy Efficiency. Version 2: available online. http://www.climatechange.gov.au/ncos. Accessed 15 Sep 12

Commonwealth of Australia (2012b) An overview of the carbon farming initiative. http://www.climatechange.gov.au/government/initiatives/carbon-farming-initiative/~/media/cfi/CFI-About-Overview_V2.pdf. Department of Climate Change and Energy Efficiency

CSIRO (2009) Intelligent grid, a value proposition for distributed energy in Australia, Commonwealth Scientific Industrial Research Organisation: Newcastle Australia

Dodson J (2011) The carbon devil in the detail on urban density 2011 [cited 25 November 2011]. http://theconversation.edu.au/the-carbon-devil-in-the-detail-on-urban-density-4226

Duit A, Galaz V, Eckerberg K, Ebbesson J (2010) Governance, complexity, and resilience. Global Environ Chang 20(3):363–368

Edis T (2012) Q&A: Q.Cells' Oliver Hartley – Part 2, interview by Tristan Edis climate spectator 2012 [cited 18 December 2012]. http://www.climatespectator.com.au/commentary/qa-qcells-oliver-hartley-part-2

ICLEI (2009) International Local Government GHG Emissions Analysis Protocol (IEAP). http://carbonn.org/fileadmin/user_upload/carbonn/Standards/IEAP_October2010_color.pdf. ICLEI – local governments for sustainability

Johansson N (2008) Production of liquid biogas, LBG, with cryogenic and conventional upgrading technology, Technology and Society, Lund University

Jones A (2004) Woking: local sustainable community energy. Woking

Jones A (2010a) Removal of the barriers to trigeneration. City of Sydney

Jones A (2010b) Sydney's energy revolution: building a low-carbon city. City of Sydney. http://www.cityofsydney.nsw.gov.au/podcasts/citytalks/Archive/2010/default.asp#Sydneyenergyrevolution

Jones A (2012) Interview with (Chief Development Officer, Energy and Climate Change, Sydney) by Jess Bunning. Perth, August 16

Jones A (2013) Email message to Jess Bunning, January 16

Kennedy S, Sgouridis S (2011) Rigorous classification and carbon accounting principles for low and Zero Carbon Cities. Energ Policy 39:5259–5268
Kenway SJ, Turner GM, Cook S, Baynes T (2008) Water-energy futures for Melbourne: the effect of water strategies, water use and urban form. CSIRO: Water for a Healthy Country National Research Flagship
Kinesis (2010) Decentralised energy master plan – trigeneration Sydney, City of Sydney, Sydney, Australia
Kinesis (2012) CCAP suite – mechanisms for change 2012 cited 27/6/2012. http://kinesis.org/tools
Kinesis Consortium (2012) City of Sydney: decentralised energy master plan, trigeneration 2010–2030, City of Sydney, Sydney, Australia
Leck H, Simon D (2012) Fostering multiscalar collaboration and co-operation for effective governance of climate change adaptation. Urban Stud: 1–8. http://usj.sagepub.com/content/early/2012/10/26/0042098012461675.full.pdf+html. Accessed 26 Oct 12
Low Carbon Australia (2013) Carbon neutral network. http://www.lowcarbonaustralia.com.au/business/going-carbon-neutral/carbon-neutral-network.aspx 2013. Cited 25 Jan 13
McCulloch B (2012) Interview with Brodie McCulloch (Managing Director SiiWA) by Jess Bunning. Tape recording, Perth, 4 July 2012
Murray J, Dey C (2009) The carbon neutral free for all. Int J Greenh Gas Control 3:237–248
Newman P, Matan A (2013) Green urbanism in Asia. World Scientific, Singapore
Newman P, Beatley T, Boyer H (2009) Resilient cities: responding to peak oil and climate change. Island Press, Washington, DC
Newman P, Rauland V, Holden D (2011) Creating resilient cities: how a new generation of tools can assist local governments in achieving their carbon abatement goals. Paper read at state of Australian cities national conference, Melbourne, Vic, 29 Nov–2 Dec 2011
Prendergast J (2012) Can trigen get us to 100 per cent renewable? Yes it can. Fifth Estate 20122012. http://www.thefifthestate.com.au/archives/33729/
Rauland V, Newman P (2011) Decarbonising Australian cities: a new model for creating low carbon, resilient cities paper read at MODSIM2011. In: 19th international congress on modelling and simulation, Perth, Australia, 12–16 Dec 2011
SimplyCarbon (2012) South Fremantle Senior High School: the journey to carbon neutral. http://simplycarbon.files.wordpress.com/2012/07/simply-carbon-e-brochure1.pdf. Accessed 25 Jan 13
Tanner T, Mitchell T, Polack E, Guenther B (2009) Urban governance for adaptation: assessing climate change resilience in ten Asian cities. IDS working papers 2009 (315): 1–47
Taper B, Holden D, Kranz L, Sander M, Helstroom R, Hashman C (2010) Urban planning, information technology and evidence based decision making for emission reductions in our cities. Sydney
VicUrban (2010) Revitalising central dandenong area 1 – Precinct energy project: request for proposal
Williams J (2012) Zero carbon homes: a road map. Earthscan, New York, NY, USA
Willis R (2008) The proximity principle: why we are living too far apart. Campaign to Protect Rural England (CPRE): London
Woking Borough Council (2011) Climate change strategy 2011. http://www.woking.gov.uk/environment/climate/Greeninitiatives/climatechangestrategy
WRI, WBCSD (2003) The GHG protocol for project accounting. http://www.ghgprotocol.org/files/ghgp/ghg_project_protocol.pdf. World Resources Institute. World Business Council for Sustainable Development
WRI, WBCSD (2004) The greenhouse gas protocol: a corporate accounting and reporting standard-revised edition. http://www.ghgprotocol.org/standards/corporate-standard. World Resources Institute. World Business Council for Sustainable Development
Wright M (2012) Trigeneration: Sydney's white elephant. Climate spectator. http://www.climatespectator.com.au/commentary/trigeneration-sydneys-white-elephant
Yu C, Brown R, Morison P (2011) Co-governing small-scale distributed water systems: an analytical framework. In: 12th international conference on urban drainage. Porto Alegre

The Rebirth of Distance in the Context of Urban Sustainability

Kobe Boussauw, Richard Hanley, and Frank Witlox

Abstract This chapter focuses on the interaction between mobility policies and spatial planning practice, and aims to contribute to a better understanding on this complex relationship. Our focus is on daily travel patterns of people (leaving freight out of the picture). We discuss the importance of spatial proximity related to the oil dependence of an urban system. We find that the debate regarding the promotion of compact development is far from finished, but that our reasoning leads to the conclusion that a further increase of residential density and land use mix in urban areas is perhaps the best guarantee for curbing excessive mobility and preparing for the end of cheap oil.

1 Introduction

An often-heard statement says that the interaction between mobility policy and spatial planning practice needs more coordination. Any concerned politician or citizen feels that the perceived increase in car traffic, and the growth of problems that are associated with car use, have "something" to do with unorganized urban expansion, sprawling new housing and industrial allotments, and ribbon development. Nevertheless, the exact nature of this relationship is unclear as is the question

K. Boussauw (✉)
Ghent University, Geography Department, Krijgslaan 281, S8 9000, Gent, Belgium
e-mail: kobe.boussauw@ugent.be

F. Witlox
Ghent University, Geography Department, Krijgslaan 281, B-9000, Gent, Belgium

R. Hanley
City University of New York, New York City College of Technology, Brooklyn, New York, USA

S.Th. Rassia and P.M. Pardalos (eds.), *Cities for Smart Environmental and Energy Futures*, Energy Systems, DOI 10.1007/978-3-642-37661-0_11,

how planning should be used as a tool to improve accessibility and steer mobility to a more sustainable course.

Preliminary conclusions of the existing research line on the interaction between urban form and sustainable travel could be formulated as follows: "A sustainable travel pattern can only be realized within an appropriate spatial framework, but other measures (financial and regulatory) are needed to effectively change travel behaviour. In other words, planning is necessary but not sufficient" (after Zhang 2002, p. 3).

While narrowing the focus on the daily travel of people (in contrast to freight transport issues) this chapter will discuss the importance of spatial proximity related to the oil dependence of an urban system.

2 Sustainable Mobility, Climate Change and Peak Oil

The objective of coming to a less car-dependent and, by extension, a less oil-dependent transport system can be argued from different perspectives. Local environmental and safety problems caused by transport have already been in the spotlight for several decades. The principal issues in this debate are air pollution, noise, deterioration of the livability of residential areas, accidents and ecological and landscape fragmentation.

This environmental approach is part of what is called "sustainable mobility" in the recent transport literature. Banister (2008) identifies four components that may contribute to the transition to a more sustainable transport system: (1) reducing the need to travel through substitution, (2) achieving a modal shift through transport policy measures, (3) distance reduction through land-use policy measures, and (4) efficiency increase through technological innovation.

Especially in the field of air pollution, accidents, and livability, we can say that in the western world a lot of progress has been made by a combination of the mentioned policies and technological developments. We do not elaborate on this: for an overview we refer to Gilbert and Perl (2008, pp. 189–264).

However, in terms of climate change (an environmental problem) and peak oil vulnerability (an economic problem), there seems to be much less progress. Below, we examine these two global phenomena, considering that these could constitute a major incentive to reduce the oil dependence of the transport system.

2.1 *Climate Change*

Transport is at the bottom of the class when it comes to greenhouse gas emissions. From 1990 to 2005 greenhouse gas emissions in the EU, for example, decreased in all sectors, except in (domestic) transport (+26 %) and international bunkers (fuel for aviation and shipping) (+64 %) (EEA 2008). Although efficiency improvement in vehicles is indeed enforceable in terms of regulation, it seems that traffic growth itself is directly linked with increasing prosperity. When mobility growth stagnates, this is usually due to a (temporary) economic recession. However, it is often argued that it should be possible to decouple economic growth from an increase in the adverse effects of traffic. Overall, proposed policy measures aimed at reducing greenhouse emissions from traffic can be divided into four packages:

1. Steering mobility to a less car-dependent course by encouraging a modal shift towards alternative transport modes (other than passenger car or truck), a more efficient utilization of vehicles, but also substitution of transport by telecommunication. Known examples in passenger transport are: car sharing, carpooling, the provision of public transport, encouraging cycling and walking, and teleworking and videoconferencing (Robèrt and Jonsson 2006).
2. Increasing, and possibly varying, financial charges on those types of mobility which emit most greenhouse gases, while alternatives may be supported in parallel. Examples in passenger travel are raising fuel taxes, introducing registration charges corresponding to the emission level of the vehicle, but also road tolls, "smart" congestion charge, free public transport and bicycle allowances (Chapman 2007).
3. Increasing vehicle efficiency by accelerating fleet replacement by more fuel efficient vehicles or vehicles that run on alternative fuels (including switching to renewable organic fuels, known as biofuels) (Anable and Bristow 2007).
4. Intervening in the spatial structure through land use planning with the aim of bringing potential destinations closer together and making long distance transport needless. With regard to person mobility, it is generally assumed that increasing density and developing a high degree of land use mix leads to less use of private cars and shorter daily distances travelled (Newman and Kenworthy 2006). The latter form of proposed policy is the background of further discussion in this chapter.

The objectives of these four packages are clear, and throughout the western world many examples can be found where these kinds of measures have been implemented and have had measurable successes. However, the finding that overall traffic and transport emissions continue to grow (Cervero and Murakami 2010), leaves room for some healthy scepticism. Decoupling transport-related emissions and prosperity appears not to be a technical but a social problem, for which an univocal response has not yet been found. Moreover, in the preceding argument we confined ourselves to the western world. The largest relative growth of transport-related emissions, however, is realized in developing countries.

2.2 *Peak Oil*

Today, the term peak oil is used to refer to phenomena caused by global scarcity of fossil oil. Although the report of the Club of Rome in 1972 predicted for 1992 a depletion of all oil reserves that were known at that time (Meadows et al. 1972), in practice peak oil is not to be seen as depletion, but as a mismatch between supply and demand (Campbell and Laherrère 1998). When demand for oil continues to rise by a steadily growing global economy, but production does not follow this trend, then, in principle, the result is a drastic increase of the relatively inelastic oil price.

Although the phenomenon of peak oil is much less known in common than climate change (Bardi 2009), it may be much more restrictive and may therefore have more far-reaching consequences, both for the economy in general and for the transport sector in particular. The reason is that peak oil is about physical limitation, which can virtually not be controlled by any policy.

A survey by ASPO (2010) shows that most scenarios locate the peak in world oil production somewhere between 2006 and 2012, although information is highly dependent on the source. The International Energy Agency (IEA 2008) estimates the world oil production in 2030 at 101.5 million barrels per day (some 20 % more than the production level of 2008) while an estimate by Aleklett et al. (2010) gives 75.8 million barrels per day (which is 10 % less compared to the production level of 2008). The debate seems characterized by an optimistic view of the IEA versus the pessimistic view that is dominant in the peer-reviewed literature.

But it is not just the point in time when peak oil will occur that is uncertain: the potential consequences are also not clear. Before 2008 it was generally assumed that reaching the production peak would be followed by strong price increases. Indeed, during 2007 oil prices rocketed up, to reach a top of about 147 USD per barrel in June 2008 (which is about 125 USD expressed in June 2010 USD) (Fig. 1). Then the price dropped again considerably in order to keep fluctuating between 80 and 120 USD, a price level that could hardly be qualified as the consequence of a dramatic shortage.

Hamilton (2008) presents a part of the explanation, by suggesting that a surge in oil prices is followed by recession, shrinking the demand for oil until a price level is reached that can be worn by the current economic system. This means that the oil price remains high, when expressed in terms of prosperity but not necessarily in monetary terms.

Despite the fact that conventional fossil fuels are finite is beyond dispute, peak oil theory may be viewed as a neo-Malthusian doctrine. The extent to which Malthus will play a role in the energy problem depends on the extent to which affordable alternatives to conventional oil will be available (Greene et al. 2006). The alternatives are known: non-conventional oil and gas (tar sand, oil shale, coal conversion), biofuel, and renewable or nuclear electricity. Typical of these alternatives, however, is that they are either much more expensive to extract compared to the current oil price (unconventional oil), or that they are not sufficiently widely available to be considered as a serious substitute (biofuel and electricity).

But what does this story mean for the transport system? According to IEA (2008), in 2006, 95 % of all energy used worldwide for transport comes from petroleum. IEA (2008) believes this share to drop to 93 % in 2015 and to 92 % in 2030, provided that biofuel substitution is widely implemented. This means that the transport sector remains extremely dependent on the availability, and thus the price, of oil.

Rodrigue et al. (2009) present the following possible effects of high oil prices on the transport system:

- Reduction of both the speed and the total amount of mobility
- Shifts to alternative modes that rely less on oil

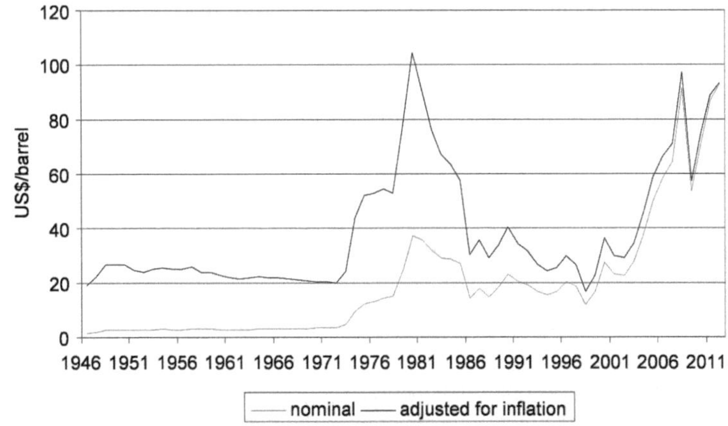

Fig. 1 Evolution of the oil price, based on annual averages (Source: Inflationdata (2012))

- Changes in the organization of transport and distribution networks in favour of more fuel- efficient vehicles and shorter total distances
- In the long-term: changes in location choices as a function of facilitating the mentioned adjustments

Although there is little evidence for long-term effects of high fuel prices on the transport system, a number of indications can be deduced from the responses to the surge in oil prices in the second half of 2007 and the first half of 2008. INRIX (2008) found a significant reduction of congestion in the first half of 2008 when the retail fuel price in the US increased by 28 % on average. The correlation between the increase in price and the decrease in automobile traffic was strongest in cities with a lot of recreational traffic (tourist destinations such as Las Vegas and Miami) and in cities where the car is dominant but public transport offers a decent alternative (e.g., Atlanta, Los Angeles). In cities where public transport was already heavily used (e.g. New York, Washington DC), the effect was less pronounced. Apparently the strongest impact was reported in the segment of tourism and recreational traffic, and in places where room for behavioural change is present.

It is possible to relate the recent volatility of oil prices and peak oil theory to the emerging literature on the theme of urban resilience. This line of research assesses which factors determine the vulnerability of an urban economy in the context of unstable oil prices and how the damage could be minimized. Transport obviously plays a major role in this theme (Dodson and Sipe 2008; Gleeson 2008).

3 The Time-Distance-Space Relationship

Based on the above-mentioned findings, it is worth taking a closer look at the mechanisms behind the continuous increase in mobility and the associated oil dependence. The annual distance covered by an average individual is increasing every year, and this is done with ever-faster means of transport. Nevertheless, most of the possible destinations where everyone is hurrying to remain relatively static: most town centres have been in the same place for decades, if not for centuries, and although cities are sprawling, they seem much less expansive than the mobility itself.

3.1 Time Distance and Travel Time Budget

Common sense says that one wants to minimize the time distance between origins and destinations, assuming that travel is a derived demand (Mokhtarian and Salomon 2001). This reasoning has been largely prompted by the fact that people, within certain financial limits, always choose the fastest available transport mode. Workers with a decent income only take public transport to commute if this is faster than the car. The market share of international rail traffic has dropped steeply since the advent of low cost airlines. Rail projects that have gained market share are invariably those whose speed is competitive with the airplane (Rodrigue et al. 2009). And biking is especially popular in dense urban centres, where it takes hours to manoeuvre a car through or to find some parking space (Verhetsel et al. 2007, p. 5).

However, the above, intuitive, argument is only a fallacy by which many urban planners and policy makers have been caught in the past. Time budget surveys, primarily those of Szalai and Converse (1972), shed new light on the matter. At the aggregate level (e.g., all inhabitants of a region) the average amount of time spent on travel appeared to be regarded as a constant, irrespective of the geographical location of the studied region. Since the US, Western Europe and Eastern Europe were involved in the investigation, we may conclude that this time budget is independent of the economic development stage or income. Szalai and Converse (1972) found an average personal travel time budget of 1 h 13 min per day, or 444 h per year.

3.2 Travel Speed

Given the constant travel time budget, the tendency to choose the fastest manner to make a trip has not led to a reduction of the time spent on travel, but to an increase in kilometrage. Within the constant travel time budget (and the financial travel budget), speed is usually maximized. The average travel speed has increased over the years, perhaps since the invention of the wheel (Ma and Kang 2011). In the economic centres of the western world, congestion and traffic regulations have somewhat constrained the increase of the average speed today, but air travel and fast public transport systems are still growing unabated. Moreover, it should not be forgotten that motorists who cause congestion just take the car because it is still faster than the alternatives. On average, the individual speed is thus higher in a congested situation than in a similar travel pattern system without congestion where more people walk or take, usually slower, public transport.

Various explanations can be found for the human inclination to travel ever faster. First, there is a simple economic explanation. Maintaining a higher speed automatically means a wider radius of action, or more precisely, a larger space-time prism (Hägerstrand 1970). The wider the radius of action, the more likely that within the associated space-time prism a suitable home, one or two nice and well-paid jobs, a decent school, a number of relatives and friends and the desired recreational and shopping facilities are located. High speed allows for spatial optimization of the utility of one's travel pattern within the constant travel time budget. But this economic approach may not explain everything. The desire for speed is as much a psychological phenomenon that is related to sensation and status.

3.3 The Combined Travel Time and Cost Budget

We may say that individual travel decisions rely on the existence of a combined travel time–cost budget, in which both the kilometre-dependent cost and the required travel time are reckoned. The ratio between the temporal and the financial component in the time–cost budget is determinant in the perception of physical distance by an individual. In large parts of Europe, the cost per kilometre has systematically declined over the years, with the introduction of the railways and tramways in the nineteenth century as a main milestone. This trend was then accelerated by the democratization of the car, which became cheaper in comparison with the general prosperity level and provided – at least until the early 1980s – definitely the fastest way to make any domestic surface trip.

Regarding the cost component, the base price of fuel is an important element. The price elasticity for fuel is highly dependent on the kind of trip one chooses. It is known that business travel and commuting are hardly affected by price increases. However, in freight transport and recreational and tourist travel, the connection is much clearer. The relationship between the amount of touristic air traffic and the price of flights is even almost linear (Litman 2007). When fuel prices rise relative to income through the peak oil phenomenon, the importance of the cost component in the combined time–cost budget will increase. Also any climate policy that is rationing the consumption of fossil fuels (e.g., through emissions trading) would significantly rise the importance of the cost component (Keppens 2006).

Since variable transport costs are almost directly linked to distance travelled,[1] the importance of physical distance in the perception of the traveller will increase when the share of the financial component in the time–cost budget is rising.

[1] Although the correlation varies with the speed: in steady-flow traffic at medium speed, an average car operates in the most efficient mode. At high speeds, but also in stop-and-go traffic, fuel consumption and emissions per kilometre go up quickly (Deakin 2007).

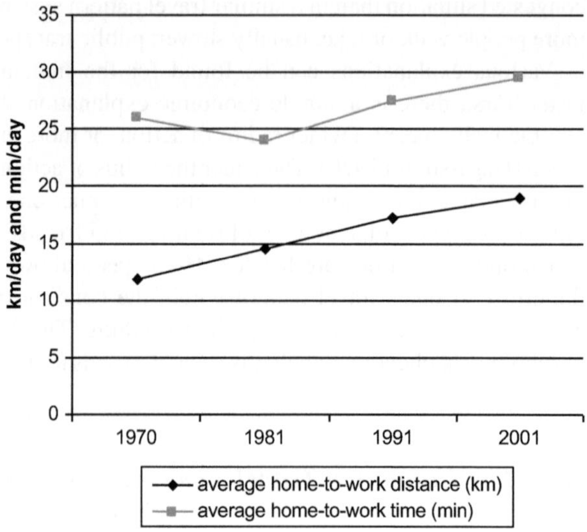

Fig. 2 Evolution of the average commuting time and distance in Belgium (Sources: Mérenne-Schoumaker et al. (1999) and Verhetsel et al. (2007))

The ratio between the cost component and the time component in the time–cost budget depends also on external factors such as the development stage of the considered region, and the general state of the world economy. In countries where the cost of motorized transport swallows, or would swallow, a large share of household income, physical distance continues to be determinant. The effort it takes to make a trip on foot or by bicycle, or even the marginal cost of fuel in the case of a motorized trip, remains largely linked to the kilometrage. In contrast, in the western world the cost of having a car, and the cost of fuel, have become basic expenses that are almost always outweighing the perceived benefit of intensifying the use of the available travel time budget.

The regional stage of development determines the generalization of car ownership and the performance of public transport, but also the congestion level and the social sensitivity to external impacts. During the 1970s, for example, in Belgium the average time spent on commuting decreased, while the length of the commute simultaneously increased. The massive shift from slow transport modes to the car allowed for a significant intensification of the use of the travel time budget. So it became possible to live further away from one's workplace, or to look for a job far from home. Since 1981, in Belgium the time spent on a particular commute is again in line with the distance. The generalization of car ownership and the saturation of the road infrastructure does not any longer allow for increases in the average commuting speed (Fig. 2) (Mérenne-Schoumaker et al. 1999; Verhetsel et al. 2007).

The external effects of transport, as an umbrella for road safety and environmental effects, play their role too. As there is more traffic on the roads, public support for government action is growing. Imposed speed limits, directly affecting the average speed, demonstrate this development. Financial policies, including fuel taxes and fixed or variable tolls, emphasize again the cost component.

4 The Rebirth of Distance

4.1 Absorption of Rising Energy Prices

In the future, potentially rising energy prices could lead to a reappraisal of physical distance. The spatial structure and the coherence and distance between different functions and services will be more pronounced in the individual perception, since one will be inclined to minimize not only travel time but also travel distance. "Proximity", a concept that is associated with physical distance (as opposed to the concept of "accessibility" that is rather associated with time distance), will gain in value, as will slower and thus more energy-efficient transport modes.

However, there is a technological buffer that may absorb for some time a considerable part of the potential price increase, since it is possible to make the fleet a lot more efficient. From 2002 to 2005 the average fuel consumption of new cars sold in Belgium decreased from 6.3 to 5.9 l/100 km (De Vlieger et al. 2007), and the most economical of all small cars get by on 3.4 l/100 km. So there is still a margin that allows reducing fossil fuel consumption and vehicle emissions significantly by using more efficient engines in smaller and lighter cars. Also, the use of alternative fuels should be taken into account, even though these will have little impact on price making.

Research on price elasticity indicates a clear link between cost and consumption of fuel, although there are noticeable differences depending on the type of trip (Bogaert et al. 2006). Long-term price elasticities are usually larger than short-term values, and are observed in terms of a reduction of kilometrage, vehicle efficiency and car ownership. However, recorded absolute values of price elasticity are usually lower than 1, meaning that demand for fuel is relatively inelastic (Brons et al. 2008). An important explanation for the inelasticity is the virtual absence of possible substitutes. There are only few, if any, fully-fledged alternative fuels available at an equal or lower price.

However, regarding the distant future, we can say that existing analyses of price elasticities do not take into account very long-term effects, such as shifts in terms of regional economic structure, real estate prices, and urban development. Brons et al. (2008) recognize that current analyses do not constitute grounds for very long-term predictions. In addition, they show that mainly travel distance is sensitive to price rises as the studied time span increases. Another important element is the *ceteris paribus* assumption, which is common in price elasticity studies. How mobility would evolve if a sharp rise in oil prices leads to an economic recession cannot be inferred from these studies.

4.2 Mobility, Accessibility and Spatial Structure

It is clear that reducing the oil dependence of the (personal) transport system has an important technological component: cars need to become more fuel efficient and alternative fuels should get adopted wherever possible. The second component, however, has to do with adapting travel behaviour: a reduction of oil dependence means less car kilometrage (both by travelling fewer kilometres, and by a modal shift), and to some extent less mobility.

In the economic approach, however, mobility is considered as a secondary activity, and not as a purpose in itself. The assumed objective is to optimize accessibility, where accessibility should be seen as a measure of the interaction potential of an individual, a family or an organization. According to Geurs and Ritsema van Eck (2001, p. 36) accessibility reflects the extent to which the "land-use transport system enables (groups of) individuals or goods to reach activities or destinations by means of a (combination of) transport mode(s)".

It is possible to realize a wide range of interaction possibilities, and thus potential destinations, within a short distance. Some of the densest urban centres in the world, such as Manhattan or Hong Kong, illustrate this. This mechanism underlies the formation of most cities, increasing accessibility based on a minimum of transport. Engwicht (1992, p. 12) puts it as follows: "Cities are an invention to maximize exchange and minimize travel."

Newman and Kenworthy (1989, 1999) were the first who managed to identify a link between the spatial structure of a variety of world cities and energy consumption for transport. They noted that population density shows an inverse relationship with energy consumption for transport. Residents of densely populated cities cover shorter distances, since their daily destinations are closer to home. In addition, they more often use public transport, compared to cities with a low population density. This phenomenon is strengthened by the higher level of congestion and parking limitations that are associated with high density.

The discourse of Newman and Kenworthy (1989, 1999) has given rise to a line of research that investigates the relationship between spatial structure and sustainability of travel patterns in more detail. Some studies seem to confirm the argument of Newman and Kenworthy (1989, 1999), while other researchers sharply criticise the methodology originally used, particularly on the demarcation of the urban areas that were examined in the original study (Mindali et al. 2004; Mees 2010). Although there is little debate about the existence of a relationship between spatial structure and travel patterns (Ewing and Cervero 2010), there is less unanimity about the sense or nonsense of conducting spatial policies aimed at sustainable mobility. Gordon and Richardson (1997) for example believe that the connection is too weak to justify interventions. According to them, as cities are sprawling spontaneously, this would be the most efficient form of urban growth within the current market economy.

However, departing from the assumption that spatial structure is a hard constraint that determines whether it is possible to shorten trip lengths or to take public transport or the bike at all, an alternative approach could be stating that compact development is a necessary precondition for anticipating urban resilience towards peak oil.

4.3 Urban Sprawl Versus the Compact City

The acknowledgement that uncontrolled and unplanned urban growth entails problems is far from new. In the peer-reviewed literature, several types of quasi-unorganized developments have been lumped together as "(urban) sprawl". In the literature, and in many policy documents, a strong link between sprawl and increased traffic is assumed (Ewing et al. 2002). This is logical bearing in mind that separation processes between origins and destinations inevitably entail larger distances to be covered. Even though there seems to be a link between the presence of sprawl and the average trip length, this reasoning does not necessarily imply a one-way causality. Sprawl is caused by increasing individual mobility, i.e. a wealth-related phenomenon that is the basis of an autonomous growth of traffic. So, we might explain the wave of suburbanization as a materialization of this increased mobility, which has itself a mutual reinforcing effect on the growth of traffic. Gilbert and Perl (2008, p. 235) formulate this phenomenon as follows: "Sprawl is believed to be facilitated by car ownership and use and also to contribute to it, in a positive feedback loop that reinforces both low-density development and motorization."

However, the extent to which spatial separation leads to an effective increase of distances that need to be covered is less clear, since this cannot be derived from local morphological characteristics. A monotonous residential lot embedded in a major employment centre could possibly lead to a more sustainable travel pattern than a compact town that is immersed in a rural area. Particularly in a context where average trip lengths, and in particular average commuting distances, have become very large in practice, it is hard to tell which kind of spatial developments are problematic in relation to mobility, and which are rather beneficial. Banister (1999) observes the issue in a non-morphological way, and suggests that it is of the utmost importance that new developments are sufficiently large and are located in or immediately subsequent to existing urban areas. Local morphology, density and spatial diversity come only in second place.

In Europe, since World War II, in a number of agglomerations tackling urban sprawl is considered an inherent part of the basic principles of what constitutes proper urban planning. The main examples that fit into this tradition are the New Towns Act (UK 1946), the Finger Plan of Copenhagen (1947), the various Notes on Spatial Planning (The Netherlands, from 1960), the Villes Nouvelles around Paris (since 1965), *et cetera.* In 1990, the European Union took a position against the further development of sprawl by endorsing the paradigm of the compact city in the Green Paper on the Urban Environment (CEC 1990). The compact city is presented as the ideal to be pursued, ensuring sustainable development. Mobility is one of the key guiding principles that form the basis of the compact city policy, since slow traffic and public transport would get more opportunities through shorter distances and higher densities.

Although the compact city policy is considered valid in a European context by most urban planners, a number of critiques have been developed. Perhaps the main criticism is the reasoning that the creation of spatial proximity by increasing density and mixing functions does not necessarily lead to a reduction in the number and the length of car trips, but merely provides the appropriate spatial framework. Moreover, the compact city policy does not take into account the size of the city. In a city like London, for example, which is by its size much more compact in a geometrical sense than a medium sized city, per capita energy consumption for transport is higher than in medium sized British cities (Breheny 1992). According to Banister (1992), in remote small towns with a limited degree of self-sufficiency, transport energy consumption per capita is easily one-third higher than average, while in rural areas this may rise up to three times the level of a regional city.

Breheny (1996) discusses the centralist and decentralist movements, which in the 1970s strongly determined the debate in the field of spatial planning. He concludes that eventually the compromise is the best solution. Compactness and diversity are important but should not lead to a loss of environmental quality. Moreover, it is not feasible to exclude greenfield developments completely, and there should also be a role for urban networks.

With regard to travel, it has become clear that the spatial structure should provide the framework that allows for a sustainable mobility pattern, with a substantial role for short-distance trips, public transport and non-motorized modes. However, it cannot be expected that travel behaviour will be steered only by the morphology of space.

5 Conclusion

Recognizing the importance of spatial proximity in relation to sustainable travel patterns is based on two tracks. First, we know from the literature that a higher population density and a better mix of functions usually accompanies shorter trip distances and a lower share of car drivers, factors indicating a more sustainable travel pattern. Although the found relationships are often weaker than would be expected intuitively, they are usually statistically significant (Boussauw and Witlox 2011). This means that strengthening spatial proximity is expected to result in a more sustainable travel pattern. How strong this effect is, depends *inter alia* on personal socioeconomic characteristics, such as income, family composition or personal preferences. In short, in this context spatial planning is used to pursue environmental and climate objectives.

Second, increasing characteristics of spatial proximity reduces the vulnerability of the spatial-economic system to increasing transport costs (Dodson and Sipe 2008). Increasing transport cost is critical for interaction possibilities in a spatial system where everything is far away from everything else. In urban structures, characterized by a high degree of spatial proximity, however, interaction possibilities remain ample, even when transport is limited. The reason why this view has hardly been addressed in the scholarly literature is the almost continuous decline in transport costs throughout history. It is only after acknowledging the sense of reality of the peak oil theory that an increase in transport cost in the long term can be seen as a likely scenario. A lack of spatial proximity may turn oil price shocks really problematic and strengthen the difference in accessibility between metropolitan and rural areas in a way that may have serious consequences for the local economy and quality of life in remote areas.

However, even if price increases would be reasonable and develop smoothly, a physical structure that is based on a high degree of spatial proximity would still provide a framework that facilitates behaviour changes in response to a vigorous climate policy.

In policy terms, this means that improving accessibility should become the paramount objective, as opposed to intensifying mobility. The ultimate goal is to achieve a high degree of accessibility based on a minimum of mobility and therefore a minimum amount of traffic. Although the debate regarding the promotion of compact development is far from finished, our reasoning leads to the conclusion that a further increase of residential density and land use mix in urban areas is perhaps the best guarantee for curbing excessive mobility and preparing for the end of cheap oil.

References

Aleklett K, Höök M, Jakobsson K, Lardelli M, Snowden S, Söderbergh B (2010) The peak of the oil age – analyzing the world oil production reference scenario in World Energy Outlook 2008. Energy Policy 38(3):1398–1414

Anable J, Bristow AL (2007) Transport and climate change: supporting document to the CfIT report. The Robert Gordon University/Loughborough University, Aberdeen/Loughborough

ASPO (2010) Association for the study of peak oil & gas. http://www.peakoil.net

Banister D (1992) Energy use, transport and settlement patterns. In: Breheny M (ed) Sustainable development and urban form. Pion, London, pp 160–181

Banister D (1999) Planning more to travel less. Town Plann Rev 70(3):313–338

Banister D (2008) The sustainable mobility paradigm. Transport Policy 15(2):73–80

Bardi U (2009) Peak oil: the four stages of a new idea. Energy 34(3):323–326

Bogaert S, De Smet L, Verdonck F, Van Biervliet K, Logghe S, Franckx L, Eyckmans J, De Jaeger S (2006) Verkennende Studie naar Prijs- en Inkomens-Elasticiteiten van Milieugerelateerde Goederen en Diensten in Vlaanderen. Flemish Environment Agency, Aalst

Boussauw K, Witlox F (2011) Linking expected mobility production to sustainable residential location planning: some evidence from Flanders. J Transport Geogr 19(4):936–942

Breheny M (1992) The contradictions of the compact city. In: Breheny M (ed) Sustainable development and urban form. Pion, London, pp 138–159

Breheny M (1996) Centrists, decentrists and compromisers: views on the future of urban form. In: Jenks M, Burton E, Williams K (eds) The compact city: a sustainable urban form? E&FN SPON, London, pp 13–35

Brons M, Nijkamp P, Pels E, Rietveld P (2008) A meta-analysis of the price elasticity of gasoline demand. A SUR approach. Energ Econ 30(5):2105–2122

Campbell CJ, Laherrère JH (1998) The end of cheap oil. Sci Am 278:78–83

CEC (1990) Green paper on the urban environment – COM(90)218. Commission of the European Communities, Brussels
Cervero R, Murakami J (2010) Effects of built environments on vehicle miles traveled: evidence from 370 US urbanized areas. Environ Plann A 42(2):400–418
Chapman L (2007) Transport and climate change: a review. J Transport Geogr 15(5):354–367
De Vlieger I, Cornelis E, Int Panis L, Schrooten L, Govaerts L, Pelkmans L, Logghe S, Vanherle K, Vanhove F, De Ceuster G, Macharis C, Festraets T, Pekin E, Turcksin L, Van Bladel K, de Jong M, Van Mierlo J, Matheys J, Timmermans J-M, De Geest C, Van Walsum E (2007) MIRA (2007) Milieurapport Vlaanderen, Achtergronddocument 2007, Transport. Flemish Environment Agency, Aalst
Deakin E (2007) The environmental consequences of sprawl. In: Transport, urban form and economic growth – Round table 137. European Conference of Ministers of Transport – OECD Publishing, Paris
Dodson J, Sipe N (2008) Shocking the suburbs: urban location, homeownership and oil vulnerability in the Australian city. Hous Stud 23(3):377–401
EEA (2008) Data and maps: EN01 energy related greenhouse gas emissions. European Environment Agency. http://www.eea.europa.eu/data-and-maps/indicators/en01-energy-related-greenhouse-gas-emissions
Engwicht D (1992) Towards an eco-city: calming the traffic. Envirobook, Sydney
Ewing R, Cervero R (2010) Travel and the built environment: a meta-analysis. J Am Plann Assoc 76(3):265–294
Ewing R, Pendall R, Chen D (2002) Measuring sprawl and its impact. Smart Growth America, Washington, DC
Geurs KT, Ritsema van Eck JR (2001) Accessibility measures: review and applications. National Institute for Public Health and the Environment, Bilthoven
Gilbert R, Perl A (2008) Transport revolutions: moving people and freight without oil. Earthscan/James & James, London
Gleeson B (2008) Critical commentary. Waking from the dream: an Australian perspective on urban resilience. Urb Stud 45(13):2653–2668
Gordon P, Richardson HW (1997) Are compact cities a desirable planning goal? J Am Plann Assoc 63(1):95–106
Greene DL, Hopson JL, Li J (2006) Have we run out of oil yet? Oil peaking analysis from an optimist's perspective. Energ Policy 34(5):515–531
Hägerstrand T (1970) What about people in regional science? Papers Regional Sci 24(1):6–21
Hamilton JD (2008) Oil and the macroeconomy. In: Durlauf SN, Blume LE (eds) The new Palgrave dictionary of economics. Palgrave Macmillan, Basingstoke
IEA (2008) World energy outlook. International Energy Agency, Paris
Inflationdata (2012) Historical crude oil prices. http://inflationdata.com/Inflation/Inflation_Rate/Historical_Oil_Prices_Table.asp
INRIX (2008) National traffic scorecard – special report. INRIX, Kirkland
Keppens M (2006) Modelling private road transport of Flemish households. The mobility and welfare effects of fuel taxes versus tradable fuel rights. Ph.D. thesis. Universiteit Hasselt, Hasselt
Litman T (2007) Transportation elasticities: how prices and other factors affect travel behavior. Victoria Transport Policy Institute, Victoria
Ma K-R, Kang E-T (2011) Time-space convergence and urban decentralisation. J Transport Geogr. doi:10.1016/j.jtrangeo.2010.06.016
Meadows DH, Meadows DL, Randers J, Behrens WW III (1972) The limits to growth. Universe Books, New York
Mees P (2010) Transport for suburbia: beyond the automobile age. Earthscan, London
Mérenne-Schoumaker B, Van der Haegen H, Van Hecke E (1999) Algemene Volks- en Woningtelling op 1 maart 1991: Werk- en Schoolpendel. Statistics Belgium, Brussels
Mindali O, Raveh A, Salomon I (2004) Urban density and energy consumption: a new look at old statistics. Transport Res Part A 38(2):143–162

Mokhtarian PL, Salomon I (2001) How derived is the demand for travel? Some conceptual and measurement considerations. Transport Res Part A 35(8):695–719

Newman P, Kenworthy J (1989) Cities and automobile dependence. A sourcebook. Gower, Aldershot

Newman P, Kenworthy J (1999) Sustainability and cities: overcoming automobile dependence. Island Press, Washington, DC

Newman P, Kenworthy J (2006) Urban design to reduce automobile dependence. Opolis 2(1):35–52

Robèrt M, Jonsson RD (2006) Assessment of transport policies toward future emission targets: a backcasting approach for Stockholm 2030. J Environ Assess Policy Manage 8(4):451–478

Rodrigue J-P, Comtois C, Slack B (2009) The geography of transport systems. Routledge, New York

Szalai A, Converse P (1972) The use of time: daily activities of urban and suburban populations in twelve countries. Mouton, The Hague

Verhetsel A, Thomas I, Van Hecke E, Beelen M (2007) Pendel in België. Deel I: De Woon-Werkverplaatsingen. Statistics Belgium, Brussels

Zhang M (2002) Conditions and effectiveness of land use as a mobility tool. Ph.D. thesis. Massachusetts Institute of Technology, Cambridge, MA

Cities for Smart Environmental and Energy Futures: Urban Heat Island Mitigation Techniques for Sustainable Cities

Konstantina Vasilakopoulou, Denia Kolokotsa, and Mattheos Santamouris

Abstract The most apparent impact of urban development on the environment is the rearrangement of its biophysical attributes. By altering the nature of the surface and generating large amounts of heat, urbanized areas modify the microclimate and air quality. The urban heat island phenomenon, which serves as a trap for atmospheric pollutants, deteriorates the quality of life and has a socio-economic impact in the urbanised areas, has been a research subject at least for the last 100 years.

Mitigation strategies have been proposed to alleviate the negative effects of the summertime urban heat. In particular these strategies take advantage of insights gained from study of the urban energy balance. Specifically, they seek to reduce the solar radiation absorbed by the surface or increase the latent heat flux away from the surface. The physical implementation of these strategies involves use of highly reflective (high albedo) roofing and paving materials, extensive planting of urban vegetation, shading etc. The aim of the present chapter is to provide the necessary knowledge concerning the recent developments in the urban heat mitigation techniques the state of the art as well as the future research prospects.

K. Vasilakopoulou (✉) • D. Kolokotsa
Technical University of Crete, Environmental Engineering Department,
Crete GR 73100, Greece
e-mail: kvasilakopoulou@isc.tuc.gr

M. Santamouris
National Kapodistrian University of Athens, Physics Department, Group Building Environmental Studies, Athens, Greece

S.Th. Rassia and P.M. Pardalos (eds.), *Cities for Smart Environmental and Energy Futures*, Energy Systems, DOI 10.1007/978-3-642-37661-0_12,

1 Introduction

Urban areas are 'complex mix of natural elements including air, water, land, climate, flora and fauna, and the built environment that is constructed or modified for human habitation and activity, encompassing buildings, infrastructure and urban open spaces' (Hardy et al. 2001).

Moreover for the first time in history, more people live now in urban than in rural areas. In 2010, urban areas are home to 3.5 billion people, or 50.5 % of the world's population. In the next four decades, all of the world's population growth is expected to take place in urban areas, which will also draw in some of the rural population through rural to urban migration. Moreover, most of the expected urban growth will take place in developing countries, where the urban population is expected to double, from 2.6 billion in 2010 to 5.2 billion in 2050. In developed countries, the number of urban dwellers will grow more modestly, from 0.9 million in 2010 to 1.1 billion in 2050. During the same period, the world's rural population will decline by 0.6 billion (United Nation 2011).

The metabolism of urban activities has become a threat to the global environment and climate change. While examining a city's environmental performance, the quality of the local environment and its impacts on the regional and global environments should be considered. Cities affect and are affected by natural systems beyond their physical boundaries through (a) the conversion of land and transformation of habitats; (b) the extraction and depletion of natural resources; and (c) the release of emissions and wastes.

One of the critical impacts of urban development on the environment is the rearrangement of its biophysical attributes. By altering the nature of the surface and generating large amounts of heat, urbanized areas modify the microclimate and air quality. The urban heat island phenomenon, which serves as a trap for atmospheric pollutants, deteriorates the quality of life and has a socio-economic impact in the urbanized areas, has been a research subject in Europe at least for the last 100 years (Santamouris et al. 2007; Santamouris 2007; Tso 1996; Taha et al. 2004; Taha 2008).

The urban heat island phenomenon is governed in large part by significant differences between the energy budgets of cities and the countryside (Landsberg 1982; Oke 1982). Important heat island studies have been performed in Europe during the last 20 years (Santamouris 2007). The reported analysis is based either on statistics of temperature differences between pairs or groups of urban and rural stations or on results obtained by networks of fixed or mobile stations across an urban area. Most of the studies are organized around the main generalization offered by (Oke 1982). The impact of the wind, cloud cover and generally of cyclonic or anticyclonic conditions on the intensity of heat island is reported (Mihalakakou et al. 2002, 2004). Also, the time period the heat island is presented, and the season where urban heat gets its maximum are also discussed in many studies. Heat island has a very important impact on the energy consumption of buildings (Geros et al. 2005; Kolokotroni et al. 2006). Increased urban temperatures exacerbate the cooling load of buildings, increase the peak electricity demand for cooling and decrease the efficiency of air conditioners. In parallel, high urban temperatures considerably decrease the cooling potential of natural and night ventilation techniques and increase pollution levels. Epidemiological studies of deaths during the heat waves worldwide have found that the elderly are more vulnerable to the combined stress of high temperatures and air pollution (Besancenot 2002; Johnson and Wilson 2009; Devi 2006). Japan and southern Europe have rapidly aging urban populations, so one can expect these regions to become more sensitive to higher temperatures and their related consequences (Zheng et al. 2010; Wong et al. 2011).

Mitigation of urban heat islands, or simply cooling urban areas, is a strategy aimed at countering the negative effects on energy demand and the environment. From a portfolio of several heat-island mitigation strategies, the ones that have attracted the researchers' attention are:

- The use of advanced materials for the urban environment, able to amortize, dissipate and reflect heat and solar radiation.
- Strategic landscaping of cities including appropriate selection and placing of green areas.
- Use of green roofs and green facades.
- Solar control systems.
- Dissipation of the excess heat in low temperature environmental heat sinks like the ground, the water and the ambient air.

The aim of the present chapter is to provide an overview of the necessary knowledge concerning the recent developments in the urban heat mitigation techniques through an analysis of the proposed methods and tools, the state of the art as well as the future research prospects.

2 Urban Heat Island Mitigation for Sustainable Cities

2.1 Cool Pavements

Pavements contribute highly to the development of urban heat islands. Various experimental and theoretical studies have shown that paved surfaces in the urban environment increase urban temperatures and deteriorate the ambient quality, (Sailor 1995; Rosenzweig et al. 2006; Menon et al. 2010; Gaitani et al. 2007). Pavements cover a high part of the urban area. Various studies in USA indicate that paved surfaces cover represent almost 29 % of the cities (Rose et al. 2003), while other studies indicate that roads, sidewalks and parking areas cover 29–39 % when view above the urban canopy and 36–45 % viewed under the canopy (Akbari and Rose 2001). It is characteristic that parking lots in USA occupy almost 37.2 billion square meters and are mainly covered by black asphalt (AASHTO 2010).

Several experimental studies carried around the world have concluded that pavements are among the major contributors of the heat island phenomenon,

(Asaeda et al. 1996; Doll et al. 1985). Their thermal balance depends on the amount of the absorbed solar radiation, the absorbed and emitted infrared radiation, the convective flux to and from the atmosphere, the conductive flux to the ground, the energy stored to the mass of the material and possible latent heat fluxes.

To mitigate the thermal problem caused by the pavements, it is necessary to decrease their sensible heat flux to the atmosphere. This can be achieved through the reduction of their surface temperature. Several techniques have been proposed to decrease the surface temperature of pavements. Paved surfaces presenting a lower surface temperature are known as 'cool pavements'. The following systems, procedures and techniques are mainly used to develop cool pavements:

(a) Surfaces presenting a higher albedo to the solar radiation. The decrease of the absorbed solar radiation contributes to lower the surface temperature of pavements. These pavements are known as 'reflective pavements'. According to the latest literature (Cambridge Systematics 2005; Nam and Buzz 2009; Ferguson et al. 2005; Levine 2011), the possible techniques to increase the albedo of pavements include: The use of conventional cement concrete pavement, the application of white topping and ultra-thin white topping techniques, the use of concrete additives like slag cement and fly ash, the use of light aggregates in asphalt concrete surfaces, the use of chip or sand seals with light aggregates, the use of roller compacted concrete pavement, the painting of the surfaces with a light color using or not microsurfacing techniques the application of color pigments and seals and the use of colorless and reflective synthetic binders,, the use of sand/shot blasting and abrading binder surfaces, resin based pavements, etc.
(b) Permeable or water retaining surfaces presenting a lower surface temperature because of evaporation processes Permeable and water retaining pavements present a lower albedo that the reflective pavements, and are appropriate for humid climates where the availability of water is not a problem. Permeable pavements may be vegetated or not. Non vegetated permeable pavements include, porous and pervious concrete, porous or rubberized asphalt, permeable interlocking concrete pavers, concrete and plastic grid pavers filled with gravels. Vegetated permeable pavements includes grass pavers, reinforced turf and concrete grid pavers use lattices of different types that allow grass to grow in the interstices, (Hunt and Collins 2008).
(c) Surfaces presenting a high or artificially increased thermal storage capacity. This may be achieved by using ingredients of high thermal capacitance or phase change materials.
(d) Artificial mechanical systems that extract the excess heat of pavements and decrease their surface temperature.
(e) Well shaded paved surfaces using natural or artificial solar control devices.

Pavements are composed by a binder together with aggregates. The albedo of the pavement may increase by using a high reflectivity surface coating, or aggregates of light color or a proper binder or a combination of the above.

The use of reflective paints as a cover of the pavements surface has been studied by various authors. In (Synnefa et al. 2006) 14 pavement surfaces covered by different types of high reflective paints were tested during the summer period. The reflectance of all the tested paints varied between 80 % and 90 %. It is reported that the use of the highly reflective coatings on the surface of the pavements reduces the daily surface temperature of a white concrete plate under hot summer conditions by 4 K and by 2 K during the night. Pavements covered with aluminum based paints presented a higher surface temperature than the other tiles mainly because of the low emissivity value they present.

When colored pavements have to be produced, infrared reflective pigments may be employed to increase their global solar reflectivity. In (Synnefa et al. 2007) ten cool colored pavement surfaces covered with infrared reflective paints were tested against conventional pavement materials of the same color. It is reported that the black colored reflective pavements presented almost 10 K lower than the standard black surface, while the cool blue presented almost 4.5 K lower surface temperature. A linear correlation between the surface temperature of the materials and their reflectivity is found.

The use of color changing coatings applied on the surface of pavements has been proposed by (Karlessi et al. 2009). In particular, it is proposed to use thermochromic coatings able to respond thermally to the environment and change reversibly their color and reflectivity from lower to higher values, as temperature rises. Almost 11 tiles covered with thermochromic paints have been prepared and tested. It is reported that the thermochromic covered presented a much lower surface temperature than pavements of the same color covered with infrared reflective coatings. The main problem of the thermochromic coatings has to do with the rapid loss of their optical characteristics

A real scale application where colored highly reflective pavements are used is described in (Santamouris et al. 2012). The project covered a total area of 4,500 m^2 and was situated next to the cost in Athens, Greece. Concrete pavements colored with infrared reflective cool paints presenting an albedo of 0.60 were installed in the park area. The project has been monitored before and after the end of the constructions and by using a combination of measurements and simulations. It was found that the achieved reduction of the average peak ambient temperature was close to 1.9 K, while the peak surface temperature in the area has been reduced by 12 K.

Unfortunately, very few real scale projects involving the use of reflected pavements are monitored in detail to document precisely their expected benefits. Aspects related to a possible deterioration of thermal comfort because of the increased reflected solar radiation have to be considered as well.

2.2 *Green Urban Spaces*

Greenery is perhaps the most important factor for improving the urban environment. Urban agglomeration, philosophy and practice has displaced vegetation from the urban areas and adversely affected the natural environment within the city. Green areas have the potential, to mitigate significantly the urban heat island phenomenon if it is inserted in the urban planning and fabric (Wong et al. 2010; Sfakianaki et al. 2009).

The rapid building construction, the incentives provided to the construction sector, the commodification of housing and the demand for land exploitation prevented the establishment of free green public spaces. The effect of urban green areas in the city's environmental quality is directly related to its size and distance from the urban region. The greater the green surface area and the smaller the distance, the greater the mitigation potential. Trees and vegetation contribute to the reduction of air and surface temperatures through evapotranspiration and shading of adjacent urban surfaces. Results of computer simulations aimed at studying the combined effect of shading and evapotranspiration of vegetation on energy use of several typical one-storey buildings in US cities showed that by adding one tree per house, the cooling energy savings varied from 12 % to 24 % (Akbari et al. 1992, 1997). Moreover, three trees per house can reduce the cooling load between 17 % and 57 %. The direct effects of shading account for only 10–35 % of the total cooling energy savings. The remaining savings result from temperatures lowered by evapotranspiration.

Urban green spaces fulfill many purposes in the urban environment that improves people's quality of life. There is therefore a broad consensus about the importance and value of urban green spaces in cities towards planning and constructing sustainable or eco-cities of twenty-first century. The ecological benefits bestowed in green spaces which range from protecting and maintaining the biodiversity to helping in the mitigation of change cannot be overlooked in today's sustainable growth and planning. Urban green spaces are especially important for improving air quality through uptake of pollutant gases and particulates which are responsible for respiratory infections. Green spaces also help in reduction of the energy costs of cooling buildings effectively. Furthermore, due to their amenity and aesthetic, green spaces increase property value. However, the most sought benefits of green spaces in a city are the social and psychological benefits. Urban green spaces, especially public parks and gardens provide resources for relaxation and recreation. In order to meet social and psychological needs of citizens satisfactorily, green spaces in the city should be easily accessible and in adequately optimal in quality and quantity. Green spaces need to be uniformly distributed throughout the city area, and the total area occupied by green spaces in the city should be large enough to accommodate the city population needs. Cities are responsible for most of the consumption of the world's re-sources and are home to most of the world's citizens as well. Bringing green space to the urban landscape can promote and inspire a better relationship with the environment while supporting important services. Green space is part of and also represents habitats and ecosystems. Many field-based measurements have found that urban parks are 1–2 °C, and sometimes even 5–7 °C, cooler than their urban surroundings, forming a "park cool island" (PCI) (Jauregui 1990; Eliasson and Upmanis 2000; Chang et al. 2007; Santamouris 2012). The differences in surface and air temperatures between the cooler park and its warmer

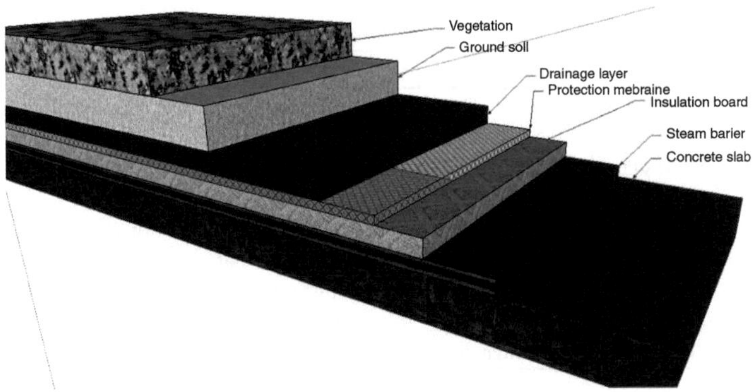

Fig. 1 Green roofs

built surroundings create a cooling oasis whose benefits can be extended beyond the park by the park's width or even farther. A stronger PCI affects a more extensive area, resulting in lower cooling energy consumption in buildings and improving the outdoor environmental conditions (Jauregui 1990; Mahmoud 2011). Considerable research efforts have been made to reveal which park characteristics are crucial to PCI formation. Previous studies have found that larger parks have stronger PCI effects but the relationship between PCI effect and park size might be non-linear (Eliasson and Upmanis 2000; Shashua-Bar and Hoffman 2003, 2004). Different park types or vegetation combinations are studied but the quantifiable effects and statistical relationships have not been quantified in detail.

Apart from urban parks, green areas can be added by the installation of green roofs and green facades.

A green or living roof is a roof or deck onto which vegetation is intentionally grown or habitats for wildlife are established. There are different types of green roof that each type functions and looks different. A typical green roof consists of a soil mixture and a drainage layer (Fig. 1). Green roofs can be categorized to intensive and extensive depending upon the use, the depth of plantation and the required maintenance. Intensive green roofs are usually traditional roof gardens that require a relatively thick soil to grow large plants and they require increased irrigation, feeding and other maintenance. They are normally accessible as a recreation space for residents and so incorporate areas of paving, seating and other architectural features. In contrast, extensive green roofs feature a lightweight growing media and self-generative plants. They are designed to be self-sustaining and should require only a minimum of maintenance, perhaps a once-yearly weeding or an application of slow-release fertilizer to boost growth. Extensive roofs are usually only accessed for maintenance. A comparison between intensive and extensive green roofs is tabulated in Table 1.

Green roofs can contribute highly to mitigate heat island as they present much lower temperatures than hard surfaces and contribute to decrease the ambient temperature through convection and evapotranspiration. Moreover green roofs are

Table 1 Comparison of green roofs

Extensive green roofs	Intensive green roofs
Low depth substrate 5–15 cm	High depth substrate 15–38 cm
Light substrate 0.55–2.1 Kg/m^2	Heavy substrate > 2.1 Kg/m^2
Short plants 2.5–61 cm	Trees, shrubs etc.
Small variety of plants (herbs, etc.)	Large variety of plants
Not accessible	Designed to be accessible
Inclination until 30°	Relatively flat
Low cost	High cost
Low irrigation demands	Irrigation is necessary
Low maintenance demands	High maintenance demands

Table 2 Comparison of green roofs

Benefits for the roof	Benefits for the building	Benefits in urban scale
Decreased U value	Reduction of thermal loads	Mitigation of urban heat island
Increased thermal mass	Improvement of comfort	Filtering of airborn pollutants
Reduction of noise pollution	Acoustic comfort	CO_2 absorption and O_2 production
Reduction of solar absorbance	Reduction of the energy consumption	Reduction of rain drainage
Increase of life	Improvement of microclimate	Improvement of urban biodiversity

characterized as passive energy technique in residential and tertiary sector as they contribute to the reduction of the demand for heating and cooling in winter and summer respectively. Niachou et al. (2001) report extensive measurements of a planted roof in Greece and they concluded that it contributes to a significant reduction of energy demand. The application of the green roof reduced the percentage of maximum indoor air temperature exceeding 30 °C to 15 % from 68 % without the green rooftop. The energy efficiency due to green roofs for a non-insulated building was 37–48 % while for an insulated buildings was quite low (5 %).

Specifically green roofs' foliage prevents gusts of cold air in the winter while during summer heat transfer is reduced as plants absorb smaller proportion of radiation compared to hard materials. Specifically evapotranspiration can achieve heat removal rates of almost 1,045,800 KJ (for a large tree) for a summer day. Moreover the green roofs can contribute to the reduction of rain drainage from precipitation (Mavroyanopoulos and Kyritsis 1986). The benefits from green roofs are tabulated in Table 2.

2.3 *Shading and Solar Control*

Shading in the urban environment is a fundamental and important technique for the improvement of the city's microclimate. Proper shading systems can contribute to a significant reduction of the radiative temperature improving simultaneously the thermal comfort conditions. The use of suitable permeable materials in open spaces can offer a significant reduction of both ambient and radiative temperature allowing the air to flow and move through them.

The efficiency of shading meshes has already been tested in the literature (Shashua-Bar et al. 2011).

Shading of the urban environment can be provided by shading devices and structures, like tents, pergolas, canopies, etc. Shaded urban surfaces receive smaller amounts of solar radiation, thus having lower temperatures. This results in lower ambient temperatures in open spaces and in reduced rate of heat convection to the building interior, when shading is applied on a building façade.

Various geometrical shapes were presented by Swaid (1992) in order to provide passive cooling by intelligent urban forms that included screening which were considered as urban shading devices. The described screens are foreseen as shading devices in the summer blocking the incoming solar radiation during day-time (that is, on-position), while being removed at night to enhance nocturnal radiative cooling (that is, off-position). In winter they are assumed to be in off-position during sunshine hours to promote the access of solar radiation and in on-position at night to obstruct the sky energy sink and reduce radiant heat losses. The efficiency of shading meshes has already been tested in the literature. Various landscape strategies including mesh shading devices are evaluated by Shashua-Bar et al. (2009), using different combinations. The effects of these treatments were tested during the summer season in two semi-enclosed courtyards located at an urban settlement in the arid Negev Highlands of southern Israel. The courtyards covered by shade trees and mesh are compared to an uncovered non-vegetated exposed courtyard, which on average reached a maximum air temperature of 34 °C in mid-afternoon. The courtyard treated with shade trees and grass yielded a daytime temperature depression of up to 2.5 K, while shading the courtyard with a fabric shading mesh, counter-intuitively, caused a relative increase of nearly 1 K. The "cooling efficiency" of these strategies was calculated as the ratio between the sensible heat removed from the space and the latent heat of evaporation, with the latter representing the amount of water required for landscape irrigation.

Movable canopies are described and proposed for Gramsci square (Scudo and Dessì 2006). The important role of shading devices which can really help to promote vitality in urban spaces through microclimatic mitigation is revealed.

Papadakis et al. (2001) reported that the radiation incident on building facades is 70–85 % reduced, when trees are planted close to the building. At the same time, temperatures in the area around the trees were significantly lower than those in non-shaded areas.

Additional case studies that incorporate shading devices and other mitigation techniques can be found in (Nikolopoulou and Steemers 2003; Nikolopoulou 2004).

2.4 Heat Sinks

Heat dissipation techniques deal with the disposal of the excess heat of the urban spaces to a sink characterized by lower temperature, like the water, the ground and the sky. Effective dissipation of the excess heat depends on two main pre-conditions: (a) The availability of a proper environmental heat sink with sufficient temperature difference for the transfer of heat and (b) the efficient thermal coupling between the urban region and the sink.

Three main processes of heat dissipation have been well studied and developed:

- Ground cooling which exploits the ground as a heat sink.
- Convective or ventilative cooling based on the use of ambient air.
- Evaporative cooling using the water as heat sink.

Ground is characterized by its ability to store the incident solar radiation as heat, conserving a constant temperature throughout the year. This way, the underground temperature is lower than the aboveground air temperature during summer and higher than the air temperature during winter. The fact that the ground has a much lower temperature than that of the environment above ground is enabling it to be used as a natural heat sink, during summer, by dissipating heat from aboveground constructions to the ground. Taking advantage of the ground temperature is achieved through ground to air heat exchangers, meaning tubes that are buried in the earth. The air enters the tubes it is circulated underground and exits the tube having a lower temperature. The designation of this system varies, as it could be referred to as "earth tubes" or as "ground-coupled air heat exchangers". Various applications of earth cooling especially for buildings can be found in the literature. Concerning cities earth cooling is used in the historic center of Tirana, in Albania (Fintikakis et al. 2011) coupled with all mitigation techniques. In the specific case study the maximum temperature drop due to all the applied techniques was 3 °C, while the maximum contribution of the earth to air heat exchangers was found to be close to 0.7 °C.

It is well known and understood that the ventilation effectiveness is strongly influenced by the urban heat island (Geros et al. 2005; Maragogiannis et al. 2011; Kolokotroni et al. 2006). Additionally, the wind velocity is reduced inside urban canyons and wind direction is significantly modified reducing the indoor air flow caused by natural ventilation techniques (Ali-Toudert and Mayer 2006; Shashua-Bar and Hoffman 2003). A comparative research on the similarities and differences of the hourly, daily and monthly variation of energy consumption in two cities with the same increase of ventilation rates is performed by (Enshen and Yimin 2005). The increase of ventilation rates within cities can have a significant impact on urban comfort especially during summer periods.

Finally water sprays and evaporative cooling in urban regions is proposed by various researchers. Fintikakis et al. (2011) showed that the contribution of the water surfaces in the urban heat mitigation potential was quite significant.

3 Strategic Spatial Planning of Cities

According to the architect and town planner Constantinos Doxiadis, the main principles that a planner should keep in mind during the process of a city or a settlement creation and at the same time the human needs that should be fulfilled by this entity, are five (Kyrtsis 2006).

First is the human need to establish contacts with other humans or elements, like nature, buildings, services, etc. in order to learn, to see, to socialize. Second is man's need for minimizing effort. This explains his effort to bring everything close to him, or the fact that he prefers to go wherever he has easy access. Third is the need for a protective space, where the impact of noise, wind, cold or other people is limited or controllable. The fourth principle is the effort or need to meliorate the relation with other humans or elements and the final principle is the effort to combine the four previous principles.

The need for shelter and the need for privacy were wisely confronted in vernacular architecture, according to the weather conditions of each area. The arrangement of buildings, open spaces (squares, parks, etc.) and streets, in relation to geomorphology and weather, is the most crucial parameter affecting environmental comfort in a city. Some of the main elements urban planning affects are: (a) protection from low/high temperatures, depending on the climate of the area, (b) protection from or enhancement of winds, (c) daylight and sunlight availability, (d) minimum air pollution sources, (e) protection from extreme noise, etc. Obtaining these requirements ensures comfortable environmental conditions both outdoors but also inside the urban buildings. The following text presents the effect on city climate of some basic spatial planning components or urban formations, like (a) Building height and urban density, (b) albedo of urban surfaces, (c) urban parks and greenery, (d) street orientation, (e) proximity to heat sinks and (f) traffic load.

The impact of urban geometry has been broadly studied for different climatic conditions. Thorsson et al. (2011) investigated the possible changes in global/regional climate when altering urban geometry in a compact mid-rise high-latitude city (Gothenburg). Results showed that urban geometry causes large intra-urban differences in T_{mrt}, on hourly, daytime and yearly time scales. In general, open areas were found to be warmer than adjacent narrow street canyons in summer, but cooler in winter. Moreover, the study confirms the potential of using urban geometry to mitigate daytime thermal stress. It was found that a densely built pattern mitigates extreme swings in T_{mrt} and PET, improving outdoor comfort conditions both in summer and in winter.

The advantages of high constructions were highlighted also by Emmanuel and Fernando (2007) who examined the sensitivity of air temperature and mean radiant temperature (MRT) of built-up urban cores to urban geometry (the density of buildings), thermal properties of human-made surfaces (albedo) and green cover (street trees), in two warm-climate cities: Pettah, Colombo (Sri Lanka) and downtown Phoenix, Arizona (USA). It was found that the best thermal comfort conditions, quantified by both air temperature and MRT, were found when all buildings on the sides of the modeled streets had the maximum height that the urban development regulations would permit. High albedo values of materials (white painted surfaces) led to low daytime temperatures in both cities and the addition of rows of trees on each side of the road appeared to have little effect.

On the contrary, in the work of Shashua-Bar et al. (2012), where the interaction of tree canopy coverage, traffic load, walls surfaces albedo, and H/W (H: building height; W: street width) at the level of a street with low aspect ratio H/W, in the city of Athens was investigated, it was found that all the examined scenarios were associated with air temperature decrease and improvement of thermal comfort in the shade, especially during the day's hottest hours, however the trees thermal effect was the dominant factor, followed by the increase of H/W and of the walls surfaces albedo. A scenario combining all the above techniques was calculated and, as expected, it offered the greatest results.

All of the abovementioned studies argue that high density urban areas (high buildings and narrow streets) lead to lower summer temperatures and better comfort conditions. However, the increase of building height, combined with inappropriate street orientation towards the prevailing winds, might have negative effects. Heat that is stored in the urban structural elements during a warm day is released during the night to the environment. In a dense urban environment, radiation is reflected or absorbed by buildings and by higher levels of air pollution. When the arrangement and the size of buildings prevents air flow and decreases wind speed, nocturnal cooling is inhibited (Smith and Levermore 2008). Thus, high density urban environments might enhance the urban heat island phenomenon.

In urban environments where heat island phenomenon is present, the height of the buildings and the surface properties of buildings and other urban structures are of crucial importance. Tall buildings create "canyon effects" in the cities, affecting wind patterns and surface and air temperatures. In a study by Georgakis and Santamouris (2006) it was found that the wind speed inside a deep canyon in Athens caused airflow rate for single side and cross ventilation configurations, to be reduced by 82 % and 68 %, respectively, compared to an undisturbed location. Furthermore, tall and dense structures reduce the ability for ventilation and increase pollution levels (Vardoulakis et al. 2003).

A strategic location for new urban formations would be close to heat sinks, like the sea, lakes and rivers and forests. Water surfaces are cooler than land during day hours, thus causing an air movement of the relatively cooler sea breeze towards the inland (Littlefair et al. 2000). Various studies argue that sites close to the coast do not suffer the heat stress of inland sites due to the sea breeze (Lopes et al. 2011; Papanastasiou et al. 2010). This leads to the conclusion that the high urban buildings should not be constructed near the shoreline and streets should allow for promotion of sea breeze towards the inland of cities.

Forests and urban parks, as it has already been mentioned in previous paragraphs, help to cool down a city, while diverting or slowing down winds and trapping pollutants. Many studies have shown that the effect of green spaces highly depend on the size and the height of the plants (Chen et al. 2009; Cao et al. 2010). Strategic planning of cities would imply the use of green spaces of adequate size between major pollution sources and the city, in order to filter air pollutants before they reach the city air and in an arrangement and place where they might block, diverse or guide the prevailing winds.

Anthropogenic heat is another parameter affecting the city climate and mainly includes heat released from the building sector, the transportation sector, and metabolism. The reduction of anthropogenic heat can primarily be succeeded by limiting the transportation needs and traffic density (Sharpe 1978). In order for this to be succeeded, city planning should provide easy and quick access to basic services and efficient public transportation. Limited traffic, combined with all the mitigation techniques analysed in previous paragraphs lead to reduced building energy consumption. Cities with low anthropogenic heat levels will have lower temperatures and urban heat island phenomenon will be less intense.

The efficiency of a plan for confronting urban heat island depends on the appropriate choice of mitigation techniques analysed in the previous paragraphs, according to the climate and the geomorphology. However, the town planning in areas under construction and the corrective actions in spatial arrangement of existing settlements are the policies that have the greatest effect and can activate the various mitigation techniques.

4 Carbon Offset of Sustainable Cities

Greenhouse gases are gases that trap infrared radiation that is released from earth into the atmosphere, thus causing earth to be warmer. The main greenhouse gases (GHGs) are water vapor (H_2O), carbon dioxide (CO_2), and methane (CH_4). The process is commonly known as the "greenhouse effect", the major cause of climate change in the planet.

The extreme growth of cities and the increase of urban population that began during the Industrial Revolution and still continues nowadays, have contributed substantially to climate change by adding CO_2 and other heat-trapping gases to the atmosphere. Carbon dioxide is the primary greenhouse gas (it accounts for 81.6 % of EU-27 greenhouse gas emissions in 2009, according to Eurostat) which is emitted both by natural processes, as part of the carbon cycle (ex. through animal and plant respiration) and by human activities. Anthropogenic CO_2 emissions come primarily from combustion of carbon based fuels, principally wood, coal, oil, and natural gas. According to the United States Environmental Protection Agency (EPA) the current CO_2 level is higher than it has been in at least 800,000 years, as CO_2 concentrations in the atmosphere have increased by almost 40 % since pre-industrial times (source: EPA). Methane is also produced through natural (natural wetlands) and human activities (agricultural activities, fossil fuel extraction, transport, etc.). Nitrous oxide is mainly produced through agricultural activities and natural biological processes (source: European Commission, Eurostat).

The latest data shows that the main sources of GHG emissions are: energy (excluding transport) with a 59.1 % share of total EU-27 greenhouse gas emissions, transport, which was the source where emissions were increasing at their fastest pace (20.2 % of the EU-27s greenhouse gas emissions in 2009), agriculture accounted for 10.3 % of all greenhouse gas emissions in the EU-27 in 2009, industrial processes accounted for a slightly lower share at 7.2 %, while emissions from waste (which includes disposal, landfill sites and waste water treatment) accounted for the remaining 3.2 % of the EU greenhouse gas emissions in 2009 (source: European Commission, Eurostat).

An important proportion of the energy that creates GHG emissions is generated by the continuously growing cities. According to Kennedy et al. (2009) the GHG emissions of cities depend on many parameters, including climate, location, the means of power generation, population density, transportation types and load, age of the city, citizens' income, etc. Many cities' administrations worldwide try to reduce emissions, both as a way to comply with international commitments and as an opportunity to improve residents' lives. This procedure firstly includes the existence of emission inventories. Inventories serve locating the major GHG emission sources and calculating the emissions per source. The Intergovernmental Panel on Climate Change (IPCC) has developed a GHG emissions inventory methodology for country emissions. Most countries report national GHG inventories through the UNFCCC (United Nations Framework Convention on Climate Change).

On the other hand, there is no internationally recognized methodology for estimating or calculating city level GHG emissions. However, during the past decade, many organisations have created methodologies for community level inventories. The main difference between country level and community level GHG emission inventories is the emissions attribution, according to Ibrahim et al. (2012). In the country level, all emissions released inside the country geographical borders are counted; however, in city level there are emissions released outside the boundaries of the city that are produced for consumption purposes inside the city (electricity generation, production of urban construction materials, agriculture, etc.) that are not usually counted.

The GHG emission inventories are used by city planners and decision makers in order to identify the primary sources of emissions, to address drivers of climate change and take climate action plans. The mitigation techniques analysed in the previous paragraphs, like the enhancement of urban vegetation, the increase of urban surfaces' albedo, the enhancement of energy efficiency in buildings, etc., are policies that contribute to the minimization of emissions from energy production and are the ones related to the urban heat island effect. The exploitation of these techniques and their combination with emission reduction techniques for other sectors (agriculture, transport, industry, etc.) in many cities around the world may well lead to lower climate change rates.

5 Conclusions

Heat island is the more documented phenomenon of climate change. It is responsible for the important increase of the cooling energy consumption in cities, increase of the pollution and decrease of the thermal comfort conditions. Development and application of successful mitigation techniques is the key to fight the temperature increase in cities and improve quality of life.

Existing mitigation techniques focus mainly on the use of additional green spaces in cities involving urban parks and green roofs and the use of highly reflective materials for roofs and pavements that are able to reflect solar radiation back to the space. Demonstration in large scale projects has shown that the application of reflective and green roofs as well as the use of green spaces and the exploitation of natural heat sinks like the ground and the water, can contribute highly to reduce urban temperatures and fight heat island.

Important research is actually carried out on this topic aiming to improve the performance of the existing mitigation techniques and to develop new efficient strategies. Materials of very high reflectivity were developed and tested in roofs and pavements with high success. New technologies on permeable pavements are also proposed and tested in humid climates.

However, it is well accepted that research on heat island mitigation techniques has to be intensified in order to offer more efficient systems, techniques and materials and finally advance the global knowledge on the topic.

References

AASHTO (2010) Rough roads ahead. American Association of State Highway and Transportation Officials, Washington, DC

Akbari H, Rose LS (2001) Characterizing the fabric of the urban environment. A case study of metropolitan Chicago, Illinois. http://www.osti.gov/bridge/servlets/purl/900694-BlMLq4/900694.pdf

Akbari H, Davis S, Dorsano S, Huang J, Winnett S (1992) Cooling our communities: a guidebook on tree planting and light colored surfacing. US Environmental Protection Agency, Washington, DC

Akbari H, Kurn D et al (1997) Peak power and cooling energy savings of shade trees. Energy Build 25:139–148

Ali-Toudert F, Mayer H (2006) Numerical study on the effects of aspect ratio and orientation of an urban street canyon on outdoor thermal comfort in hot and dry climate. Build Environ 41(2):94–108, http://www.scopus.com/inward/record.url?eid=2-s2.0-26444614659&partnerID=40&md5=ae6531f93d6dd40e09d7f30989ff0ec4

Asaeda T, Ca VT, Wake A (1996) Heat storage of pavement and its effect on the lower atmosphere. Atmos Environ 30(3):413–427, http://www.scopus.com/inward/record.url?eid=2-s2.0-0030087830&partnerID=40&md5=830cd1e721591866f3f74bdbb365206e

Besancenot J-P (2002) Heat waves and mortality in large urban areas. Vagues de chaleur et mortalité dans les grandes agglomé rations urbaines 1(4):229–240, http://www.scopus.com/inward/record.url?eid=2-s2.0-0141748484&partnerID=40&md5=7f7631581798dd4a05f6c5ebea0c48b5

Cambridge Systematics, Inc (2005) Cool pavement report, EPA cool pavements study – Task 5. Prepared for Heat Island Reduction Initiative

Cao X et al (2010) Quantifying the cool island intensity of urban parks using ASTER and IKONOS data. Landscape Urban Plann 96(4):224–231, http://www.scopus.com/inward/record.url?eid=2-s2.0-77953127933&partnerID=40&md5=322d3bf5705eedcc1abe60e2e0bb0821

Chang C-R, Li M-H, Chang S-D (2007) A preliminary study on the local cool-island intensity of Taipei city parks. Landscape Urban Plann 80(4):386–395, http://linkinghub.elsevier.com/retrieve/pii/S0169204606002076. Accessed July 30, 2012

Chen H et al (2009) Analysis on relations between the pattern of urban forests and heat island effect in Chengdu. Shengtai Xuebao/Acta Ecologica Sinica 29(9):4865–4874, http://www.scopus.com/inward/record.url?eid=2-s2.0-70350145316&partnerID=40&md5=dd7ff78f6f653eb66d49bee60aa7edfe

Department of Economic and Social Affairs Population Division (2011) Population distribution, urbanization, internal migration and development: an international perspective. United Nations, New York

Devi SS (2006) Urban heat islands and environmental impact. In: 86th AMS annual meeting. http://www.scopus.com/inward/record.url?eid=2-s2.0-77949289643&partnerID=40&md5=9112fe53c15c7ad3caab07026d99b2e8

Doll D, Ching JKS, Kaneshiro J (1985) Parameterization of subsurface heating for soil and concrete using net radiation data. Boundary-Layer Meteorol 32(4):351–372, http://www.scopus.com/inward/record.url?eid=2-s2.0-0022220928&partnerID=40&md5=a9cd20f7a213e25dc8c63e097d5a1a61

Eliasson I, Upmanis H (2000) Nocturnal air fow from urban parks-implications for city Ventilation. Theor Appl Climatol 107:95–107

Emmanuel R, Fernando HJS (2007) Urban heat islands in humid and arid climates: role of urban form and thermal properties in Colombo, Sri Lanka and Phoenix, USA. Clim Res 34(3):241–251, http://www.scopus.com/inward/record.url?eid=2-s2.0-44349095912&partnerID=40&md5=8cf1a20357dc5daf9fee202008fd1a4f

Enshen L, Yimin X (2005) Identifications: the relative increase rates of cooling or heating energy consumption are approximate in different cities with the same increase of ventilation rate. Build Environ 40(4 SPEC. ISS.): 497–505. http://www.scopus.com/inward/record.url?eid=2-s2.0-11944257548&partnerID=40&md5=4b1ae7df2434f2f2d6789f2349c60938

Ferguson B et al (2005) Reducing urban heat islands: compendium of strategies cool pavements. http://www.epa.gov/hiri/resources/pdf/CoolPavesCompendium.pdf

Fintikakis N et al (2011) Bioclimatic design of open public spaces in the historic centre of Tirana, Albania. Sust Cities Soc 1(1):54–62, http://www.scopus.com/inward/record.url?eid=2-s2.0-79952553201&partnerID=40&md5=5108e968b71772b9264493c6d1c2ffdd

Gaitani N, Mihalakakou G, Santamouris M (2007) On the use of bioclimatic architecture principles in order to improve thermal comfort conditions in outdoor spaces. Build Environ 42(1):317–324, http://linkinghub.elsevier.com/retrieve/pii/S0360132305003409

Georgakis C, Santamouris M (2006) Experimental investigation of air flow and temperature distribution in deep urban canyons for natural ventilation purposes. Energy Build 38(4):367–376, http://www.scopus.com/inward/record.url?eid=2-s2.0-31144461378&partnerID=40&md5=2926d2a2f1743b4ca7de59b1fd5ffaa3

Geros V et al (2005) On the cooling potential of night ventilation techniques in the urban environment. Energy Build 37(3):243–257, http://www.scopus.com/inward/record.url?eid=2-s2.0-10044289512&partnerID=40&md5=9ac1b130305556d2d39430c7f9437acf

Hardy JE, Mitlin D, Satterthwaite D (2001) Environmental problems in an urbanizing world. Erthscan Publications, Sterling

Hunt WF, Collins KA (2008) Urban waterways. Permeable pavement: research update and design implications. North Carolina Cooperative Extension Service

Ibrahim N et al (2012) Greenhouse gas emissions from cities: comparison of international inventory frameworks. Local Environ 17(2):223–241, http://www.scopus.com/inward/record.url?eid=2-s2.0-84858391319&partnerID=40&md5=760f660b790d60f9147bc63a4faa34e3

Jauregui E (1990) Influence of a large urban park on temperature and convective precipitation in a tropical city. Energy Build 16:457–463

Johnson DP, Wilson JS (2009) The socio-spatial dynamics of extreme urban heat events: the case of heat-related deaths in Philadelphia. Appl Geogr 29(3):419–434, http://www.scopus.com/inward/record.url?eid=2-s2.0-67649191759&partnerID=40&md5=b21a8fc6ff5958099e055dc1dd46a066

Karlessi T et al (2009) Development and testing of thermochromic coatings for buildings and urban structures. Solar Energy 83(4):538–551, http://www.scopus.com/inward/record.url?eid=2-s2.0-62249150228&partnerID=40&md5=a01e0a43a867d5589acdf88af93b089c

Kennedy C et al (2009) Greenhouse gas emissions from global cities. Environ Sci Technol 43(19):7297–7302, http://www.scopus.com/inward/record.url?eid=2-s2.0-70349617486&partnerID=40&md5=37442767d9f4778060f56bc8e97f98c2

Kolokotroni M, Giannitsaris I, Watkins R (2006) The effect of the London urban heat island on building summer cooling demand and night ventilation strategies. Solar Energy 80(4):383–392, http://www.scopus.com/inward/record.url?eid=2-s2.0-33645992480&partnerID=40&md5=76cc21e137420bd46b5a72abfff856c4

Kyrtsis A (2006) Constantinos A. Doxiadis: texts, design drawings, settlements. IKAROS, Athens

Landsberg HE (1982) The urban climate. http://www.scopus.com/inward/record.url?eid=2-s2.0-0020404164&partnerID=40&md5=9128f6d5bfc173ecccb43b8f52781a02

Levine KK (2011) Cool pavements research and technology. Institute of Transportation Studies Library

Littlefair PJ et al (2000) Environmental site layout planning: solar access, microclimate and passive cooling in urban areas. Building Research Establishment (BRE), London

Lopes A et al (2011) The influence of the summer sea breeze on thermal comfort in funchal (Madeira). A contribution to tourism and urban planning. Meteorologische Zeitschrift 20(5):553–564, http://www.scopus.com/inward/record.url?eid=2-s2.0-84555191281&partnerID=40&md5=6da8803d038cfaa483f6dda80934ae9e

Mahmoud AHA (2011) Analysis of the microclimatic and human comfort conditions in an urban park in hot and arid regions. Build Environ 46(12):2641–2656, http://www.scopus.com/inward/record.url?eid=2-s2.0-79960910262&partnerID=40&md5=5c61ea7c97f5446df4145fad92a263f1

Maragogiannis K, Kolokotsa D, Maria EA (2011) Study of night ventilation efficiency in urban environment: technical and legal aspects. In: The 52nd international scientific conference of Riga Technical University Energy Resources 2011, Riga

Mavroyanopoulos GN, Kyritsis S (1986) The performance of a greenhouse heated by an earth-air heat exchanger. Agric Forest Meteorol 36(3):263–268, http://www.scopus.com/inward/record.url?eid=2-s2.0-0006720891&partnerID=40&md5=bec6f6c8f1d63bccac46604649498911

Menon S et al (2010) Radiative forcing and temperature response to changes in urban albedos and associated CO_2 offsets. Environ Res Lett 5(1). http://www.scopus.com/inward/record.url?eid=2-s2.0-77950656112&partnerID=40&md5=1163d091cb8a98373fb16b09790ffd9a

Mihalakakou G et al (2002) Application of neural networks to the simulation of the heat island over Athens, Greece, using synoptic types as a predictor. J Appl Meteorol 41(5):519–527, http://www.scopus.com/inward/record.url?eid=2-s2.0-0036574086&partnerID=40&md5=f54a4919b8dbf7b14a534198d13af435

Mihalakakou G et al (2004) Simulation of the urban heat island phenomenon in Mediterranean climates. Pure Appl Geophys 161(2):429–451, http://www.scopus.com/inward/record.url?eid=2-s2.0-0842344455&partnerID=40&md5=8f1121ee3f1cd0d8087baab9c3f5b371

Nam T, Buzz P (2009) Strategies for design and construction of high-reflectance asphalt pavements. TRB Report No. 2098

Niachou A et al (2001) Analysis of the green roof thermal properties and investigation of its energy performance. Energy Build 33(7):719–729, http://www.scopus.com/inward/record.url?eid=2-s2.0-0035451948&partnerID=40&md5=a131f98bc9bc4561afcf027b8f525462

Nikolopoulou M (2004) RUROS project. http://www.cres.gr/kape/education/1.design_guidelines_en.pdf

Nikolopoulou M, Steemers K (2003) Thermal comfort and psychological adaptation as a guide for designing urban spaces. Energy Build 35(1):95–101, http://www.scopus.com/inward/record.url?eid=2-s2.0-0037210644&partnerID=40&md5=12693db4a888634e7cf17cd557015fd5
Oke TR (1982) The energetic basis of the urban heat island (Symons memorial lecture, 20 May 1980). Q J R Meteorol Soc 108(455):1–24, http://www.scopus.com/inward/record.url?eid=2-s2.0-0020371056&partnerID=40&md5=123d5c173aefbd2a0ab4fca4968559c8
Papadakis G, Tsamis P, Kyritsis S (2001) An experimental investigation of the effect of shading with plants for solar control of buildings. Energy Build 33(8):831–836, http://www.scopus.com/inward/record.url?eid=2-s2.0-0035480254&partnerID=40&md5=c9a9ed6f052db07b4b90a500b9fa85e6
Papanastasiou DK et al (2010) Temperature, comfort and pollution levels during heat waves and the role of sea breeze. Int J Biometeorol 54(3):307–317, http://www.scopus.com/inward/record.url?eid=2-s2.0-77951765486&partnerID=40&md5=45d2220e57705f050b9e95d9f0e4f9bd
Rose LS, Akbari H, Taha H (2003) Characterizing the fabric of the urban environment. A case study of Greater Houston, Texas
Rosenzweig C et al (2006) No title. http://www.giss.nasa.gov/research/news/20060130/103341.pdf
Sailor DJ (1995) Simulated urban climate response to modifications in surface albedo and vegetative cover. J Appl Meteorol 34(7):1694–1704, http://www.scopus.com/inward/record.url?eid=2-s2.0-0029528239&partnerID=40&md5=e18caeb6f1f54f862b61b35e9e715d9f
Santamouris M (2007) Heat island research in Europe – the state of the art. Adv Build Energy Res 1:123–150
Santamouris M (2012) Cooling the cities – a review of reflective and green roof mitigation technologies to fight heat island and improve comfort in urban environments. Solar Energy. http://www.scopus.com/inward/record.url?eid=2-s2.0-84864191264&partnerID=40&md5=cc70150812073a2eb2cd4f3c3688d86e
Santamouris M, Paraponiaris K, Mihalakakou G (2007) Estimating the ecological footprint of the heat island effect over Athens, Greece. Climatic Change 80(3–4):265–276, http://www.springerlink.com/index/10.1007/s10584-006-9128-0
Santamouris M et al (2012) Using cool paving materials to improve microclimate of urban areas-design realization and results of the Flisvos project. Build Environ 53:128–136, http://dx.doi.org/10.1016/j.buildenv.2012.01.022
Scudo G, Dessì V (2006) Thermal comfort in urban space renewal. In: The 23rd conference on passive and low energy architecture, Geneva, pp 6–8
Sfakianaki A et al (2009) Theoretical and experimental analysis of the thermal behaviour of a green roof system installed in two residential buildings in Athens, Greece. Int J Energy Res 33(12):1059–1069, http://www.scopus.com/inward/record.url?eid=2-s2.0-75949095299&partnerID=40&md5=6adf0f34477159dfecd13eee846f6b20
Sharpe R (1978) The effect of urban form on transport energy patterns. Urban Ecol 3(2):125–135, http://www.scopus.com/inward/record.url?eid=2-s2.0-49349121981&partnerID=40&md5=b3131711878101f55fb2d26d67710a2a
Shashua-Bar L, Hoffman ME (2003) Geometry and orientation aspects in passive cooling of canyon streets with trees. Energy Build 35(1):61–68, http://www.scopus.com/inward/record.url?eid=2-s2.0-0037210760&partnerID=40&md5=681afb98d05ba747fd55ecf9db771f94
Shashua-Bar L, Hoffman ME (2004) Quantitative evaluation of passive cooling of the UCL microclimate in hot regions in summer, case study: urban streets and courtyards with trees. Build Environ 39(9):1087–1099, http://www.scopus.com/inward/record.url?eid=2-s2.0-1764424315&partnerID=40&md5=5c0ab63a3da38498ae98fff33506cba1
Shashua-Bar L, Pearlmutter D, Erell E (2009) The cooling efficiency of urban landscape strategies in a hot dry climate. Landscape Urban Plann 92(3–4):179–186, http://www.scopus.com/inward/record.url?eid=2-s2.0-68349146541&partnerID=40&md5=3cecb3776f273dcbf6af4283ecfc8e20

Shashua-Bar L, Pearlmutter D, Erell E (2011) The influence of trees and grass on outdoor thermal comfort in a hot-arid environment. Int J Climatol 31(10):1498–1506, http://www.scopus.com/inward/record.url?eid=2-s2.0-79960251591&partnerID=40&md5=c20549af540bdec1eecfc39ccb7673b6

Shashua-Bar L, Tsiros IX, Hoffman M (2012) Passive cooling design options to ameliorate thermal comfort in urban streets of a Mediterranean climate (Athens) under hot summer conditions. Build Environ 57:110–119, http://www.scopus.com/inward/record.url?eid=2-s2.0-84861330012&partnerID=40&md5=64be082826140179a8189e46231c55db

Smith C, Levermore G (2008) Designing urban spaces and buildings to improve sustainability and quality of life in a warmer world. Energy Policy 36(12):4558–4562, http://www.scopus.com/inward/record.url?eid=2-s2.0-55249118566&partnerID=40&md5=73faecd26692228a31189924ec178e69

Swaid H (1992) Intelligent urban forms (IUF) a new climate-concerned, urban planning strategy. Theor Appl Climatol 191:179–191

Synnefa A, Santamouris M, Livada I (2006) A study of the thermal performance of reflective coatings for the urban environment. Solar Energy 80(8):968–981, http://www.scopus.com/inward/record.url?eid=2-s2.0-33746324228&partnerID=40&md5=28b2daba12ba5c6d8500ce5be16f76a9

Synnefa A, Santamouris M, Apostolakis K (2007) On the development, optical properties and thermal performance of cool colored coatings for the urban environment. Solar Energy 81(4):488–497, http://www.scopus.com/inward/record.url?eid=2-s2.0-33947495516&partnerID=40&md5=6f8337fcf9cccb429858576c403981b8

Taha H (2008) Meso-urban meteorological and photochemical modeling of heat island mitigation. Atmos Environ 42(38):8795–8809, http://www.scopus.com/inward/record.url?eid=2-s2.0-55549112186&partnerID=40&md5=b0bc0fad82dfd842c42d00938aac39d2

Taha H et al (2004) The potential of urban environmental control in alleviating heat-wave health effects in five US regions. In: 16th conference on biometeorology and aerobiology, pp 147–154. http://www.scopus.com/inward/record.url?eid=2-s2.0-22144447519&partnerID=40&md5=b38637950778703d438c17d6aabdf4a5

Thorsson S et al (2011) Potential changes in outdoor thermal comfort conditions in Gothenburg, Sweden due to climate change: the influence of urban geometry. Int J Climatol 31(2):324–335, http://www.scopus.com/inward/record.url?eid=2-s2.0-78751692009&partnerID=40&md5=9f692069bcbb4c9a75a11058a1afad2d

Tso CP (1996) A survey of urban heat island studies in two tropical cities. Atmos Environ 30(3):507–519, http://www.scopus.com/inward/record.url?eid=2-s2.0-0030072291&partnerID=40&md5=1c9d41e8295dc79a639959c26a1bfea0

Vardoulakis S et al (2003) Modelling air quality in street canyons: a review. Atmos Environ 37(2):155–182, http://www.scopus.com/inward/record.url?eid=2-s2.0-2242483771&partnerID=40&md5=6d41e2037a2ff3aa43d35ee4a38aac12

Wong NH et al (2010) Perception studies of vertical greenery systems in Singapore. J Urban Plann Develop 136(4):330–338, http://www.scopus.com/inward/record.url?eid=2-s2.0-79251552559&partnerID=40&md5=080851edbc77c13695f41be83fe0a691

Wong NH, Kardinal Jusuf S, Ignatius M (2011) Urban heat island study on building morphology related with micro-climate condition and energy consumption within Singapore commercial area. In: PLEA 2011 – architecture and sustainable development, conference proceedings of the 27th international conference on passive and low energy architecture, pp 445–450. http://www.scopus.com/inward/record.url?eid=2-s2.0-84864123607&partnerID=40&md5=ae9589bc19daf0c7688dd0f2d2a7989b

Zheng J, Wu G, Li Y (2010) Characteristics of an urban heat island of different underlying surfaces and the assessment of a human comfort index for Zhengzhou. In: 2010 4th international conference on bioinformatics and biomedical engineering, iCBBE 2010. http://www.scopus.com/inward/record.url?eid=2-s2.0-77956144911&partnerID=40&md5=ef7d3b281ae1e42041c2a07349d2c4c5

Building Conservation Towards a Sustainable Future: Use of GPR

Zeynep Aygen and Gökhan Kılıç

Abstract According to the website of Mary Robinson Foundation for Climate Justice, 2.7 Million people worldwide, most of them being extremely poor and 1.3 billion of them with no access to electricity, rely on solid fuels such as traditional biomass and coal for cooking and heating.[1] In this world where a number of experts are concentrating their research on finding alternative and environmentally friendly energy sources, masses of existing buildings are being demolished causing energy loss and polluting the environment with demolition waste. These demolitions are justified through the conventional scientific belief introduced by modernity that older buildings are of poor quality.

This chapter aims to look to the debate in regard of the heath loss issues in existing buildings and point out to new research and alternative methodological approaches to provide proof of the current underestimation of the thermal transmissivity of traditionally built walls, the main reason leading to their demolition.

> Colonel Johnson, Chief Constable of Middleshire might be of the of the opinion that nothing could beat a wood fire, but Hercule Poirot was of the opinion that central heating could and did every time[2]

---

[1] http://www.mrfcj.org/about/our_work/access-to-energy.html?gclid=CMO7vOin9LYCFQm33godgUYAzQ, accessed 01/05/2013.

[2] Agatha Christie, *Hercule Poirot's Christmas*, Twentieth Edition, Fontana & Collins: Glasgow, 1988, p. 66.

Z. Aygen (✉)
Mimar Sinan Fine Arts University, Istanbul, Turkey
e-mail: Aygen_zeynep@hotmail.com

G. Kılıç
Izmir University of Economics, Izmir, Turkey
e-mail: gokhankilic79@gmail.com

S.Th. Rassia and P.M. Pardalos (eds.), *Cities for Smart Environmental and Energy Futures*, Energy Systems, DOI 10.1007/978-3-642-37661-0_13,

Agatha Christie's famous character Hercule Poirot, who always complains of the draught, dampness and cold in the historic houses of Britain, stands for the firm belief of European modernity, empirical development being a remedy for all problems in a continuously improving world supported by new, groundbreaking technologies. This belief is also reflected in scientific views regarding the environmental feasibility of historic or traditional architecture being of poor quality. As Anne Power demonstrates in her paper assessing the policy of demolition of older buildings with the aim to help to energy efficiency, the UK based Environmental Change Institute's proposal in 2005 meant three million demolitions in the country until 2050; however the proposal did not include an assessment of the environmental impact of these demolitions at all.[3] Demolitions cause physically energy loss, demolition waste to be disposed and pollute the environment and demolition costs in addition to the new building costs; socially demolitions cause displacement and destruction of neighbourhood relationships. Social sustainability should have the same value as physical sustainability, but unfortunately in a number of countries including UK this is not the case.

On the other hand Germany and Netherlands have a different approach by preferring refurbishment and building conservation to demolitions. The German Federal Government's programme covering 17 million blocks of pre-1984 homes dating back to 2007[4] is the result of a long time respect policy dating back to 1980s the focus of which was saving the pre-war building stock in the run-down areas of Berlin and other industrial cities of Germany. In Berlin the rehabilitation of the nineteenth century factory worker neighbourhood Kreuzberg in Berlin by the governmentally funded institutions IBA/ STERN in the late 1980s is a pioneering example how a sustainable rehabilitation of the existing building stock can support environmental sustainability by saving energy and money. There is evidence that the rehabilitation and upgrading costs of the low quality nineteenth century buildings in Kreuzberg including the installation of environmentally friendly heating systems an eco-waste disposal equipments has amounted for each case to 40–65 % of the costs of a new building of the same environmentally friendly category.[5] Moreover the Kreuzberg Project is an iconic case study for social sustainability, as it was run with community participation principles in order to integrate both owners and tenants to the project, with the tenants being mostly guest workers and immigrants from non-European countries.[6] Since then Germany continued to apply these principles based on sustainable rehabilitation of the existing building stock in preference to new buildings. A recent Federal Government project in Germany has been

[3]Anne Power, Does Demolition or Refurbishment of Old and Inefficient Homes Help to Increase our Environmental, Social and Economic Viability?, *Energy Policy*, Nr 36, 2008, pp. 4487–4501.

[4]ibid

[5]*Behutsame Stadterneuerung* in Kreuzberg, Paper 5.6, STERN: Berlin, 1987.

[6]For more on this aspect of the project see Zeynep Aygen (ed.), *Bürger statt Städter: Bürgerbeteiligung als Strategie der Problemlösung und der sozialen Integration*, Arbeitskreis das Moderne und Islam/ das Arabische Buch: Berlin, 1999.

introduced under the name 'Energetische Stadtsanierung', which invites users in urban neighbourhoods to have a joint heating system by grouping their buildings together by considering the aspects of the relevant building culture, historic building conservation issues, building economy and social structure in each entity.[7] The federal credits provided by the KfW bank group dedicated to increase the number of energy efficient homes in Germany, funded in the period 2006–2012 2.9 million home refurbishment projects, which in turn means 5, nine million tonnes reduction of CO-2 emissions for a period of 30 years.[8] 1.5 Milliards Euro has been dedicated to the period 2012–2014 for energy efficient refurbishment projects.

The main reason for UK and other countries to have prejudices against the existing building stock is the categorisation of the buildings predating 1920 as buildings of a solid wall construction as opposed to buildings of cavity wall construction with damp proof membranes. The use of modern insulation materials in traditional buildings by inadequate interventions mainly in 1970s and 1980s changed in many cases airflow in traditional walls and caused dampness leading to damage and in the same time to bad reputation for these buildings. Any historic building built with a technology no more widely available need to be examined properly before it is deemed for demolition as poorly insulated. Caroline Rye, who investigated the U-values (thermal transmissivity) of traditionally built walls using two different quantification methods; an in situ method which measures a U-value from the wall itself and a U-value produced by a widely used standard method of calculation, discovered a discrepancy between the two sets of U-values. Her research shows that the standard U-value calculation led to the underestimation of the thermal performance of 75 % of her case studies.[9] Moreover she cites also interesting examples of the injustice traditional home owners have been subject to during EPC (Energy Performance Certificate) assessments as a consequence of the limitations of the used software and gap of knowledge about historic construction methods by the assessors.[10] According to her further research results on behalf of the Society for the Protection of Ancient Buildings (SPAB), in some instances it now appears that heat loss through vernacular materials can be up to three times lower than expected.[11]

The use of special methods in energy assessment of historic buildings and introduction of alternative insulation methods will lead to a more sustainable and financially viable future for smart cities. One of the technologies with a future for the

[7]http://www.bmvbs.de/SharedDocs/DE/Artikel/SW/energetische-stadtsanierung.html, accessed 13/ 12/ 2012.

[8]http://www.bmvbs.de/SharedDocs/DE/Artikel/SW/co2-gebaeudesanierung-energieeffizient-bauen-und-sanieren-die-fakten.html?nn=78052, accessed 13/ 12/ 2012.

[9]Carline Rye, *The Energy Profiles of Historic Buildings: A Comparison of the* In Situ *and Calculated U-Values of Traditionally Built Walls*, MSc Thesis supervised by Zeynep Aygen, University of Portsmouth: Portsmouth, 2011.

[10]Ibid, p.15-16.

[11]http://www.spab.org.uk/advice/energy-efficiency/, accessed 15/ 12/ 2012.

Fig. 1 St Mary's Church (1884–1887), Chatham (Kilic 2012)

assessment of 'solid' walls is GPR (Ground Penetrating Radar) survey. In order to demonstrate that a solid wall contains much more than only being solid, the authors chose the St Mary's Church in Chatham, UK as a case study specially investigated for this chapter.

1 Survey of St Mary's Church, Chatham

1.1 St Mary's Church

Chatham, which had historically a famous Dockyard, closed in the 1980s, is also well known by Charles Dickens fans, as he lived there during two different periods of his childhood, in Ordnance Terrace, and in The Brook. Chatham was in the past also home to The Royal Marines, whose barracks were situated; there is the Lloyds of London building is now, on Dock Road. Lloyd's moved out several years ago, and their old building is now Medway Council's headquarters.[12]

St Mary's Church (Fig. 1) was Chatham's original parish church. It stands in Dock Road, which must have been a busy road during the time of the Royal Marines. The site is believed to have been used as a place of worship since pagan times. The church was used as a heritage centre for some time, but is now closed. It is apparently awaiting conversion to offices, which makes it interesting for our case study, as it will need central heating in this case.

[12]http://www.medway.gov.uk/, accessed 15/ 12/ 2012.

2 Objectives

- To assess the condition of the voids/cavity on the St Mary's Church masonry wall (wall condition).

3 Methodology

The Ground Penetrating Radar (GPR) survey was performed using the 2,000 MHz Antenna. Designed specifically for the inspection of concrete/masonry application, this high frequency array is lightweight and manoeuvrable yet provides high quality, densely sampled data. Denser sampling produces higher quality tomography and 3D images which assist considerably in the interpretation of data (Fig. 2).

The system is composed of an array of eight horizontally polarized 2GHz channels spaced at 10cm, mounted on a lightweight and highly manoeuvrable trolley and powered by a large, 24 Ah 12 V battery.

3.1 GPR Survey

A GPR survey is performed in by marking a grid on the wall using chalk or temporary paint and pushing the radar across the grid in straight lines. The location of the grid is referenced by recording the coordinates with respect to a fixed location. Due to the frequency of the antenna and the size of the array, it is recommended that multiple surveys are performed covering moderate areas, rather than one large survey. For optimum results it is recommended to push the radar in both the T and L axis (Fig. 3).

Each area surveyed will be defined with a zone name and saved in a folder on the acquisition laptop. Further processing of the acquired data takes place back in the office.

The process for performing a GPR survey is the following:

1. Mark out an 8 cm grid for each area with respect to a fixed location (take a photograph if possible).
2. Perform each GPR acquisition by recording the start coordinates of scans and pushing the radar across each of the marked, grid lines.
3. The data will be saved and further processing takes place offsite, back in the office.

It is recommended to take a photo of the survey area including the marked grid and the fixed reference point. This will help during the data processing stage which takes place offsite, back in the office.

Fig. 2 St Mary's Church wall surveyed area by GPR

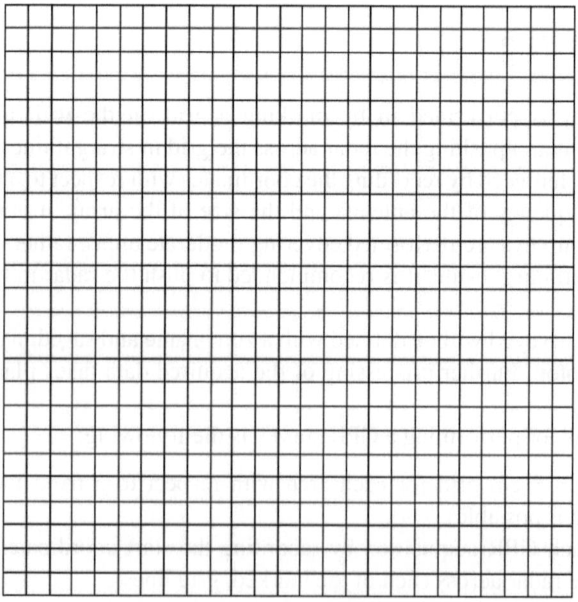

Fig. 3 T/L grid

Fig. 4 GPR Operation

4 Results

This example shows us that there are alternative methods to read traditionally built walls. GPR not only demonstrates which parts of a solid wall have cavities; it also helps us to understand the existence of different layers so that we have evidence of traditional thermal insulation methods such as clay or horse hair and clay mixture. English Heritage, the government institution in charge of England's historic and traditional buildings suggests that one way of minimising the energy loss through dampness is to use vapour permeable insulation materials and counts in this category sheep's wool, wood fibre board and hemp batts. They also acknowledge the fact that these materials can be more expensive altough they are more suitable.[13] On the other hand these materials are only expensive because their production is limited. If the preservation of traditional structures becomes more common, the costs of maintanance will sink, too (Figs. 4, 5, and 6).

The application of GPR will ensure that traditionally built walls are not dismissed as performing poorly in terms of their energy loss. Moreover the use of this methodology will help to reveal craks and damaged structural elements hidden otherwise in the wall. The assessment of a 8 m^2 wide area as in this case took only a day, of which 1 h was used for marking the area, 3 h for the surveying operation, 1 h for the procession and 2 h for the interpretation. If one considers the fact this

[13] http://www.english-heritage.org.uk/your-property/saving-energy/carrying-out-the-work/, accessed 15/12/2012.

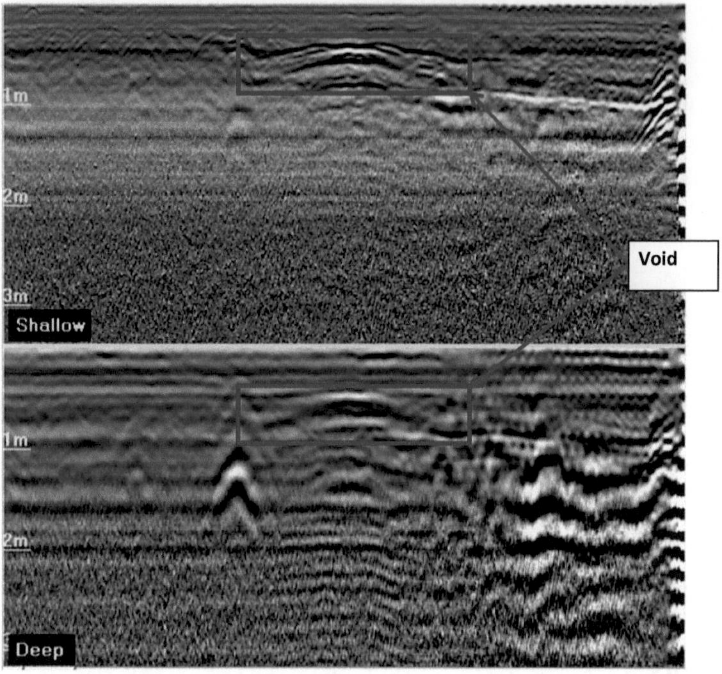

Fig. 5 Voids

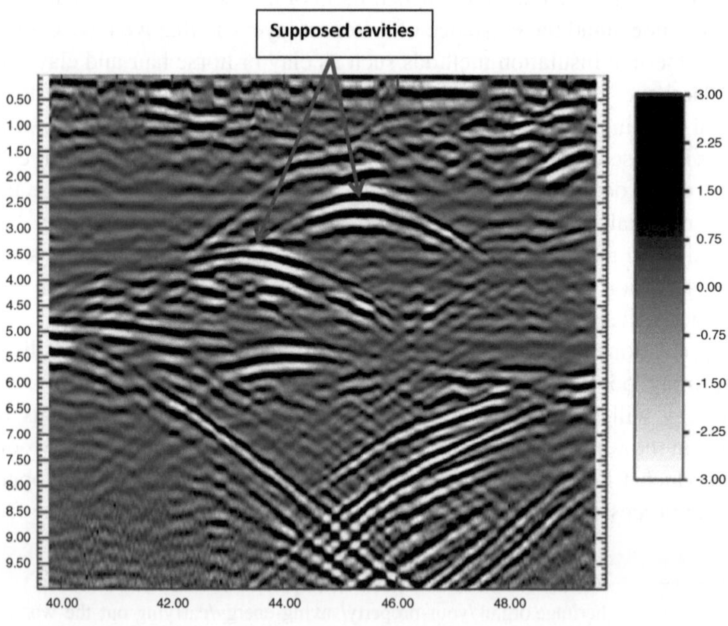

Fig. 6 Supposed cavities

exercise would be done professionally with team work, the resullts can be obtained much more quickly.

There is really very little research on the energy saving performance of historic buildings and further environmental sustainablity characteristics of them such as daylight provision mechanisms, passive ventilation and natural disaster resistance. Most of the existing research is also either available in general guides or as part of scientific papers about the adaptation of historic buildings.[14] Scientific research with empirical results, specific technological suggestions in detail and most importantly on realistic assessment methodologies related to historic structures is extremely limited. As there is no space to discuss all the aspects of this subject in this limited chapter, it is hoped that it at least will encourage further researchers to concentrate and improve energy assessment methods used currently poorly in terms of historic structures and dismiss the prejudices of the industry that they should be demolished for the sake of a greener future. On the contary, it is not possible to establish smart cities on the ashes of the past.

[14]For example see for the first category Sue Roaf, David Crichton and Fergus Nicol, *Adapting Buildings and Cities for Climate Change: A 21st Century Guide*,Elsevier & Architectural Press: London and for the second category a good example is Craig Langston, Francis K. W. Wong, Eddie C. M. Hui, Li-Yin Shen, Strategic Assessment of Building Adaptive Reuse Opportunities in Hong Kong, *Building and Environment*, Volume 43, Issue 10, October 2008, pp. 1709–1718.

Evaluation of the Shading Efficiency of the Shading Devices Installed in the Tram Stations in Athens

Christos Gousis and Ioannis Tzouvadakis

Abstract The purpose of this study is to evaluate the shading devices installed in the stations of the Athenian tram, in regard to the shading efficiency they provided during summer months and for the time period from 9:00 to 17:00, which is the time period of the day with the greatest solar intensity. The existing network of stations of the Athenian tram was recorded after conducting in situ visits, where the characteristics of the installed shading devices, such as orientation, dimensions and symmetries, were registered. Based on the registered data, the shading devices were categorized according to their symmetries in four categories and also they were drawn in three dimensional designs in ACAD. For the evaluation of the provided shading efficiency, two methods were used. The first one was based on the Ecotect program, a program analysing bioclimatic behaviour of structures, which in this case provided overshadowing percentages, while the second one was based on a program created specifically for this study in MATLAB source code. The shading efficiency was evaluated per orientation for the two most common types of shading devices (type A and C) and also for all of the existing shading devices installed in the tram stations. The overshadowing percentages of the shading devices are affected by the orientation of their major axis. The orientations providing the higher percentages are on the West – East axis and around it, while on the North – South axis and around it, the percentages become smaller.

C. Gousis (✉) • I. Tzouvadakis
National Technical University of Athens, Athens, Greece
e-mail: euchristos@gmail.com; itzouvad@central.ntua.gr

S.Th. Rassia and P.M. Pardalos (eds.), *Cities for Smart Environmental and Energy Futures*, Energy Systems, DOI 10.1007/978-3-642-37661-0_14,

1 Introduction

Tram is a tested transportation solution, capable to address the traffic jam problem and can be found in the majority of modern big cities, assisting to minimize the time travel and the emissions in urban areas. Athens is included among the cities listed to have tram as a mean of Mass transportation, with a network that spreads up to 27 km, and serves 65,000 passengers on a daily basis.

Athens climatic conditions are characterized by frequent sunshine, especially during the summer months. Sunshine together with high ambient temperatures in the summer months impose an important need for shading in outdoor spaces, like outdoor stations of Means of Public Transportation. Outdoor thermal environment factors, such as air temperature, relative humidity, wind speed and solar radiation, affect evaluation of thermal comfort (Lin et al. 2010). Human thermal comfort is defined as a condition of mind that expresses satisfaction with the thermal environment (American Society of Heating and Refrigerating Engineers [ASHRAE] 1997).

According to different studies four are the main physical parameters that characterize the thermal environment and determine the thermal comfort sensation. These are ambient air temperature, wind speed, relative humidity and mean radiant temperature (Givoni 1976; Gagge et al. 1976; Berger 1993; Gaitania et al. 2007).

The purpose of this study is to evaluate the shading devices installed in the stations of the Athenian tram, in regard to the shading efficiency they provided during summer months and for the time period from 9:00 to 17:00, which is the time period of the day with the greatest solar intensity.

2 About the Athenian Tram

The first tram started its operation in Athens in 1882 but after World War II, its descending started, and lead to its gradual abolition from 1953 to 1977. The new Athenian tram started its operation in July of 2004 and its network consists of 48 stations, of which "SYNTAGMA", "SEF" and "ASKLIPIO VOULA" are terminal stations. In 16 stations the boarding platforms are located in the centre of the station, in 30 stations the boarding platforms are located sideways, while in two of the terminal stations both types of boarding platforms exist. Tram operates 20 h per day on weekdays and 24 h a day on weekends. The average waiting time for passengers is estimated to be around 7.5–10 min long (http://www.tramsa.gr/).

Every station is equipped with shading devices, seats for the passengers, audiovisual informative system, automatic machines for buying tickets, ticket validation machines, large metallic cupboards for the hardware, projection screens for information (e.g. maps), clocks, waste bins and closed TV circuit (Fig. 1).

Fig. 1 Shading device on a side-way boarding platform, in the station "2nd Agiou Kosma"

3 Categorization of Shading Devices of the Athenian Tram

The shading devices installed in the tram stations consist of a metallic frame, on top of which semi-transparent panels are located, with an angle of +8° from the horizontal level. These shading devices cover part of the platform, providing to the passengers protection from weather conditions (i.e. sunshine, rainfall).

In the present study as "orientation of a shading device" is considered the orientation of the major axis of the shading device. The platform and as results the shading devices always have identical orientations, following the pattern of the rail's network.

For the needs of the present study a 3D model of the shading devices has been designed on computer in ACAD. In respect to their morphology, two types of symmetries have been identified. The first type regards symmetry about the vertical level and the second one regards symmetry about the horizontal level. In the first type of symmetry the shading device is coded as "Y shaped" for the shading devices that are symmetrical about the vertical level and as "Γ shaped" (Γ: gamma[1]) for the shading devices that are not symmetrical about the vertical level. "Γ shaped" shading devices are found in side-way platforms and "Y shaped" in central platforms. In the second type of symmetry the shading device is coded as "S" when it is symmetrical about the horizontal level and as "A" when it is asymmetrical about the horizontal level. In the network of the tram stations four combinations of their composition exist in the platforms, which have been coded as types A, B, C and D as shown in Table 1.

Table 2 gives an overview of all stations of the tram network together with their main characteristics (name of the station, number and type of shading devices, and orientation) as obtained from the in situ visits in each of the 47 stations.

[1] Γ: pronounced as "gamma", is a letter of the Greek alphabet.

Table 1 Categorization of the shading devices of the tram network in four types, based on the morphological combinations of their symmetries

	Y shaped **A**symmetrical	
Type of shading device of central platform (Y shaped) 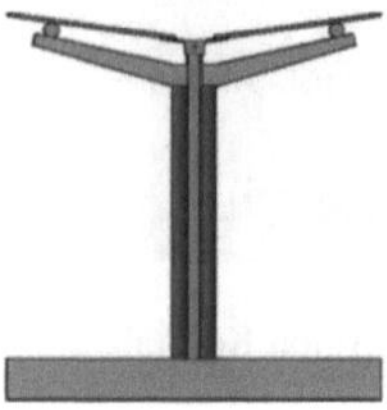	**Type A (YS)** **Y** shaped **S**ymmetrical	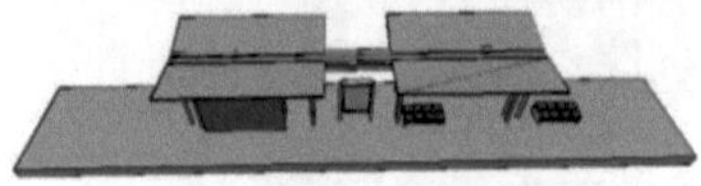
	Type B (YA) **Y** shaped **A**symmetrical	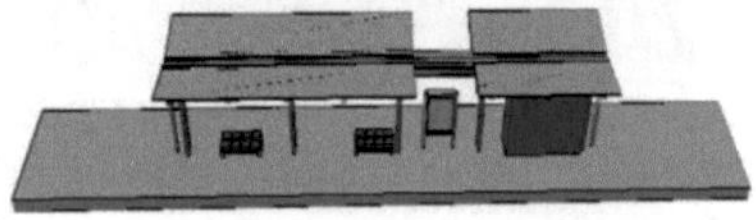
Type of shading device of side-way platform (Γ shaped) 	**Type C (ΓS)** **Γ shaped** **S**ymmetrical	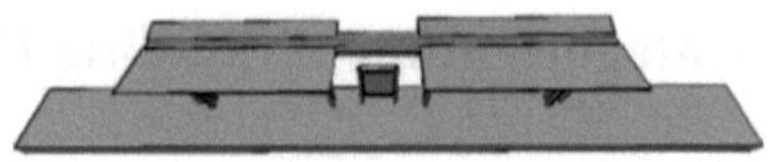
	Type D (ΓA) **Γ** shaped **A**symmetrical	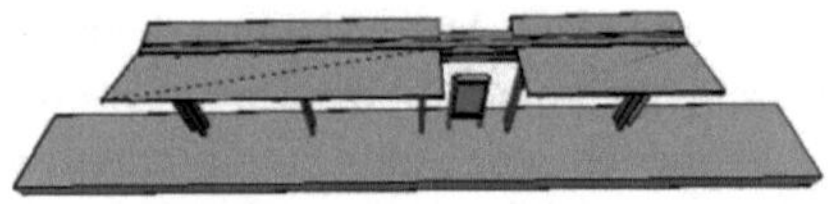

In Fig. 2 a breakdown of the stations of the tram network according to the type of the installed shading devices is presented. As it can be seen the most common shading devices are of type C and A, with the 57 % of the stations having type C shading devices and 21 % of the stations having type A. Stations having type B shading devices follow with 17 % and finally stations with type D with 4 %.

In Fig. 3, a breakdown of the tram stations per orientation and per type of shading device, is presented. As shown, type C shading devices, which are the most common ones in the tram network, are oriented mainly in SSE and SE directions, while type A shading devices are oriented mainly in SSW directions.

4 Shading Analysis

The sun path alters the shading morphology during the day-time and during different seasons in a year, resulting in different morphology of shades covering the platform, in relation to time. The evaluation of the shading efficiency is based on the overshadowing percentages of the platforms. As "overshadowing percentage" is defined, the ratio of the shadowed surface to the total surface of the platform.

Table 2 Overview table of the 47 stations of the Athenian tram, presenting their main characteristics and the overshadowing percentage for the 21st of July

No	Station name	No of shading devices	Shading device type	Platform orientation
1	SYNTAGMA	1	D	SW
2	ZAPPEIO	2	C	E
3	VOULIAGMENIS	1	B	SW
4	FIX	2	C	SSW
5	KASOMOULI	2	C	SSW
6	NEOS KOSMOS	2	C	S
7	BAKNANA	1	A	S
8	EGAIOU	1	B	SW
9	AG. FOTINIS	1	B	SW
10	M. ALEXANDROS	1	A	SE
11	AGIA PARASKEUI	1	B	SSW
12	MIDEIAS MIKALIS	1	A	SSW
13	EUAGGELIKI SXOLI	1	B	SSW
14	AXILLEOS	1	A	SSW
15	AMFITHEAS	2	C	WSW
16	PANAGITSA	1	A	SSW
17	MOUSSON	1	A	S
18	BATIS	2	C	WNW
19	FLOISVOS	2	C	WNW
20	PARKO FLOISVOU	2	C	NW
21	TROKANTERO	2	C	NE
22	AGIA SKEPI	2	C	NNE
23	DELTA FALIROU	1	A	WNW
24	TZITZIFIES	1	A	NW
25	KALLITHEA	2	C	WNW
26	MOSXATO	2	C	E
27	NEO FALIRO	2	C	W
28	SEF	2	B	W
29	EDEM	2	C	SE
30	PIKRODAFNI	2	B	SE
31	MARINA ALIMOU	2	C	S
32	KALAMAKI	2	C	SE
33	ZEFIROS	2	C	SE
34	LOUTRA ALIMOU	2	C	SSE
35	ELLINIKO	2	C	SSE
36	1st AGIOU KOSMA	2	C	SSE
37	2nd AGIOU KOSMA	2	C	SSE
38	AGIOS ALEXANDROS	2	C	SSW
39	EL. OLIMPIONIKON	2	C	SSE
40	KENTRO ISTIOPLOIAS	2	C	SE
41	PLATEIA VERGOTI	2	C	SE
42	PARALIA GLIFADAS	2	C	ESE
43	PALAIO DIMARXEIO	2	C	ESE
44	PLATEIA V.KATRAKI	2	D	E
45	AGGELOU METAXA	1	A	ESE
46	PLATEIA ESPERIDON	1	A	SSE
47	KOLIMBITIRIO	1	B	SSW

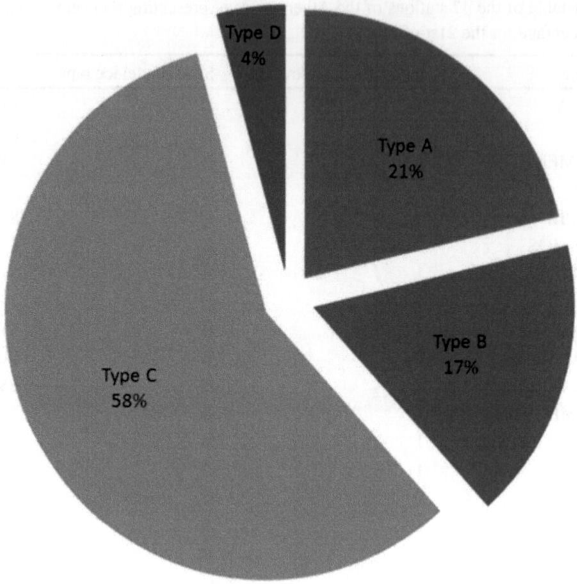

Fig. 2 Breakdown of the tram stations per type of shading device installed

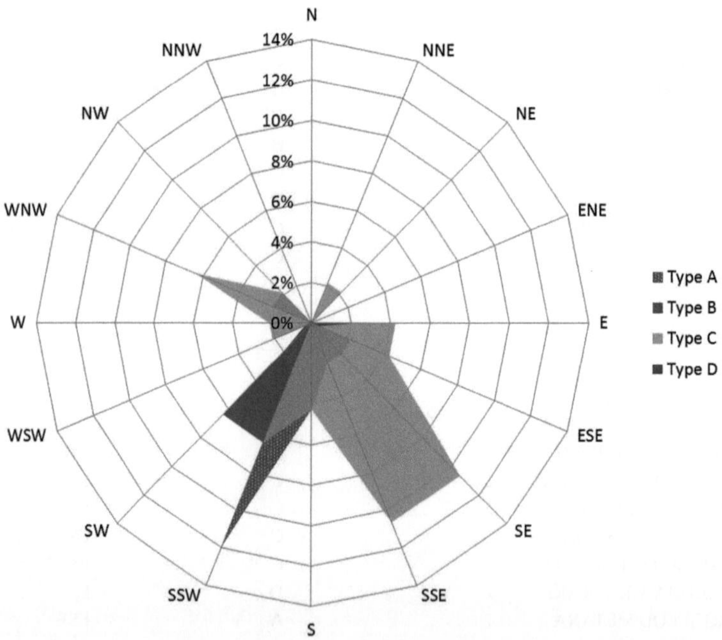

Fig. 3 Breakdown of the 47 tram stations per orientation and per type of shading device

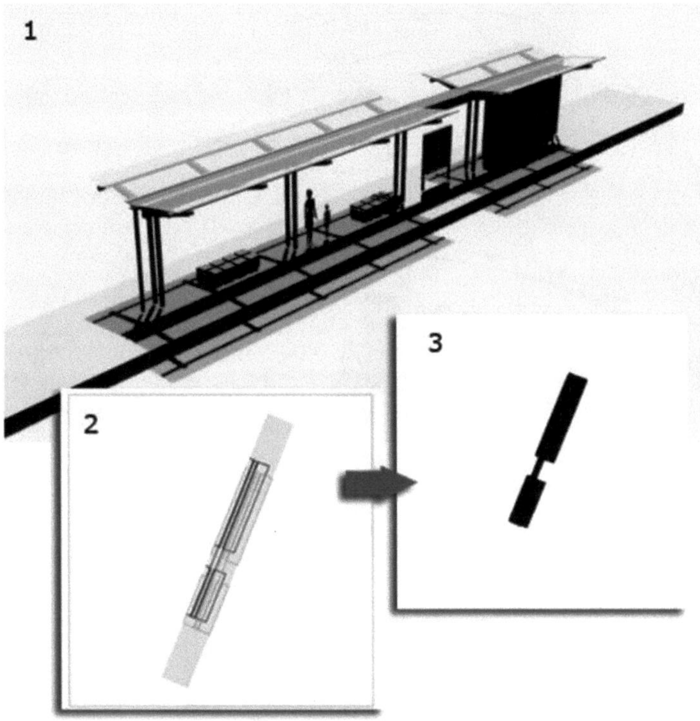

Fig. 4 Schematic overview of the first methodology

For example when the overshadowing percentage is 100 % the whole platform is shaded, while when it is 0 % there is no shade covering the platform.

4.1 Two Alternative Methodologies for Calculating Overshadowing Percentages

There are several methods used to calculate an overshadowed surface. In the present paper the calculation is based on the two following methods.

The first methodology is based on measuring the area of the overshadowed surfaces of the platforms by using images created by 3D computer graphics software, simulating the sun path. The majority of 3D computer graphics software have the ability of simulating the sun path and producing images of shaded 3D objects. For the needs of the present study the three dimensional models of the stations shading devises were designed in AutoCAD from which their shaded images derived. The pictures then where processed through Photoshop in order the overshadowed surfaces to appear as black polygons on white background. The area of the polygons was calculated by a program that was created specifically for this case in MATLAB source code (Fig. 4).

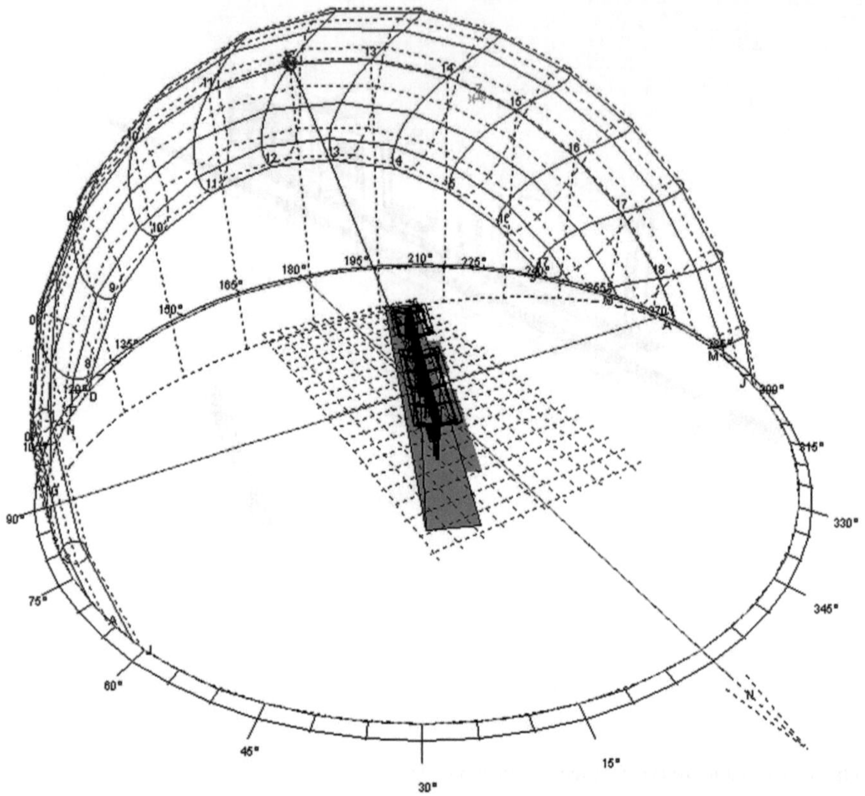

Fig. 5 Autodesk Ecotect analysis annual sun-path diagram over a shading device of the Athenian tram. For every sun position an overshadowing percentage is calculated

Another way to determine overshadowing surfaces is by the use of environmental analysis tools that allow designers to simulate building performance. In this study, as second methodology the software Autodesk Ecotect Analysis was used. This methodology has in fact proved to be less time consuming than the first one and its results were relatively accurate in comparison to the results given from the first methodology (Fig. 5).

4.2 Assumptions of the Shading Analysis

The shading analysis was performed on an hourly basis for the 21st day of each month and for 16 main orientations of the major axis of the platforms. In specific the shading analysis was performed from 9:00 to 17:00, corresponding to the 8 h of the day which bear the highest solar intensity and are around the solar noon, which is approximately at 13:30 for every day studied, given that daylight saving system is in use.

Athens geographical coordinates are 37°58′N 23°43′E.

Greece is in the Eastern European Time Zone. Eastern European Standard Time (EET) is 2 h ahead of Greenwich Mean Time (GMT + 2). Like most states in Europe, Summer (Daylight-Saving) Time is used in Greece, where the time is shifted forward by 1 h; 3 h ahead of Greenwich Mean Time (GMT + 3). After the summer period the time in Greece is shifted back by 1 h to Eastern European Time (EET) or (GMT + 2).

The overshadowing percentages refer to the below mentioned platform areas per type of shading device:

- Type A: platform area 118.9 m^2
- Type B: platform area 139.5 m^2
- Type C: platform area 81 m^2
- Type D: platform area 95 m^2

4.3 *Weight Factor*

For the 21st day of each month, eight different overshadowing percentages derived, one for each of the 8 h of the day which are examined. In order to be able to compare the overshadowing percentages of the different orientations and the different days of the year, it was considered necessary to sum them up to 1 % representing each day. The average of the hourly percentages could be one way to address this matter, but this way it would not be taken into consideration that the percentages do not have equal weights. This is due to the fact that the need for shading becomes more important during the hours with the most intense solar radiation. Thus, a weighted average of the overshadowing percentages was used, weighting as more important the percentages that account to the hours closer to the solar noon.

The determination of the weight factors of the overshadowing percentages of each hour is based on the solar radiation intensity of each hour.

The solar radiation (Arzogloy-Kontiza 1989) reaching a horizontal surface outside of the atmosphere ($G_o(d)$), for a random day (d) in the year and in a specific geographical latitude, in relation to the vertical solar radiation $G_{on}(d)$ is given by the following equation:

$$G_o(d) = G_{on}(d) \times \cos\theta_z \tag{1}$$

Angle θz, also called zenith angle, is the angle between zenith and the sun, while its complementary angle, $a = 90° - \theta z$, gives the altitude angle of the sun.

For calculating the weight factor of each hour (W_{hour}), for the different months, firstly we calculate the solar radiation intensity for each one of the 8 of a day and then each one of them is divided by their sum. The equation for calculating the

Table 3 Weight factors for summer months June, July and August

	10:00	11:00	12:00	13:00	14:00	15:00	16:00	17:00
June	0.101	0.121	0.135	0.142	0.142	0.135	0.121	0.101
July	0.100	0.121	0.136	0.143	0.143	0.136	0.121	0.100
August	0.097	0.120	0.137	0.145	0.145	0.137	0.120	0.097

weight factor is given below (Eq. 2), while the values of the weight factors for the summer months, June, July and August are presented in Table 3.

$$W_{hour} = \frac{G_o(d)_{hour}}{\sum_1^8 G_o(d)_t} = \frac{\cos\theta_{z\,hour}}{\sum_1^8 \cos\theta_{zt}} \tag{2}$$

where t takes values from 1 to 8, referring to the 8 h that are examined.

5 Evaluation of the Shading Efficiency of the Shading Devices

5.1 Shading Devices of Type A (YS) and C (ΓS)

In this section the results of the shading analysis for the shading devices of types A and C are presented, for the 16 main orientations of their major axis. Focus is given to the results corresponding to the summer period, when the solar intensity and ambient temperature are higher and thus the needs for shading are more important.

In the framework of this study, type A shading device is centered to a platform with dimensions 31.3 m × 3.8 m, and of total area 118.9 m^2, while type C shading device is centered to a platform with dimensions 31.3 m × 2.6 m and of total area 81 m^2. Thus, the overshadowing percentages, given in Table 4 correspond to the above mentioned areas respectively; and are the weighted averages of the overshadowing percentages for the basic orientations of the major axis of the shading devices for months June, July and August.

From Figs. 6 and 7 it can be seen that the pattern of change per month, for the overshadowing percentages of the two types of shading devices is rather similar. Specifically, it is observed that both for type A and C shading devices, the shading efficiency is better for orientations near the East–West axis (E-W, ESE-WNW, ENE-WSW), as opposed to the orientations near the North -South axis (N-S, NNW-SSE, NNE-SSW), where the overshadowing percentages present their minimum values.

For type A shading devices, as shown in Fig. 6, the overshadowing percentages of June and July range at the same levels, with those of July being slightly higher for several orientations. The maximum values ranges between 30 % and 35 %, while the minimum between 21 % and 24 %. The overshadowing percentages of August

Table 4 Weighted averages of the overshadowing percentages for the basic orientations of the major axis of the shading devices for months June, July and August

	Type A (YS) shading devices			Type C (SΓ) shading devices		
	June (%)	July (%)	August (%)	June (%)	July (%)	August (%)
N	23	21	21	16	16	16
NNE	24	22	21	18	17	16
NE	29	27	24	22	21	18
ENE	32	30	24	29	29	23
E	35	35	26	35	32	26
ESE	33	33	28	29	28	22
SE	28	28	23	20	20	18
SSE	23	24	22	18	18	17
S	23	21	21	17	16	15
SSW	24	22	21	18	21	17
SW	29	27	24	22	23	19
WSW	32	30	24	30	28	23
W	35	35	26	34	32	28
WNW	33	33	28	30	29	24
NW	28	28	23	24	22	19
NNW	23	24	22	17	16	17

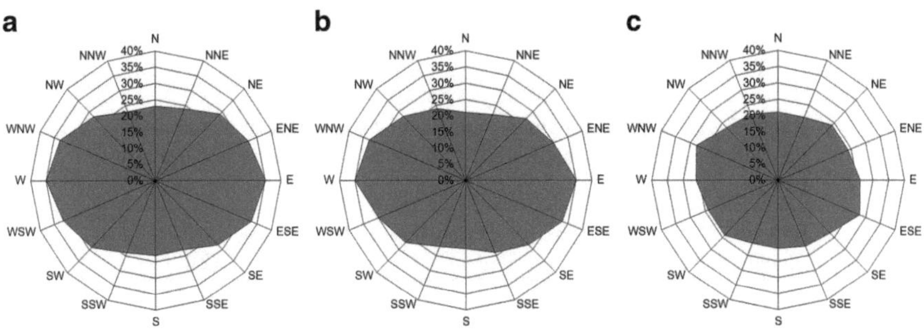

Fig. 6 Overshadowing percentages per orientation for type A shading devices for months: (**a**) June, (**b**) July, (**c**) August

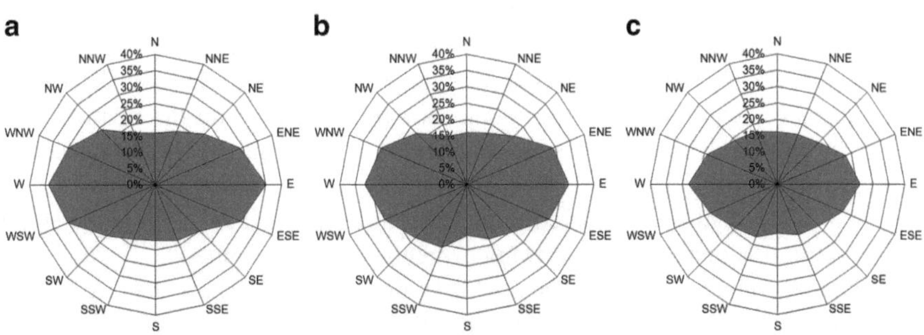

Fig. 7 Overshadowing percentages per orientation for type C shading devices for months: (**a**) June, (**b**) July, (**c**) August

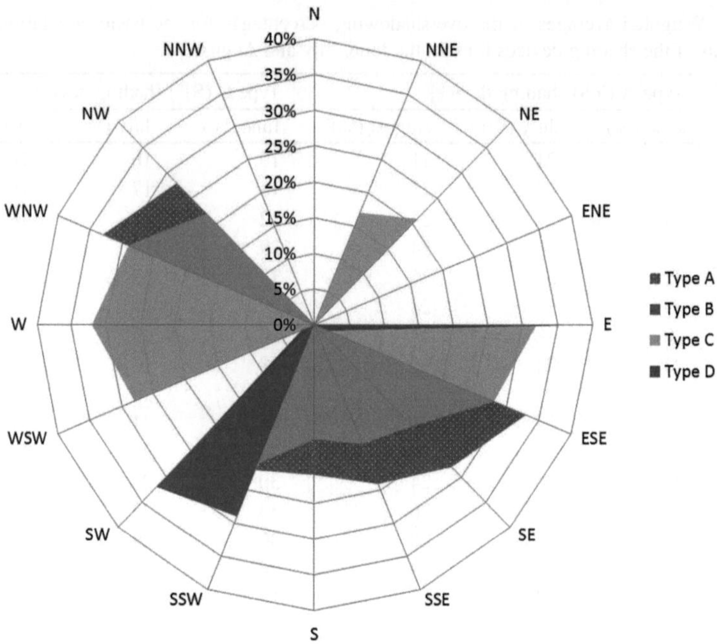

Fig. 8 Overshadowing percentages for the 47 stations per type of shading device for the 21st of July

Table 5 Average, maximum and minimum values of the overshadowing percentages per type of shading device for the 47 stations and for the 21st of July

Category *Sample size (N)*	Type A (10)	Type B (8)	Type C (27)	Type D (2)
Average	25.4 %	31.5 %	22.7 %	31.5 %
Min-max	(21–33 %)	(29–38 %)	(16–32 %)	(28–35 %)

are slightly lower (by 9 %) compared to those of June and July. Furthermore for August the effect of the orientation on the overshadowing percentages is limited.

The overshadowing percentages for type C shading devices (Fig. 7) follow a similar pattern of change as the one of type A. In specific, the overshadowing percentages of June and July have values at similar levels with the relative ones of type A, with maximum values ranging between 35 % and 29 %, and minimum values between 16 % and 21 %. For August, the overshadowing percentages also present a 9 % reduction, compared to the values of June and July, while the effect of the orientation on the overshadowing percentages is also limited.

It should be noted that due to the double symmetry of the type A shading device, the values of the overshadowing percentages for a random orientation are equal to the ones for the opposite orientation (corresponding to a 180 °C turn). Type C

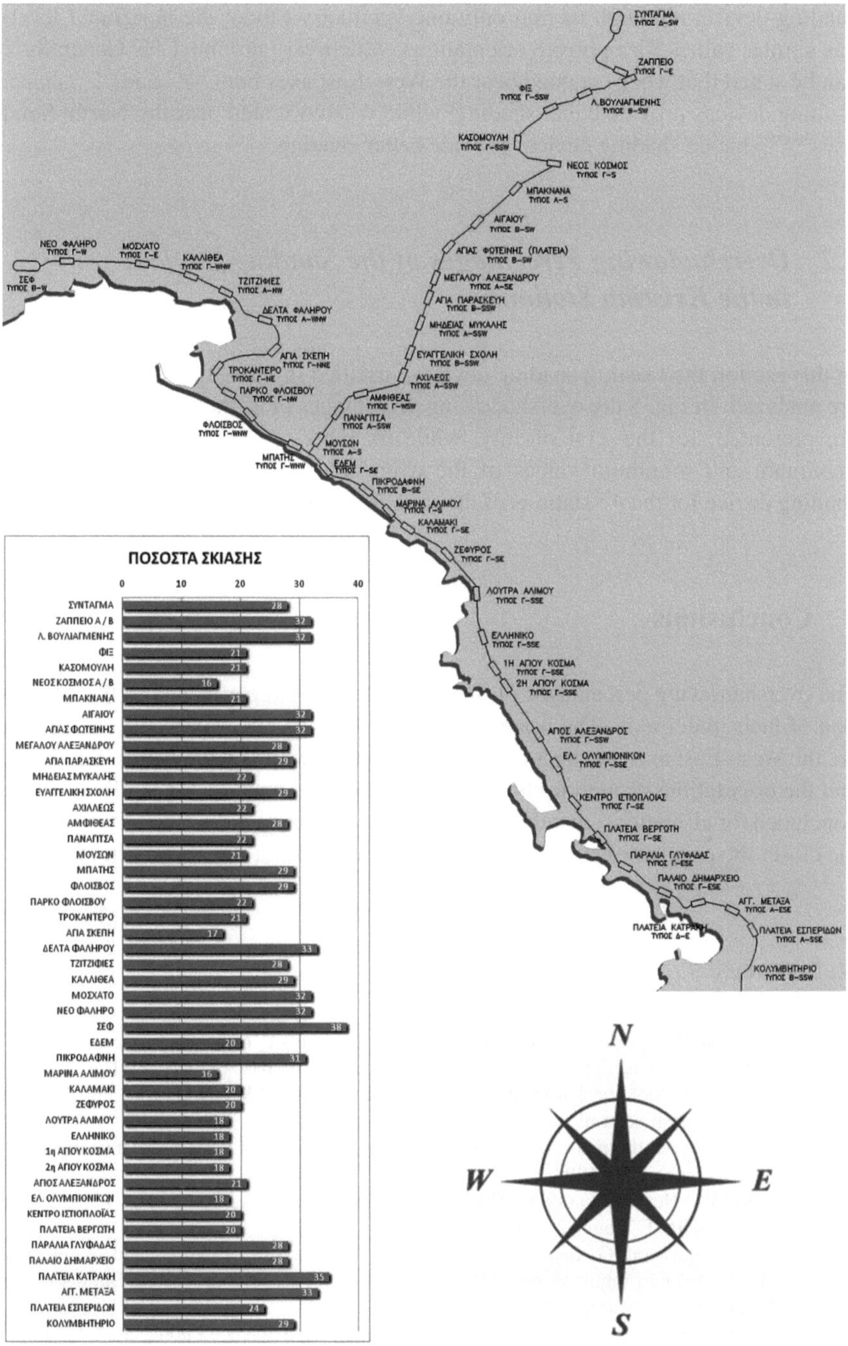

Fig. 9 Overshadowing percentages of the existing tram stations for the 21st of July and a graphical depiction of the network on the map

shading device, although having only one symmetry (about the horizontal level), has similar values for opposite orientations, which vary around 1 %. Generally, it can be stated that when oriented near the West–East axes both "Y" and "Γ" shaped shading devices provide equal shading, while when oriented near the North–South axes "Y" shaped shading devices provide better shading.

5.2 *Overshadowing Percentages of the Shading Devices in the Existing Stations*

In this section the existing shading devices, installed in the Athenian tram network are analyzed. In Fig. 8 the overshadowing percentages of all stations per orientation are presented, for the 21st of July, while in Table 5 is presented the average, maximum and minimum values of the overshadowing percentages per type of shading device for the 47 stations of the tram network.

6 Conclusions

The overshadowing percentages of the shading devices are affected by the orientation of their major axis. The orientations providing the better percentages for July are the West–East axis and the orientations around it, while on the North–South axis and the orientations around it, the percentages become smaller. Generally it can be concluded for all stations, that the higher the value of the overshadowing percentage, the closer the orientation of the shading device to the West–East axis (Fig. 9).

References

Arzogloy-Kontiza M (1989) Introduction in astrophysics. University of Athens, Athens

American Society of Heating and Refrigerating Engineers [ASHRAE] (1997) Handbook of fundamentals: physiological principles, comfort, health. ASHRAE, New York http://isbndb.com/d/publisher/ashrae.html

Berger X (1993) About thermal comfort. In: Proceedings of the third PASCOOL meeting. CMF Subgroup, Florence, May 1993

Gagge AP, Nishi Y, Nevins RG (1976) The role of clothing in meeting FEA energy conservation guidelines. ASHRAE Trans 82(Part II):234

Gaitania N, Mihalakakou G, Santamouris M (2007) On the use of bioclimatic architecture principles in order to improve thermal comfort conditions in outdoor spaces. Build Environ 42:317–324

Givoni B (1976) Man, climate and architecture, 2nd edn. Applied Science Publishers, London

http://www.tramsa.gr/. Accessed 5 Oct 11

Lin T-P, Matzarakis A, Hwang R-L (2010) Shading effect on long-term outdoor thermal comfort. Build Environ 45:213–221

Modeling and Control of Large and Flexible Wind Turbines in Variable Speed Mode

Wei Lin and Xinghua Liu

Abstract As wind turbines are increased in size and power, flexible structure of wind turbines is becoming increasingly important and cannot be neglected. In this paper a four-mass drive-train model is presented by considering flexible structure modes of wind turbines. Due to the complexity of the four-mass model of the drive train, the focus of this work is on the design of nonlinear variable speed controllers for a reduced-order, two-mass model of the drive train. Both the state feedback controller and the measurement feedback controller (i.e., using the information of rotor speed only but not the torsional angle) are proposed to achieve asymptotic tracking control for prescribed rotor speed reference signals, thus capturing maximum wind power. Simulation results are provided to demonstrate the effectiveness of the proposed control schemes. Generalizations of our control system design methods to larger and flexible wind turbines with high-order drive train models are currently under investigation and will be reported elsewhere.

W. Lin (✉)
Case Western Reserve University, Department of Electrical Engineering and Computer Science, Cleveland, Ohio, USA
e-mail: linwei@case.edu

X. Liu
Shenzhen Graduate School, Harbin Institute of Technology, HIT Campus of Shenzhen Town, Xili, Shenzhen China
e-mail: leo.xinghua.liu@gmail.com

S.Th. Rassia and P.M. Pardalos (eds.), *Cities for Smart Environmental and Energy Futures*, Energy Systems, DOI 10.1007/978-3-642-37661-0_15,

1 Introduction

Due to environmental pressures and concerns, wind power, as one of promising and green renewable energy, has received a great deal of attention and been developed significantly worldwide. Over the past decade, both the rotor size and structural weight of wind turbines have increased considerably in order to get the wind turbine running, maintaining efficiency and producing more electricity. For example, Enercon E-126, which is the largest wind turbine nowadays, has a hub height of 135 m, rotor diameter of 126 m, total weight of 6,000 t and can generate up to 7.5 MW power. All of this raises a challenging issue on how to model and control such larger and more flexible wind turbines.

A horizonal-axis wind turbine typically consists of three blades, hub, low-speed shaft, gearbox, high-speed shaft, double-fed induction generator (DFIG), nacelle and tower. Contrary to the conventional turbine generators such as steam, hydro and diesel, a wind turbine has a relatively low shaft stiffness when viewed from the generator side. In fact, the shaft stiffness of a wind turbine generator is only 1–1.5 % of the afore-mentioned turbine generators in per unit, and is even lower than the electrical stiffness (Hinrichsen and Nolan 1982). For this reason, a simplified, one-mass drive-train model has been commonly used to describe wind turbines (Beltran et al. 2008; Boukhezzar and Siguerdidjane 2005; Johnson et al. 2008), by treating the drive-train dynamics as a rigid connection and neglecting the flexibility of the shaft. Although the one-mass drive-train model is intuitive and convenient for the control systems design of wind turbines, it cannot capture the dynamics of torsional effects, which has a significant influence on the power fluctuations and the interaction of the wind turbine with the grid. Moreover, oversimplifying the models of wind turbine could introduce significant errors in dynamic behavior and stability (Salman and Teo 2003).

The flexibility of the low-speed shaft or blades has been considered, for instance, in (Ramtharan et al. 2007) and (Wang and Weiss 2009), and was shown to have a significant impact on the dynamic performance of the DFIG in transient stability studies. In the afore-mentioned papers, more accurate yet complex models for wind turbines, such as two-mass and three mass drive-train models, were developed and used for the purpose of control, by not neglecting flexible structural modes of wind turbines (Trudnowski et al. 2004; Wang and Weiss 2009). In the case when both the flexibilities of the shaft and blades are taken into account, a three-mass drive-train model was derived (Ramtharan et al. 2007). In the work (Papathanassiou and Papadopoulos 1999), the authors considered the flexibility of high-speed shaft instead of the flexibility of blades, resulting in another type of three-mass drive-train model. In the subsequent paper (Papathanassiou and Papadopoulos 2001), the authors further presented a six-mass drive-train model that includes six masses (i.e., three blades, a hub, a gearbox and a generator) and the associated five springs, by extending the three-mass drive-train model (Papathanassiou and Papadopoulos 1999) to a more general situation. In the popular wind turbine simulation software FAST, a typical horizontal-axis wind turbine (HAWT) with three blades has been modeled with 24 degrees of freedom (DOFs), by considering most, if not all, possible combinations of flexible structural modes in wind turbines (Jonkman and Buhl 2005). While these higher-order models give more accurate descriptions for larger and flexible wind turbines, they also increase the complexity substantially. On one hand, the model complexity makes the design of control systems of wind turbines complicated and difficult. On the other hand, it results in more time-consuming computation in simulations, particularly, in the situation when performing simulations of wind farm, power systems and wind turbines with the grid (Perdana 2008).

In the existing literature, most of wind turbines control systems are designed based on the linearization of the two-mass or higher-order drive-train models. An obvious reason is that the linearized models make the design of controllers simple and easy to be implemented in practical applications. On the other hand, a system model of wind turbines is highly nonlinear due to the character of aerodynamics and the flexibility of blades and tower. One of the objectives of this paper is to develop a nonlinear model with moderate order, in particular, a four-mass drive-train model, which incorporates and generalizes the commonly used two-mass drive-train model. This will be done by taking into account flexible structural modes of wind turbines such as the flexibility of the shaft and blades. The other goal of this paper is to investigate the question of how to control a large and flexible wind turbine described by a higher-order nonlinear model when it operates in variable speed mode. In this direction, several papers have studied the problem of variable-speed control for wind turbines, primarily based on the lower-order linearized model. For instance, adaptive control of wind turbines was studied in (Song et al. 2000), while a robust variable-speed control scheme was proposed in (Sivrioglu et al. 2008). In (Beltran et al. 2008) and (Thomsen and Poulsen 2007), sliding mode control and disturbance rejection schemes have been presented, respectively. The work (Boukhezzar and Siguerdidjane 2005) has developed a nonlinear control method without using wind speed measurement.

In this paper, we present some preliminary results on modeling and control of a large and flexible wind turbine in variable speed mode. The modeling of the flexible wind turbine system is presented in Sect. 2, where a four-mass drive-train model is given. Due to the complexity and the nonlinear nature of the model thus derived, it is rather difficult to carry out a nonlinear, variable speed control design for the four-mass drive-train model directly. To simplify the problem, we discuss in Sect. 3 how to control a reduced-order, variable-speed wind turbines. Specifically, we employ the commonly used two-mass drive-train model, which can be viewed as a reduced-order model of the four-mass drive-train model obtained in Sect. 2, to conduct a nonlinear feedback design. Nonlinear controllers are introduced to achieve asymptotic tracking of a prescribed rotor speed so as to maximize wind power capture. In Sect. 4. the effectiveness of our nonlinear controllers is demonstrated via simulations. Concluding remarks are given in Sect. 5.

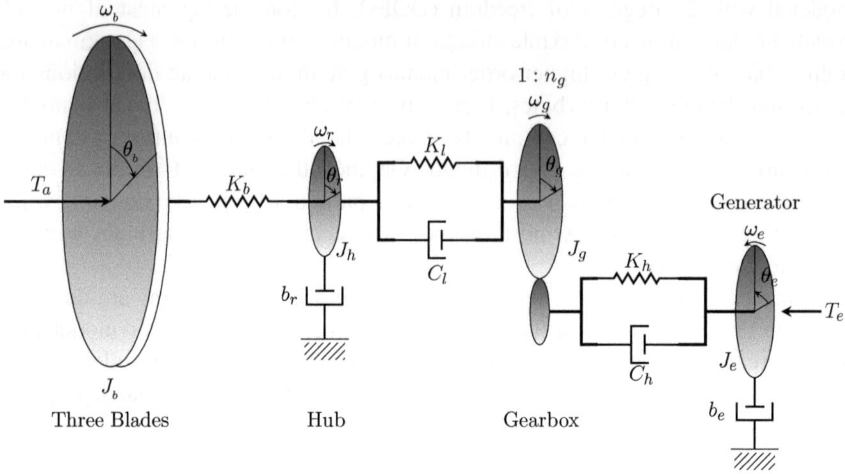

Fig. 1 The four-mass drive-train model

2 Two Nonlinear Models of Wind Turbines

In this section, a nonlinear model with appropriate order is developed for a horizonal-axis wind turbine by using the Lagrange's equation. In particular, we derive a four-mass drive-train model that generalizes and incorporates the well-known two-mass drive-train model. This is accomplished by treating the key components of wind turbine as mass-spring systems, as shown in Fig. 1, where J_b, J_h, J_g and J_e are the inertia of the blades, hub, gearbox and generator, respectively. The stiffness factor K_b indicates the effective flexibility of three blades, K_l represents the flexibility of the low-speed shaft accompanied by the damping factor C_l, and the high-speed shaft stiffness factor is denoted by K_h associated with the damping factor C_h. The parameters b_r and b_e are the damping factors of the rotor and generator, respectively.

In what follows, a mechanical model of the wind turbine system shown in Fig. 1 is derived based on the Lagrange's equation. To this end, define

$$L = T - V, \tag{1}$$

where T is the kinetic energy of the system, V the potential energy and L the Lagrangian. From Fig. 1, it is not difficult to see that

$$T = \frac{1}{2}J_b\dot{\theta}_b^2 + \frac{1}{2}J_h\dot{\theta}_r^2 + \frac{1}{2}J_g\dot{\theta}_g^2 + \frac{1}{2}J_e\dot{\theta}_e^2 \tag{2}$$

$$V = \frac{1}{2}K_b(\theta_b - \theta_r)^2 + \frac{1}{2}K_l(\theta_r - \theta_g)^2 + \frac{1}{2}K_h(\theta_g n_g - \theta_e)^2. \tag{3}$$

Using of the general form Lagrange's equation with conservative and nonconservative forces, i.e.

$$\frac{\mathrm{d}}{\mathrm{d}t}\left(\frac{\partial L}{\partial \dot{q}_i}\right) - \frac{\partial L}{\partial q_i} + \frac{\partial D}{\partial \dot{q}_i} = Q_i = \frac{\delta W}{\delta q_i} \tag{4}$$

with

$$D = \frac{1}{2}C_l(\dot{\theta}_r - \dot{\theta}_g)^2 + \frac{1}{2}C_h(\dot{\theta}_g n_g - \dot{\theta}_e)^2 \tag{5}$$

$$\delta W = T_a \delta\theta_b - b_r \dot{\theta}_r \delta\theta_r - b_e \dot{\theta}_e \delta\theta_e - T_e \delta\theta_e, \tag{6}$$

being the Rayleigh's dissipation function and the virtual work, respectively, we can obtain the following differential equations that describe the dynamics of the four-mass drive-train model (including four general coordinates and three energy storage devices)

$$\dot{\theta}_{K_b} = \omega_b - \omega_r \tag{7}$$

$$\dot{\theta}_{K_l} = \omega_r - \omega_g \tag{8}$$

$$\dot{\theta}_{K_h} = \omega_g n_g - \omega_e \tag{9}$$

$$\dot{\omega}_b = \frac{1}{J_b}[-K_b\theta_{K_b} + T_a] \tag{10}$$

$$\dot{\omega}_r = \frac{1}{J_h}[K_b\theta_{K_b} - K_l\theta_{K_l} - C_l(\omega_r - \omega_g) - b_r\omega_r] \tag{11}$$

$$\dot{\omega}_g = \frac{1}{J_g}[K_l\theta_{K_l} - K_h n_g\theta_{K_h} + C_l(\omega_r - \omega_g) - C_h n_g(\omega_g n_g - \omega_e)] \tag{12}$$

$$\dot{\omega}_e = \frac{1}{J_e}[K_h\theta_{K_h} + C_h(\omega_g n_g - \omega_e) - b_e\omega_e - T_e] \tag{13}$$

where $\theta_{K_b} = \theta_b - \theta_r$, $\theta_{K_l} = \theta_r - \theta_g$, $\theta_{K_h} = \theta_g n_g - \theta_e$ are the torsional angles of the shafts, and (7)–(9) are the corresponding dynamic equations.

As explained in (Burton et al. 2001) and (Heier and Waddington 1998), the power captured by the wind turbine is

$$P_m = \frac{1}{2}C_p(\lambda, \beta)\rho\pi R_w^2 V_w^3 = k_w\omega^3 = T_a\omega \tag{14}$$

where ρ is the air density, R_w is the blade tip radius, V_w is the wind speed and C_p is the power coefficient, which is a function of the tip-speed ratio λ and the pitch angle β. The tip-speed ratio $\lambda = \frac{R_w\omega}{V_w}$, where ω is the rotor speed. Clearly,

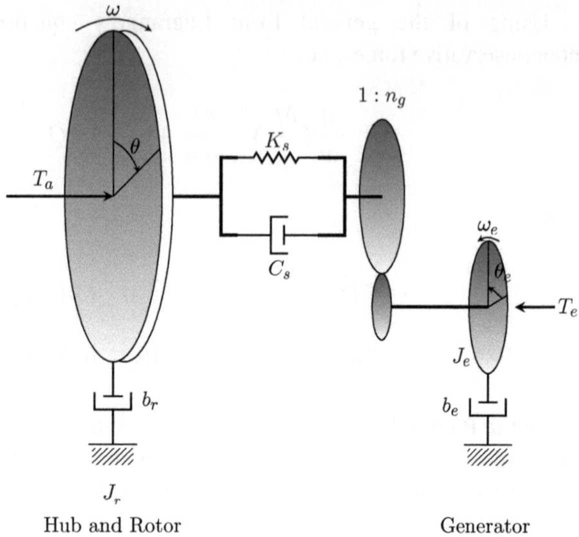

Fig. 2 The two-mass drive-train model

$$T_a = \frac{P_m}{\omega} = k_w \omega^2 \tag{15}$$

where $k_w = \frac{1}{2} C_p \rho \pi \frac{R_w^5}{\lambda^3}$.

Using the order-reduction method introduced in (Ramtharan et al. 2007), or ignoring the flexibility of the blades and the high-speed shaft, we can deduce from the four-mass drive-train model the following two-mass drive-train model shown in Fig. 2. The dynamic equations of the two-mass drive-train model are given by

$$\dot{\theta}_k = \omega - \frac{1}{n_g}\omega_e, \quad \theta_k := \theta - \frac{1}{n_g}\theta_e \tag{16}$$

$$\dot{\omega} = \frac{1}{J_r}(-K_s\theta_k + k_w\omega^2 - C_s\omega - b_r\omega + \frac{C_s}{n_g}\omega_e) \tag{17}$$

$$\dot{\omega}_e = \frac{1}{J_e}(\frac{K_s}{n_g}\theta_k + \frac{C_s}{n_g}\omega - \frac{C_s}{n_g^2}\omega_e - b_e\omega_e - T_e) \tag{18}$$

3 Nonlinear Controllers for Variable-Speed Wind Turbines

In this section, we briefly discuss how to design nonlinear tracking controllers for the variable-speed wind turbine described by, instead of the complex four-mass drive-train model (7)–(13), the reduced-order two-mass drive-train model (16)–(18). As shown in Fig. 3, the control objective is to design a generator control torque T_e

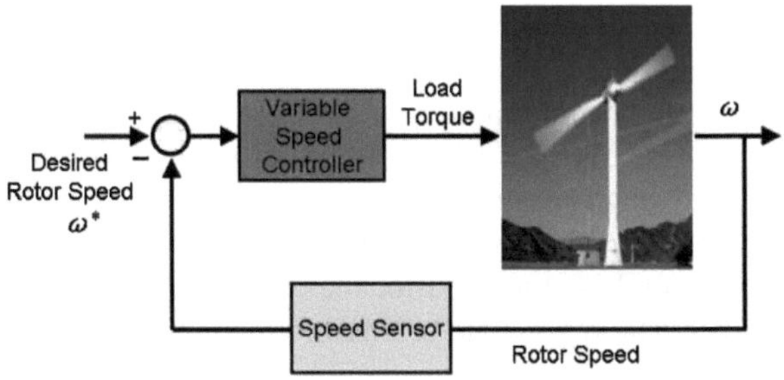

Fig. 3 A tracking control system of wind turbine

such that the rotor speed ω of the wind turbine asymptotically tracks the desired speed ω^*.

Notably, when a wind turbine is operating at variable speed mode and keeps the pitch angle at a constant, if the rotor speed is adjusted continuously according to the wind speed variation, the tip-speed ratio can be maintained at an optimal point, thus enabling the wind turbine to capture the maximum wind power. In other words, the wind turbine maximum power capture problem boils down to the tracking problem described in the previous paragraph.

To solve the tracking control problem for the wind turbine represented by the two-mass drive train model (16)–(18), we introduce the change of coordinates,

$$z = \theta_k, \quad \xi_1 = \omega, \quad \xi_2 = \frac{C_s}{J_r n_g}\omega_e - \frac{K_s}{J_r}\theta_k$$

which is a global diffeomorphism, to transform the system (16)–(18) into the following equivalent form

$$\dot{z} = -\frac{K_s}{C_s}z + \xi_1 - \frac{J_r}{C_s}\xi_2 \tag{19}$$

$$\dot{\xi}_1 = \xi_2 + \phi_1(\xi_1) \tag{20}$$

$$\dot{\xi}_2 = v + \phi_2(z, \xi_1, \xi_2) \tag{21}$$

$$y = \xi_1 \tag{22}$$

where $v = -\frac{C_s}{J_e J_r n_g} T_e$ and

$$\phi_1(\xi_1) = -\frac{C_s + b_r}{J_r}\xi_1 + \frac{k_w}{J_r}\xi_1^2$$

$$\phi_2(z, \xi_1, \xi_2) = \left(\frac{K_s^2}{J_r C_s} - \frac{b_e K_s}{J_e J_r}\right)z + \left(\frac{C_s^2}{J_r J_e n_g^2} - \frac{K_s}{J_r}\right)\xi_1 + \left(\frac{K_s}{C_s} - \frac{C_s}{J_e n_g^2} - \frac{b_e}{J_e}\right)\xi_2$$

Clearly, the nonlinear system (19)–(21) is in the normal form with a triangular structure and minimum-phase (Isidori 1995). However, the fact that both the states ξ_1 and ξ_2 enter the zero dynamics makes the problem of global asymptotic tracking non-trivial. Nevertheless, using a feedback design method inspired by (Lin and Gong 2003), we can explicit construct a tracking controller that solves the problem, as summarized in the following theorem. Due to the limit of the space, a detailed proof and the Lyapunov-based method for the design of a tracking controller are omitted in this paper. The readers are referred to the paper (Lu and Lin 2011) for details.

Theorem 1. *For the wind turbine system described by the two-mass drive train model (19)–(22), there is a state feedback nonlinear controller* $v = \beta(z, \xi_1, \xi_2, y_r, \dot{y}_r, \ddot{y}_r)$, *with* $y_r(t) = \omega^*$ *being a smooth and bounded speed signal, such that the rotor speed* $\xi_1 = \omega$ *globally asymptotically tracks the desired speed* $y_r(t) = \omega^*$.

Remark 1. It is known that global asymptotic tracking can be achieved for nonlinear systems in a lower-triangular form with trivial zero dynamics. In the presence of zero dynamics which involves more than one component, namely, the output of the system ξ_1 and its derivative such as ξ_2, even the global stabilization problem (not mentioning the more difficult tracking problem) may not be solvable (Isidori 1995). Fortunately, for the two-mass drive train model (19)–(22), the zero dynamics is linear and our state feedback design takes full advantage of such a linear property, yielding a solution to the tracking problem, i.e., force the rotor speed $\xi_1 = \omega$ to follow asymptotically the desired speed $y_r(t) = \omega^*$.

The tracking controller developed in Theorem 1 (see Lu and Lin 2011 for further details) requires the information of all the states θ_k, ω and ω_e of the system (16)–(18). However, the torsional angle θ_k is usually difficult to measure in practice. For this reason, the tracking controller obtained in Theorem 1 is not practically implementable. In what follows, we present an observer-based nonlinear controller by using the information of ω only, to achieve asymptotic tracking for the prescribed reference speed signal $y_r(t)$.

For the wind turbine system described by (19)–(22) in (z, ξ_1, ξ_2) coordinates, which is equivalent to the two-mass drive model (16)–(18), ξ_2 is not measurable because it involves θ_k. The only measurable signal is the state $\xi_1 = \omega$ that can be used for feedback design. With this in mind, the following result can be established.

Theorem 2. *For the wind turbine system described by (19)–(22), there exists a nonlinear output feedback controller of the form*

$$\dot{\varsigma} = \alpha(\varsigma, \xi_1), \qquad \varsigma = [\hat{z}\ \hat{\xi}_2]^T, \tag{23}$$

$$v = \beta(\varsigma, \xi_1, y_r, \dot{y}_r, \ddot{y}_r), \tag{24}$$

such that the rotor speed $\xi_1 = \omega$ *globally asymptotically tracks the desired speed* $y_r(t)$*, where* $y_r(t)$ *is a prescribed reference speed signal, twice differentiable and bounded.*

The theorem can be proved by using the state feedback controller derived from Theorem 1, together with the design of a reduced-order observer that estimates the unmeasurable states z and ξ_2 of the system (19)–(22). It has been shown that global stabilization of nonlinear systems with a triangular form is solvable by an output dynamic compensator if the linear growth condition is satisfied and the zero dynamics is exponentially stable (Qian and Lin 2003). Employing the idea of (Qian and Lin 2003) with appropriate modifications, one can build a reduced-order observer to estimate the states z and ξ_2 of the system (19)–(22), as done in (Lu and Lin 2011). A new ingredient of our output feedback design is the estimation of the zero dynamics state z, which was not considered in (Qian and Lin 2003). Thanks to the minimum-phase property of the nonlinear system (19)–(22), the estimation of the state z can be carried by our design method, even though both the states ξ_1 and ξ_2 enter the zero dynamics. A thoroughly proof of Theorem 2 can be found in (Lu and Lin 2011) and is omitted here due to the limit of space.

Remark 2. Both the state and output tracking control schemes introduced in this section are applicable to wind turbines operated in region 2, i.e., from cut-in to rated wind speed region, so as to realize the maximum wind power capture. Our nonlinear tracking control methods for the two-mass drive-train model are expected to be extended to the variable speed control of the three-mass drive-train model or even the four-mass drive-train model. The research in this direction is currently on going and a progress will be reported elsewhere.

4 Simulation Results

In this section, we present some simulation results to demonstrate the effectiveness of the nonlinear tracking control schemes introduced in the previous section. The wind turbine under consideration is with 1.7 MW rated power and its diameter of the blades is 35 m (Garcia-Sanz and Elso 2009). The physical parameters are given in Table 1. The rotor speed is adjusted to track the following desired trajectory asymptotically:

$$y_r(t) = \omega^* = 1.1 + 0.5\big[1 + \sin(\frac{\pi}{10}t)\big]$$

The simulation results using both the state and output feedback controllers for the two-mass drive-train model are shown in Fig. 4a, b, respectively. Obviously, it can be seen from the simulations that the proposed control schemes are able to achieve speed tracking asymptotically.

Table 1 Wind turbine's parameters

J_r	$4 \times 10^6\,\mathrm{kg \cdot m^2}$	J_e	$20\,\mathrm{kg \cdot m^2}$
b_r	$980\,\mathrm{N \cdot m/s}$	b_e	$0.2\,\mathrm{N \cdot m/s}$
n_g	38.06	k_w	$1.14 \times 10^5\,\mathrm{kg \cdot m^2}$
K_s	$10 \times 6\,\mathrm{N \cdot m}$	C_s	$500\,\mathrm{N \cdot m/s}$

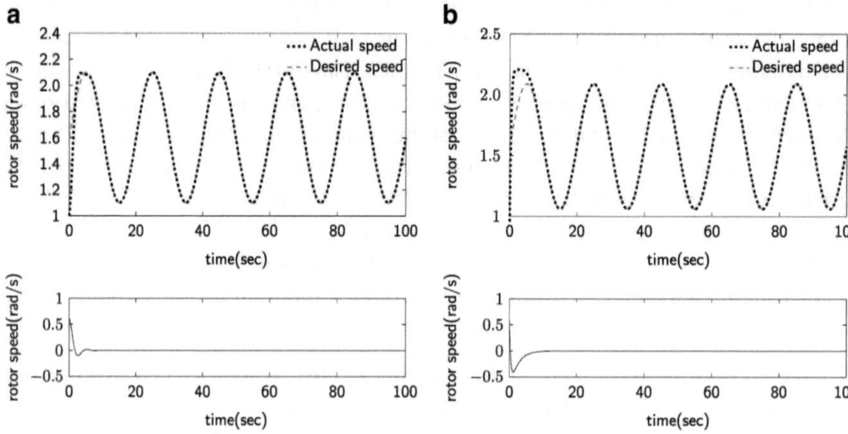

Fig. 4 Variable speed control simulation of the two-mass drive-train model. (**a**) Rotor speed tracking with state feedback control. (**b**) Rotor speed tracking with output feedback control

5 Conclusions

Variable speed control of wind turbines is one of the effective ways to increase the efficiency of wind power generation. In this paper, modeling and asymptotic tracking control for large and flexible wind turbines have been investigated. By taking into account certain flexible structural modes of wind turbines, we modeled the wind turbine system as a four-mass drive-train model and the dynamic equations of the system were derived from the Lagrange's equation. Consequently, a reduced-order, two-mass drive-train model can be deduced from the four-mass drive train model. For the wind turbine described by the two-mass drive train model, we discussed briefly how to design both state and output feedback nonlinear controllers to achieve global asymptotic tracking for the rotor speed. Analysis and simulations have demonstrated that the proposed nonlinear control schemes are capable of achieving satisfactory rotor speed tracking. The variable speed control of wind turbines represented by the higher-order nonlinear model, such as the three-mass or four-mass drive-train model, is under investigation currently. The work reported in this paper is indeed a good starting point for such investigations, and is expected to be extended to certain higher-order drive-train models.

Acknowledgements This work was supported in part by 973 Program (2009CB320601), 111 Project (B08015), Shenzhen Key Lab on Wind Energy and Smart Grids (CXB201005250025A), Shenzhen Key Fundamental Research Projects (JC201005260110A, JC201105160551A) and Oversea Talents Innovation fund (KQC201105300002A).

References

Beltran B, Ahmed-Ali T, El Hachemi Benbouzid M (2008) Sliding mode power control of variable-speed wind energy conversion systems. IEEE Trans Energy Convers 23:551–558

Boukhezzar B, Siguerdidjane H (2005) Nonlinear control of variable speed wind turbines without speed measurement. In: 44th IEEE conference on decision and control, and the European control conference 2005, Seville, 12–15 Dec 2005, pp 3456–3461

Burton T, Sharpe D, Jenkins N, Bossanyi E (2001) Wind energy: handbook. Wiley, New York

Garcia-Sanz M, Elso J (2009) Dynamic modeling of wind turbines. A handout of Department of EECS, Case Western Reserve University

Heier S, Waddington R (1998) Grid integration of wind energy conversion systems. Wiley, London

Hinrichsen EN, Nolan PJ (1982) Dynamics and stability of wind turbine generators. IEEE Trans Power Apparatus Syst PAS-101:2640–2648

Isidori A (1995) Nonlinear control systems, 3rd edn. Springer, New York

Johnson KE, Pao LY, Balas MJ, Fingersh LJ (2008) Control of variable-speed wind turbines: standard and adaptive techniques for maximizing energy capture. IEEE Control Syst Mag 26:70–81

Jonkman JM, Buhl ML Jr (2005) FAST user's guide, NREL/EL-500-29798. National Renewable Energy Laboratory, Golden

Lin W, Gong Q (2003) A remark on partial state feedback stabilization of cascade systems by small-gain theorem. IEEE Trans Autom Control 48:497–500

Lu Z, Lin W (2011) Asymptotic tracking control of variable-speed wind turbines. In: Proceedings of the 18th world congress, IFAC, Milano, 28 Aug–2 Sep 2011, pp 8457–8462

Papathanassiou SA, Papadopoulos MP (1999) Dynamic behavior of variable speed wind turbines under stochastic wind. IEEE Trans Energy Convers 14:1617–1623

Papathanassiou SA, Papadopoulos MP (2001) Mechanical stresses in fixed-speed wind turbines due to network disturbances. IEEE Trans Energy Convers 16:361–367

Perdana A (2008) Dynamic models of wind turbines: a contribution towards the establishment of standardized models of wind turbines for power system stability studies. PhD thesis, Chalmers University of Technology

Qian C, Lin W (2003) Output feedback control of a class of nonlinear systems: a non-separation principle diagram. IEEE Trans Autom Control 47:1710–1715

Ramtharan G, Jenkins N, Anaya-Lara O, Bossanyi E (2007) Influence of rotor structural dynamics representations on the electrical transient performance of FSIG and DFIG wind turbines. Wind Energy 10:293–301

Salman SK, Teo ALJ (2003) Windmill modeling consideration and factors influencing the stability of a grid-connected wind power-based embedded generator. IEEE Trans Power Syst 18:793–802

Sivrioglu S, Ozbay U, Zergeroglu E (2008) Variable speed control of wind turbine: a robust backstepping approach. In: Proceedings of the 17th world congress, IFAC, Seoul, 6–11 July 2008, pp 1183–1188

Song YD, Dhinakaran B, Bao XY (2000) Variable speed control of wind turbines using nonlinear and adaptive algorithms. J Wind Energy Ind Aerodyn 85:293–308

Thomsen SC, Poulsen NK (2007) A disturbance decoupling nonlinear control law for variable speed wind turbines. In: IEEE mediterranean confrence on control and automation. doi:10.1109/MED.2007.4433869

Trudnowski DJ, Gentile A, Khan JM, Petritz EM (2004) Fixed-speed wind-generator and wind-park modeling for transient stability studies. IEEE Trans Power Syst 19:1911–1917

Wang C, Weiss G (2009) Stability analysis of the drive-train of a wind turbine with quadratic torque control. Int J Robust and Nonlinear Control 19:1886–1895

Sustainable Design for Campus Residential Housing

Smita Rakshit and Anthony Filipovitch

Abstract Alarmed by continuing natural disasters around the world every industry is adopting sustainable practices to lessen the environmental damage to the planet. Academic institutions are also making efforts to make the campus sustainable and environment-friendly. Along with other academic sectors, campus housing is considered an important element of enhancing the academic experience. Although in recent times campus housing has focused on energy efficiency, recycling and using 'green' materials, only a limited number of universities are incorporating the design principles of new urbanism to create mixed-use residential areas for students. Mixed-use developments are focused on increased density, pedestrian-friendly environment, reduced ground cover accompanied by other design principles. This essay considers the advantages of adopting new urbanist principles for residential housing in higher education institutions.

1 Introduction

Alarmed by continuing natural disasters around the world, planning professionals have raised their voices to encourage environmentally friendly development. Every industry is adopting sustainable practices to lessen the environmental damage to the planet. This has been reflected in profiles that include endorsements such as LEED certification and energy star ratings. Colleges and universities have not stood apart from these efforts. They play an important role preparing students to be future leaders in environmental awareness. Campus housing is a vital element for enriching the experience, not just of the residential students who live there but of everyone who comes on a campus. Although many institutions have already adopted some

S. Rakshit • A. Filipovitch (✉)
Minnesota State University, Urban & Regional Studies Institute, Mankato, USA
e-mail: Anthony.Filipovitch@mnsu.edu

S.Th. Rassia and P.M. Pardalos (eds.), *Cities for Smart Environmental and Energy Futures*, Energy Systems, DOI 10.1007/978-3-642-37661-0_16,

measures to develop environment-friendly housing for students, there is ample scope for further work in this area. To this point, most of the efforts toward sustainable student housing has focused on recycling, using recycled and environmentally responsive materials, installing 'green' building components, and reducing energy use. But only a few institutions have thought "outside the box" (of individual residence halls) and applied the principles of new urbanism to create mixed-use residential areas for students. The principles of new urbanism are a modern trend guiding the development of built space by addressing issues such as ground-cover, increased density, and transit and pedestrian oriented development. This essay considers the advantages of adopting new urbanist principles for residential housing in higher education institutions.

2 Organizations Promoting Sustainability on Campus

Following the publication of *Our Common Future* by the Brundtland Commission (1987), 20 university presidents and chancellors formed an association to carry the principles of environmental sustainability into the academic environment. The result was the formation of the Association of University Leaders for a Sustainable Future (ULSF), which issued the Talloires Declaration in 1990. The ULSF grew to more than 350 members 20 years later (ULSF 2012). The Declaration is a ten-point call to action and includes such principles as "practicing institutional ecology," "creating an institutional culture of ecology," and "educating for environmentally responsible citizenship." In the United States, the American College and University Presidents' Climate Commitment (ACUPCC) in 5 years has gathered commitments from more than 650 higher education institutions to develop plans to reduce greenhouse gas emissions and, in the long run, to achieve climate neutrality (ACUPCC 2012). They also commit to publishing their action plan, inventory, and progress reports on the web-based ACUPCC Reporting System.

Campus housing has the potential to be an important element of an educational institution's contribution to institutional sustainability. The Association of College and University Housing Officers- International (ACUHO-I) has declared that sustainability is "deeply embedded expectative for campus operations and buildings" (Torres-Antonini and Dunkel 2009). To address this, a growing number of educational institutions are adopting environment-friendly residences which reflect the institution's commitment to encouraging campus sustainability. The Society for College and University Planning (SCUP) organizes webcasts and seminars to help promote sustainability in institutions (ACUHO-SCUP 2008). ACUHO-I and SCUP has organized webcast on December 3, 2008 on Trends I Campus Housing: Data and Core Concepts from Design Innovations. This allowed listening to site via telephone and web during the question and answers period (ACUHO-SCUP 2008). The presenters included Planner Sue Gott from University of Michigan, Cynthia Parish Bologh, Principal investigator ABUHO-I Construction and Renovation Survey, and Jim Curtin, principal architect of Solomon Cordwell Buez Architects.

3 Traditional Campus Housing

It was not until nineteenth century that the importance of campus housing was recognized by American colleges. Although Oxford and Cambridge long had a residential college structure, it was Thomas Jefferson who designed the University of Virginia as an "academical village." Soon other university presidents began to observe the educational and political advantages of campus housing (Dober 1996). By the 1950s, the impact of low-quality student housing on the ability to attract students was seen as an educational crisis. The Higher Education Facilities Act of 1963 and subsequent legislations allowed educational institutions to construct buildings as per their requirements. But often campus housing was compelled to compromise program objectives and design quality due to inadequate financing (Dober 1996). This led to construction of typical high-density multi-storied buildings with limited amenities. Most of them have double-loaded corridors connecting single or double rooms. Dober describes this as an architectural mistake that needed to be addressed. For example, Washington State University was compelled to build 6–8-story high-density shelter-model dormitories, most of them with no design relationship to other campus buildings, and were placed along the perimeter of the university. Some institutions labored under this problem for 30 or more years of limited maintenance and an administration concerned about filling the structures.

In reaction to this situation, institutions began constructing better-planned housing for students on their campuses in the 1980s. But they did this by focusing more on rehabilitation and improvement of existing buildings and sites. For example, Harvard spent $75 million, Brown University spent $35 million and University of Connecticut spent $25 million to rehabilitate and modify existing structures (Dober 1996). In the mid-1990s universities began to realize the importance of campus housing as an important academic experience. Campuses such as University of Miami, Florida were making efforts to make the campus housing more domestic in scale by reducing density and eschewing barrack-style housing. Energy conservation, ADA (American Disability Act) compliance, and safety and security of the housing were also being addressed by the universities in this era.

Some students chose to live off-campus to take advantage of greater variety in housing choice and the perception that off-campus housing would be cheaper than on-campus housing. As a result, housing officials now are focusing on diversifying the choices in dormitories to satisfy a wider range of students. For example, Cabrini College designed its student housing to look like single-family homes (but different in material, plan and siting) and they are close both to an adjacent residential area and to the campus recreational facilities. This permitted the college to rent space to the community should enrollment decline. Kutztown University built an additional wing to its existing 400-bed dormitory, a 20-bed unit that looks like a house. Harvard University converted a motel into a law school dormitory and Massachusetts Institute of Technology transformed an industrial building into a graduate student housing.

Dober, in his book *Campus Architecture* (1996), categorizes current campus housing as falling into three models:

3.1 Shelter Model

The baseline "shelter model" is a unit that is less than 150 sq. ft. of area per student with limited amenity and minimal space for social gathering.

3.2 Campus-Life Model

A campus-life model is based on units of 150–200 sq. ft. per students with amenities like laundry, snack bar, game-room, fitness center, isolated group-study spaces etc. and a few spaces for social gathering.

3.3 Academic Model

The academic model is based on larger units of 200 sq. ft. per student with all the amenities and a combination of formal and informal gathering spaces for students. It may also have space for faculty in residence, tutor offices, a library, and multi-purpose spaces. This model is particularly popular for graduate students and executives who enroll for short-term training.

Dober also notes that smaller cafeterias with or without outdoor sitting space are becoming popular since they provide not only greater dining choices but also create opportunities for informal learning through social interaction.

4 Categories of Sustainable Campus Housing

While Dober describes shifts in campus housing due to market pressures and pedagogical objectives, Maruja Torres-Antonini and Norbert W. Dunkel (2009) look at changes in design and use of campus housing in response to environmental and sustainability concerns. They studied 87 self-identified and self-reported sustainable campus housing initiatives created by colleges and universities in the United States. They identified three categories of sustainable campus housing:

4.1 *Green Campus Housing*

These are the energy-efficient housing structures that uses energy, water and materials most efficiently and ensures elimination of negative impacts on environment throughout the life-cycle of the structure. They have confirmed the quality of their efforts through LEED certification, energy star ratings, and other environmental standards. These campuses are focused on the environmental impacts of their structures and do not necessarily have an educational program focused on sustainability.

4.2 *Sustainability-Themed Living-Learning Communities*

These are the communities that meet the requirements of "residential learning communities" – "a residential education unit in a college or university that is organized on the basis of an academic theme or approach and is intended to integrated academic learning and community living" (Midden 2008). In this case, the learning communities are focused on issues of sustainability, both as learned in concept and as lived in practice. They may or may not be associated with formal academic programs, but they encourage environment-friendly lifestyle.

4.3 *Campus Housing Sustainability Hubs*

These are the residential complexes that combine the characteristics of both green campus housing and sustainability-themed living-learning communities. They provides ample opportunity to experience sustainability principles and lifestyle in their living environment. They also encourage interaction and participation in the community to obtain knowledge and results in positive attitudinal and affective change (Cross 1998).

This categorization marks a transition to an era of campus housing that is environment-friendly and sustainable. Many universities are also trying to integrate New Urbanist principles to achieve highest performance.

5 Principles of New Urbanism

In 1993, a group of architects came together to form the Congress for New Urbanism (CNU). Their goal was to counter the post-World War II emphasis on sprawl and low density with high quality design that emphasized neighborhood, community, pedestrian-oriented places that conserved the natural and built environments (CNU 2012). One of the definitive works of this movement was Peter Katz's

The New Urbanism (1993). While the CNU Charter lists 29 principles, they have been consolidated into 10 main principles, 6 based on design, 3 expressing the implications of the design principles (increased density, green transportation, and sustainability), and the final one—quality of life—which is the net effect of the 9 previous principles. The six design principles (New Urbanism 2012) are:

5.1 Walkability

For a place to be walkable, most of the necessities for daily living should be within a 10-min. walking radius (1,000–1,200 yards; about 1,000 m). Designs should be "pedestrian friendly."

5.2 Connectivity

The space should be stitched with an interconnected street grid, with a hierarchy of roads from boulevards to streets to alleys/walkways.

5.3 Mixed-Use and Diversity

Within the neighborhood, there should be a mix of shops, offices, apartments, and houses.

5.4 Mixed Housing

The mix of housing in a neighborhood should include a range of sizes and prices to encourage a diverse mix of people living in and using the neighborhood.

5.5 Quality Architecture and Urban Design

The physical space should generate a sense of place and a feeling of beauty.

5.6 Traditional Neighborhood Structure

The space should be designed so there is a recognizable center and edges, with public space at the center and a range of densities within a 10-min walk. The highest densities should be toward the center, and natural habitats should be connected and integrated into all of the spaces.

6 "New Urbanist" Campus Residences

Federally funded Urban Renewal projects in the United States in the 1960s resulted in sprawl in suburban areas, depopulation in the city cores, and blight in traditional cities. The Federal government also funded highways nationwide that drove people out of the city cores and encouraged them to commute long distances every day to work from home.

Yet some universities in these older cities are successfully adopting new urbanist principles for student housing. Examples include Ohio State University (Burnett 2005), University of Pennsylvania (Allen 2002), and Washington State University, St. Louis (Madsen 2001). They are at the forefront of rethinking campus housing for students.

6.1 Ohio State University-South Campus Gateway

Ohio State University has inititated a mixed-use residence for students to revitalize the areas around the perimeter of the campus. 'High Street,' which is an imortant street on the east edge of the campus, was dilapidated along with the surrounding area. This in turn led students move further away to find 'safe' off-campus housing. Desiring to bring the students back close to the university, the university decided to create a mixed-use development called South Campus Gateway to spruce up an undesireable neighborhood. Master Planner David Dixon in 2005 recognized the problem and decided to go for public-private partnership to build the proposed mixed-use development.

The project extends to four blocks on each side of the High Street. Designed by renowned architectural firm Elkus Manfredi, it is comprised of 890,000 sq. ft. of which 580,000 sq. ft. are devoted to residential space for graduate students, faculty and visiting professors; retail, and entertainment space and a parking structure account for the remaining 310,000 sq. ft (Manfredi 2012a,b). It provides 184 apartments dedicated to the students and affiliates, 12 restaurants, one 8-screen cinema, a university bookstore, and a 1,200-car parking garage to support retail businesses (Martin and Allen 2009).

The university's Campus Partner's president Terry Foegler described this project as a 'Signature Project' for revitalizing the campus area and the corporation invested $151 million for this redevelopment. The president also indicated that shortly after the initiation of the project, it was successful in attracting students and (perhaps more importantly) retail businesses in an area surrounding the university that had been facing decline (Wolf 2005). A number of both local and chain restaurants with diversified flavor are now in operation. Student enrollment in the university has gone up and a large number of those students have chosen South Campus Gateway for housing. One of the most important consequences is that the project has improved safety in the area (Gebolys 2010).

Ohio is an outstanding example of incorporating new urbanism design concepts for campus housing and successfully implementing it. It has been able to attract a number or retail businesses, both small and large, within a confined area of four blocks. This implies that the students living in that area are getting a number of facilities within 10 min walking distance. All the restaurants have included outdoor seating and one 2-story restaurant even has a balcony that enhances the indoor-outdoor relationship (Wolf 2005). One of the main objectives of the project has been to create a center of activity with diverse uses in the area . South Campus Gateway is an exemplary redevelopment project that successfully executes new urbanism principles.

6.2 University of Pennsylvania-Sansom Commons

The University of Pennsylvania in Philadelphia was founded by Benjamin Franklin in 1749 and is located in an inner-city neighborhood of West Philadelphia, known as 'University City'. Although one of the most prestigious institutions in the United States, the University of Pennsylvania experienced decline in student population and deterioration in its surrounding neighborhood beginning in the 1950s through 2000 (Rodin 2005). Alarmed by the worsening situation, the university began West Philadelphia initiatives beginning in 1994. The principle strategies were:

- Stimulating the housing market
- Clean, safe and attractive neighborhood
- Attracting retail development
- Encourage economic development
- Improving public schools

The Sansom Commons was developed as part of a strategy for attracting retail development in the area adjacent to the university. The 300,000 sq. ft. project encompasses six city blocks and includes 37,000 sq. ft of retail and restaurants, a 190,000 sq. ft. hotel which is named as The Inn, and a 56,000 sq. ft. Penn Bookstore. At the center of the project is a public square, developed as a vital connection between the campus and the neighborhood (Burnett 2005).

The university has played the role of both a developer and subsidizer, stimulating the housing market by renovating abondoned properties in that specific neighborhood and selling them to public (Rodin 2005). The targeted retail activities were apparel, groceries, dining and entertainment. Both local and chain retailers are running their businesses profitably. The success of Sansom Commons have led to $370 million of private investment in West Philadelphia including a mixed-use complex of 282 market-rate apartments for mixed-income people including students, new retail businesses and banks (Rodin 2005).

University of Pennsylvania's Sansom Commons is also an example of how new urbanist design principles can transform a deteriorated area into an urban center. The project has encouraged pedestrians to walk around in the area in a safe, healthy environment. It has successfully attracted students, visitors and residents with its retail and cultural amenities. the design also incorporates Fredrick Law Olmsted's idea of integrating public space with retail space to create enjoyable streets and provide opprotunities for diversified people to come together (Rodin 2005).

6.3 Washington University in St. Louis-South 40 Village

South 40 Village of Washington University in St. Louis is another example that has recognized the advantages of mixed-use development and brought together uses such as residence halls, new food services, retail shops, auditorium, student activity space and a fitness center. The complex was designed by Mackey Mitchell Architects and has created an area incorporating all the residence halls in the university. The complex encompasses an area of 40 acre.

The project is located at the intersection of two major streets of Washington University's residential community. Mackey Mitchell architects describe the project as creating a European-style streetscape that would encourage a pedestrian-friendly environment. The pedestrain 'spine' connects the South 40 residential campus to the central academic area of the university. The residence halls are arranged on both sides of this spine. Lower levels of these halls consist of uses such as café, retail businesses, shops and entertainment with outdoor dining areas. The upper stories are dedicated for students' residences.

A significant feature of this complex is the green roof over the commissary kitchen and the loading dock. The 7,500 sq. ft. roof enhances the quality of the area (Clayprop Construction 2012). The project is LEED certified. The architects were aiming to create a space that would encourage social interaction and group gathering (Madsen 2001).

South 40 Village is an attempt to incorporate New Urbanist design principles for the university's redevelopment project. The project created a link between the academic and the residential parts of the univesity. The outdoor spaces have successfully increased the communication among the students and have encouraged pedestrians to use the link.

7 Conclusion

Mixed-use, new urbanist development for campus residential housing is an emerging trend, and an area that is ripe for further research. There are only a few articles published on mixed-use campus residential housing, and no rigorous assessment of their success in meeting their intended outcomes. Regardless, more and more universities are starting to realize its advantages and are adopting new urbanism principles for campus residential housing. Several university-owned mixed-use housing projects are under construction as this review is being written. For example University of Southern California's University Park redevelopment project will include student housing, retail, hotel and academic spaces, and has already been approved by City Council (University of Southern California 2012). The University of Nebraska has approved a mixed-use development in which parking will be owned by university and student housing and retail space will be owned by the private sector (UNL 2012). In short, new urbanist mixed-use developments for campus residential housing, along with energy conservation and recycling, is an emerging sustainability trend for university campuses.

References

ACUHO-SCUP (2008) Trends in campus housing data and core concepts from design innovations. http://www.scup.org/page/events/online_program/2008/trends-in-campus-housing-december. Accessed 25 Nov 2012

ACUPCC (2012) Implementation guide. American College & University Presidents' Climate Commitment, Boston. http://www.presidentsclimatecommitment.org/. Accessed 1 Dec 2012

Allen PV (2002) Penn extends retail push. Phila Bus J 21:27

Brundtland GH (1987) Report of the world commission on environment and development: our common future. United Nations, New York. http://www.un-documents.net/our-common-future.pdf. Accessed 1 Dec 2012

Burnett L (2005) All mixed up. Contract 47(3):48

Clayprop Construction (2012) South 40 village at Washington University. http://www.claycorp.com/p/264/washington-university-south-campus-redevelopment/. Accessed 2 Dec 2012

CNU (2012) Charter. http://www.cnu.org/charter. Accessed 12 Jan 12

Cross KP (1998) Why now? Why learning communities? About Campus 3(3):4–11

Dober RP (1996) Campus architecture: building in the groves of academe. McGraw-Hill, New York

Gebolys D (2010) National Planning Excellence Award for completion. Am Plan Assoc 76(4):31

Madsen J (2001). A space divided: newest addition at Washington University campus offers contrasts in style. Buildings. Gale Group, Farmington Hills, Michigan

Manfredi E (2012a) The Inn at Penn. http://www.elkus-manfredi.com/upload/attachment/INNPENN_Broch.pdf. Accessed 27 Nov 2012

Manfredi E (2012b) THe Ohio State University South Campus Gateway. http://www.elkus-manfredi.com/upload/attachment/OHIOST_Broch.pdf. Accessed 27 Nov 2012

Martin J and Allen M (2009) Students in my backyard: housing at the campus edge and other emerging trends in residential development. Plann High Educ 37(2):43–44

Midden R (2008) The residential learning communities international clearinghouse. http://pcc.bgsu.edu/rlcch/index.php. Accessed 12 Jan 2012

Peter K (1993) The new urbanism: toward an architecture of community. McGraw-Hill, New York

Rodin J (2005) The 21st century urban university: new roles for practice and research. J Am Plann Assoc 71(3):237–249

Torres-Antonini M, Dunkel NW (2009) Green residence halls are here: current trends in sustainable campus housing. Journal of College and University Student Housing. Volume 36, No. 1. April/May 2009

ULSF (2012) Association of university leaders for a sustainable future. http://www.ulsf.org/index.html. Accessed 1 Dec 2012

University of Southern California (2012) Village at USC. http://village.usc.edu/. Accessed 11 Dec 2012
UNL (2012) Regents approve mixed-use parking/ housing facility. http://newsroom.unl.edu/announce/todayatunl/1764/9883. Accessed 11 Dec 2012
Wolf B (2005) South Campus Gateway rapidly approaches completion. Real Estate Forum 60:6

House Biographies: Housing Studies on the Smallest Urban Scale

Eveline Althaus and Marie Antoinette Glaser

Abstract This article discusses 'house biographies' as a research approach for analyzing the dynamics of residential buildings' built and lived space over time. Focusing on the smallest urban unit – the house within its local, historical, social and cultural context - is not only a way of researching the qualities and problems of a specific urban residential setting but can also make a fruitful contribution to a more sustainable planning that considers both technical and social issues. Thinking about future cities requires one to take into account what makes some buildings, including their premises, endure and which factors allow them to be appreciated over a long period. Doing 'house biographies' enables one to gain such differentiated knowledge, taking the small scale as a starting point for higher-level thoughts and actions. By placing the emphasis on methodological principles as well as applied research methods, this article provides an insight into the logics and ways of doing a house biography.

Buildings that endure have a history of decisions and actions taken, uses made and people involved throughout their career of alternating periods of crisis and consolidation. Buildings that endure are interesting objects of study because they carry multiple traces of life over time: stories, conflicts, myths, images, as well as objects, documents, repair work, additions, and so forth.

Enduring buildings can be sustainable because they persist over time; in a sense, they are sustainable because of their potential for durability. It is precisely these perspectives of sustainability that could be advantageous to planners and investors who have long-term objectives related to housing. If a building is loved, it will be cared for. If it is cared for continually, it will endure.

E. Althaus • M.A. Glaser (✉)
ETH CASE, Centre for Research on Architecture, Society & the Built Environment, Zurich, Switzerland
e-mail: glaser@arch.ethz.ch

S.Th. Rassia and P.M. Pardalos (eds.), *Cities for Smart Environmental and Energy Futures*, Energy Systems, DOI 10.1007/978-3-642-37661-0_17,

In a current research project,[1] we aim to identify the processes, structures and practices that lead certain residential buildings to successfully endure over time, from their erection until present. In this article, we discuss the analytical approach of the 'house biography' that was developed by our interdisciplinary team. To understand this process one has to examine the physical, the cultural as well as the social qualities of the enduring building within its environment. This means asking about not only the conditions of construction but also the cultural context, ideals, concepts, and how the building is dealt with from a practical standpoint (by the residents, the owner, the manager and the public). In this way, a new, more integrated view of sustainability emerges that is neither purely eco-technically oriented nor greatly bound to economics. When thinking about the qualities of a place, it is important to apply a dynamic perspective: our built environment has a history, consisting of manifold stories, memories and narratives of people.

To make use of the term sustainability, which threatens to become both enigmatic and empty due to ubiquitous use, we particularly focus on the level of the social and cultural dimension, which refers to social values, norms and ideas (Hugentobler et al. 1997; Hugentobler and Gysi 2000). A house, then, is seen as sustainable when it is appreciated and loved for a long period of time by the respective tenants or when it is bequeathed to subsequent generations until it enters the cultural memory of a society. A house is durable if it supports existing values and ideals and can at the same time be integrated into the following generations' ideals.

This article reflects upon ways of analysing the career or trajectory[2] of a house over time: in short, on doing house biographies. To our understanding, a house biography is a research instrument that analyses the qualities and conflicts as well as the dynamic development of the lived and built space of a house or a housing complex over time (Glaser 2011). Therefore, it focuses on the smallest urban scale: on the house. It is the unit in the middle between the large space: street, neighbourhood, town, countryside, and the smaller unit: flat, room, interior (Schlögel 2009). We use the term 'house' in this research in a more generic way, that is to say that house refers not only to the 'built space', i.e., the material and construction of the structure, but also to the cultural and historical dimension of the 'lived space'. The latter includes the way that people treat the built space: use, appropriation, relocation, modification, the tactile and optical perception, the appreciation and emotions, the conceptual and planning related discussions. The spatial term 'house' does not just comprise the incidents within the property lines where a residential building has been erected; the residential environment, with its

[1]At the ETH CASE – ETH Wohnforum, an on-going research project analyzes the house biographies of large residential complexes built in Switzerland between 1960 and 1980. The project is funded by the Swiss National Science Foundation and runs from 2011 to 2014. The primary investigator is Dietmar Eberle, researchers are Marie A. Glaser, Eveline Althaus and Claudia Mühlebach.

[2]The term 'trajectory' was introduced by Anselm Strauss to delineate social processes that are shaped by interaction (Strauss 1991).

infrastructure and its social and spatial aspects, with which the residential building and its residents are in a relationship, is part of this.

1 House Biographies and Theories of Space

The term 'house biography' was originally proposed by Johann Friedrich Geist in his fundamental monograph on the 'Berliner Mietshaus' (*Berlin rental house*) in which he studied the history of the typical urban block apartment houses of early twentieth century Berlin, focusing especially on their social history and everyday culture (Geist and Kürvers 1984).[3] Alison Blunt, used the term 'house biography' in a case study on Christadora House, a 'skyscraper settlement' in New York City, to "investigate the relationships between the built form of a settlement house and new forms of residence within it", and to show that the building itself "was not only shaped by, but also recast, embodied practices of settlement, inhabitation and domestic life" (Blunt 2008, p. 550).

Uncovering the interplay of built space, social life and action is crucial to understanding the biography of a house. We use the term 'biography' to stress that we think of a house – or a housing complex – not only as a built structure but also as a system that is full of life, a system that is dynamic and performative (see also Jacobs et al. 2012), a system that is connected to everyday experiences and narratives of people (see also Gieryn 2002; Jenkins 2002). Furthermore, a house is not an isolated entity but should be seen in relation with its urban environment and general social trends, as a building on its site with a complex differentiation of inner and outer, (semi-)public and private spaces.

A house biography is interested in both people *and* their spatial settings, in practices (action) and structure, in the visible and the invisible, in permanence and change. There are multiple forces that keep a house together or let it fall apart over the years (Jacobs et al. 2007). The aim of the house biography approach is to trace back over time the various dimensions that contribute to the durability, quality, appreciation and long-term functioning of a house. What are the material components of a house, its construction and facilities? How do people experience a specific dwelling environment in their everyday life? How do they perceive it? How do they deal with, appreciate, modify, appropriate it? How do they create an atmosphere of home? Which management and maintenance strategies are applied? How do people remember the stories of a house and which future(s) do they

[3]Since it has proven difficult to work with purely historical sources, Geist's method has been supplemented by an empirical approach. Where Geist remains historically oriented on account of the lack of people to discuss with, the opportunity arises here to establish contact with residents, owners and management and to explore their uses and experience of the housing complex (Glaser et al. 2010).

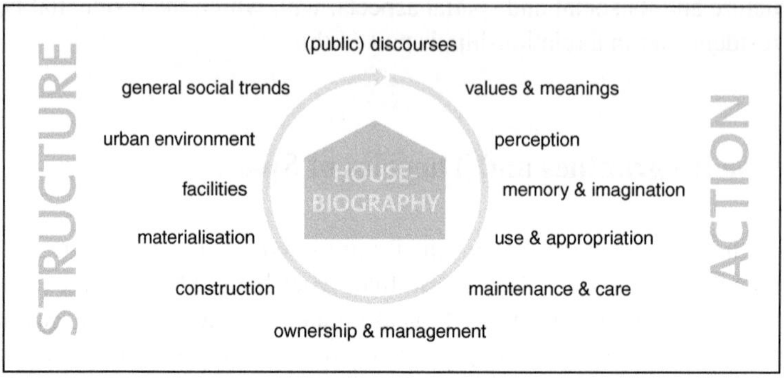

Fig. 1 Multi-dimensional approach of a house biography

imagine? What are the public and scientific discourses and which meanings and values do they portray? (Fig. 1)

Two theories of space mainly inspire this research approach. One is Henri Lefebvre's (1974) thinking of space as socially produced in an interplay of the triad 'perceived', 'conceived' and 'lived space'. Lefebvre's 'perceived space' refers to the conditions and material structures that are (re-)produced through everyday practice. This aspect might be seen for instance in the various forms of everyday perceptions that are linked to a house, but also in common routines, actions and behaviours that are taking place. The second aspect, the 'conceived space' refers to the cognitive and theoretical discourses and knowledge systems of space, that in Lefebvre's thinking always go along with power relations. Here, concepts and models of architects, scientists, experts and the media come into play. They determine how a house should be read and understood, for example by defining its built structure, by evaluating the condition of its solidity and ageing, but also by reviewing specific dwelling qualities or the social mix of its inhabitants. Lefebvre's third aspect, the 'lived space' complements the other two aspects, referring to the symbolic level and imaginary expressions, that have the (subversive) potential to give to a space other meanings and visions than the prevailing discourse and ruling order suggest. In our case, this aspect can be seen, for instance, in the way people appropriate their house and dwelling environment; how they modify, decorate and take care of it; which personal signs and individual traces they leave in public or semi-public places and so forth. Lefebvre's theory has to be read and understood in its context and remains in some points ambiguous, yet he deserves the credit for having opened up new ways of analysing space, by emphasizing the interplay of social and cultural processes that lead to its production (see Schmid 2005; Steets 2008). A more recent, very fruitful approach to analysing the social constitution of space is the relational theory of space by Martina Löw (2001, 2008). With reference to Anthony Giddens's theory of structuration (Giddens 1984), Löw conceives space in the duality of structure *and* action. As such she consequently thinks of space as a relational ordering and denies the idea of space as a container that Lefebvre also criticized but could not always avoid. Following Löw, to understand space you need to consider, on the one hand, the spatial configurations and the ways in which they are produced, the processes of 'spacing' as she calls it; on the other hand, she points out that space must be actively connected by people through processes of perception, imagination or memory. Space is constituted, as Löw points out, through such activities of 'synthesis'. Very inspiring in Löw's approach is her process-related thinking: she conceives of space as a relational ordering of bodies that are constantly in movement. And through this movement, the ordering itself is subject to change. This approach enables us to look at the transformations of space, which is also important for a house biography that tries to think of space in its temporal dimension and to tell the story of a house over time.

When reflecting upon the house biography approach in terms of theory, we could say that the particular spatial character of a house or a housing estate is produced and reproduced through the actions (or practices, in Bourdieu's language) of a variety of people. Architects, city planners and building companies defined and conceived where and how a house took shape; their thinking was embedded in the logic of their time and institutional background, for instance providing good quality dwelling space or building at low costs and making profit or installing recreational facilities to promote a lively community. Owners and property managers decided what was to happen with the ageing buildings, whose architecture and structure were not necessarily fitting current expectations and standards. Over time, residents move in and out and position themselves in various and changing constellations in a house. The way they use or appropriate their dwelling environment, but also the way they are perceived in public discourse, has a feedback effect on the spatial character of the place.

2 Doing House Biographies: Methodological Reflection

In short, a house biography is a dense portrait of a house, in which architectural, historic and ethnographic analyses are brought together (Glaser et al. 2010). In order to guarantee such a multi-disciplinary synopsis, researchers from architecture/design and the humanities (cultural anthropology/sociology and history) should collaborate in a team (Glaser 2011).

A mixed-method approach is applied: according to *historical* research procedures, research in archives and analysis of historical documents; following *ethnographic* methodology, site observations, qualitative interviews, commented walks and visual documentations[4]; and corresponding *architectural* investigations, analysis of construction plans, layouts and concepts as well as of the materialisation

[4]Our on-going research undertakes several case studies simultaneously and therefore cannot apply an extensive ethnography, which would require spending a long period in the field. This project works more with the breadth of material collection than in-depth.

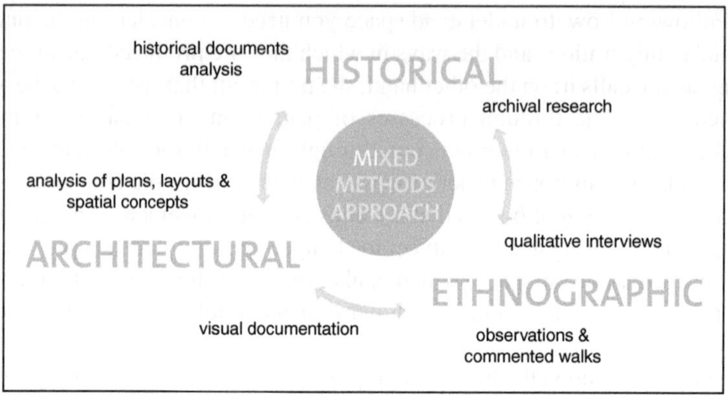

Fig. 2 Methodology sketch

of a house (Fig. 2). With this mix of methods we are trying to spot the stories of use, appropriation and appreciation of a house from various perspectives over time. Specifically, the perspectives of residents, property owners, property managers, architects and the general public are discussed.

Carrying out house biographies requires critical (self-)reflection. Since researchers bring along their personal housing experiences and their own notions of 'house' and 'home', it is important to become aware of our own subjective attitudes and assumptions, to question our prejudices, to discuss our housing practices and our ways of seeing and perceiving flats, houses, settlements and urban neighbourhoods. To enhance this process it is crucial to go on site, to walk a place and 'dive' into it, to let it 'speak' to you, to make notes and photographs and to reflect on them critically. The collaboration and discussion in an interdisciplinary research team is very fruitful for doing so.

Research in archives is a way to understand how the knowledge of a housing estate is historicized. When conducting house biographies, the knowledge of existing literature and professional archives often is limited. More material can be found in the archives of property owners and management, material such as plans, renovation concepts, documents on the on-going maintenance work, bills and accounting material, letters and reports from various people (authorities, residents, social workers, etc.), personal notes of the manager, residents' dossiers, press releases, media reports and so on. It is, however, crucial to be aware that this material can reflect the selection criteria and archiving practices of the different managers over the years. We found, for example, in an archive on a housing complex belonging to the city of Zurich, large amounts of data on a conflict in which the former property manager was involved in the 1990s. Studying the archival data that was left by him, we were sure of the importance of this issue for the biography of the house; however, when talking to the people who are currently in charge and when interviewing residents that moved in later, we realized that memory is short and this story is hardly remembered today. Integrating ethnographic methods, like interviews, field visits and observations, therefore, also has an important corrective potential to put issues and events into perspective. When researching the biography of a house, it is critical to talk to people and to listen to their stories. A very important person on the site of a housing complex for instance, is the concierge ('Hauswart'). He is the binding element between property management and the residents. He (or occasionally also she) usually knows a lot about the housing building or complex, not only the technical functioning but also the everyday stories, gossip and conflicts among residents as well as the communal spaces and the manifold specificities and qualities of the house (Strebel 2011).

Semi-structured interviews with different user groups are an important source of knowledge. These situations require sensitivity to carefully build trust in dialogue and to develop techniques to focus on the house and not on an individual biography. When dealing with space it is important not only to consider the narrative but also to work visually. Various methods may serve here: working with video in collaboration with a professional filmmaker[5] as well as working with photography and participant photo-interviews (Kolb 2001), asking residents, for instance, to portray their own house and housing. The choice of the visual approach must account for the specific context of a house or a housing estate as well as the disposition of the people that dwell there or are in charge of it.

3 Conclusions

In this article we discussed the multidisciplinary approach of house biographies in housing studies. The house biographies methodology can also be used for a comparative study of selected case studies. Doing multiple house biographies on buildings of one typology, e.g., large housing complexes built in Switzerland from 1960 to 1980, allows us to draw conclusions about the performance and the perception of that typology. By studying the changing career of crises and peaks of enduring residential complexes, we aim to identify the processes, the structures and the practices that led them to successfully persist over time. They are still in use and well maintained. As our research shows, strategies of maintenance and renting versus mixed ownership can mean as much for durability over time as social networks and an active neighbourhood, as well as the materialisation and renovation cycles of the built structure. The location of the housing site and the development of its immediate urban environment greatly affect the career of the house. Crucial to this approach is the understanding of the complex interweaving of practices, following Lefebvre, of the lived, the conceived and the perceived space that contribute to the biography and have an impact on the lifespan of a building. The synthesis of the multiple biographies in our study shows factors that can support the endurance of a house and keep it from falling apart. The understanding of a house

[5] See http://www.hausbiografien.arch.ethz.ch/

as a multidimensional setting within its local historical, social and cultural context can be one starting point when starting to plan for future cities. To make cities more liveable, we need to reflect upon them from multiple perspectives from the house to the site to the neighbourhood to the city scale. A house biography is an analytical instrument functioning on the smallest urban unit – a house on its site. As residential housing still forms the major part of our cities now and in the future, it provides an approach for analysing the dynamic interplay of the built with its social and cultural dimension and can offer a contribution to a more integrated sustainable planning.

References

Blunt A (2008) The 'skyscraper settlement': home and residence at Christodora House. Environ Plann A 40(3):550–571

Geist JF, Kürvers K (1984) Das Berliner Mietshaus. 1862–1945. Prestel Verlag, München

Giddens A (1984) The constitution of society. Outline of the theory of structuration. Polity Press, Cambridge

Gieryn TF (2002) What buildings do. Theor Soc 31:35–74

Glaser MA (2011) Durability in housing – aesthetic of the ordinary. In: Sang L, TU Delft (eds) Aesthetics of sustainable architecture. 0I0 Publishers, Rotterdam, pp 198–213

Glaser M, Adam A, Joss A (2010) Zur Karriere des Dauerhaften. Hausbiographien wertgeschätzter Wohnungsbauten aus den Jahren 1900 bis zur Gegenwart. Wissenschaftlicher Schlussbericht. ETH Wohnforum, Zürich

Hugentobler M, Gysi S (2000) Sustainable urban development in Zurich. Learning from successful projects. In: Nyströom L (ed) City and culture: cultural processes and urban sustainability. J Hous Built Environ 15(4):306–320

Hugentobler M, Henz A, Gysi S et al (1997) Nachhaltige Stadtentwicklung: ein Evaluations- und Handlungsforschungsprojekt in der Stadt Zürich. Synthesebericht, Zürich

Jacobs JM, Cairns S, Strebel I (2007) 'A tall storey ... but a fact, just the same': the red road highrise as a black box'. Urban Stud 43(11):609–629

Jacobs JM, Cairns S, Strebel I (2012) Doing building work: methods at the interface of geography and architecture. Geogr Res 50(2):126–140

Jenkins L (2002) 11, rue du conservatoire and the permeability of buildings. Space Cult 5(3):222–236

Kolb B (2001) Fotobefragung. Bilder zur Gesundheit. Dissertation an der Fakultät für Sozial- und Humanwissenschaft, Universtität Wien

Lefebvre H (1974) La production de l'espace. Editions Anthropos, Paris

Löw M (2001) Raumsoziologie. Suhrkamp Verlag, Frankfurt am Main

Löw M (2008) The constitution of space: the structuration of spaces through the simultaneity of effect and perception. Eur J Soc Theory 11:25–49

Schlögel K (2009) Im Raume lesen wir die Zeit: über Zivilisationsgeschichte und Geopolitik. Fischer Verlag, Frankfurt am Main

Schmid C (2005) Stadt, Raum und Gesellschaft. Henri Lefebvre und die Theorie der Produktion des Raumes. Franz Steiner Verlag, München

Steets S (2008) Raum & Stadt. In: Baur N et al (eds) Handbuch soziologie. VS Verlag für Sozialwissenschaften, Wiesbaden, pp 391–412

Strauss A (1991) Creating sociological awareness. Collective images and symbolic representations. Transaction Publishers, New Brunswick/London

Strebel I (2011) The living building: towards a geography of maintenance work. Soc Cult Geogr 12(3):243–262

For the Smarter Good of Cities: On Cities, Complexity and Slippages in the Smart City Discourse

Henriette Steiner and Kristin Veel

Abstract The notion of the Smart City describes the city as a system of information and flow, one that, although complex and wayward, can be controlled, manipulated and optimised to increase efficiency in sectors such as transportation infrastructures, health care, etc. This way of thinking takes for granted that there exists something like a common goal of optimisation which would benefit the larger whole of the city and which would make purposeness and meaning come together in the built environment. It thus propagates a rhetoric that echoes modernist visions from the early twentieth century of betterment of culture through technology. What remains to be understood in a cultural-theoretical perspective, however, is what consequences this way of thinking has on the urban cultural level.

This article takes a closer look at the smart city discourse as it is set forward in IBM's online exhibition site *Before Cities Got Smart*. Using strategies of close reading and analysis of the visual material presented on this site, the aim is to illuminate how art, technology and advertisement are brought together to account for and promote the Smart City model. What becomes apparent is the problematic way in which the *complexity* of such environments is dealt with in this discourse. In order to illuminate the deeper urban-cultural significance of the complexities at work in the use of automated technologies in the contemporary city, we turn to French philosopher Gilles Deleuze and his notion of 'societies of control' as well as the artworks of the Canadian artist David Rokeby. The artworks approach the implications of smart city technologies in a different and critical way, one that allows

H. Steiner (✉)
ETH Zurich, Department of Architecture, HIL G 65.1, Wolfgang-Pauli-Strasse 15, CH-8093 Zurich, Switzerland
e-mail: steiner@arch.ethz.ch

K. Veel
University of Copenhagen, Department of Arts and Cultural Studies, Karen Blixensvej, 2300 København S, 21.3.49, Copenhagen, Denmark
e-mail: Veel@cantab.net

S.Th. Rassia and P.M. Pardalos (eds.), *Cities for Smart Environmental and Energy Futures*, Energy Systems, DOI 10.1007/978-3-642-37661-0_18,

for the complexity of these environments to be enhanced rather than glossed over. In this way, we hope to develop a more differentiated cultural-theoretical platform for discussing the Smart City concept and the ideological implications embedded in its use of technology within the context of contemporary urbanism.

The question of the ideal city is as old as urban civilisation itself. So is the question of how to mediate between, on the one hand, the urban reality as it unfolds around us in all its complexity and (potential) disorder and, on the other hand, the vision of a city that is well-structured, clean and free of vice. Think of the Heavenly City of Jerusalem in the Bible as opposed to earthly cities or even to cities of sin such as Sodom and Gomorrah. Think of the built ideal cities of the Renaissance. Or, think of the well-known utopian urban projects from the twentieth century, the tabula rasa Modernism of the 1920s and 1930s as known in the work of the architect Le Corbusier;[1] or the cities of power of the fascist rulers of that century, for instance Berlin as envisioned by Albert Speer for Adolf Hitler.

An example of a current discourse on the urban condition, and one with a similarly all-encompassing vision, is the concept of the Smart City.[2] Although far less pompous, ideological or rigid in its aspirations, the ambition of the Smart City is to create a new kind of urbanity. One that is more beautiful, clean and harmonious than the urban reality as we currently know it. In this discourse, the city is described as a system of information and flow that, though complex and wayward, can be controlled, manipulated and optimised to increase efficiency in sectors such as transportation infrastructures and health care. One concrete suggestion of the Smart City is the self-reflexive computation of streetlights, which are monitored through sensors and which can constantly adapt to the ever-changing traffic flow, thus aiding people's way through the city. Such propositions by the smart city propagandists are not far removed from visions of a more beautiful, seamless and happy society. Disorder is managed and brought under control.

When scrutinising the rhetoric of many of the promotors of Smart City technologies, it becomes apparent that their aim is no less than the betterment of culture as a whole. It is a way of thinking which takes for granted that there exists something like a common goal of optimisation, which could benefit the larger whole of the city and which would make purpose and meaning come together in the built environment. In the Smart City, the relationship of meaning between the daily life-actions of individuals and the larger structures of the city is envisioned in a way that resembles the optimised and seamless workings of a computerised system. This makes us wonder: if urbanity has to do with the city as a site for social interaction, negotiation and the associated, inevitable potential for conflict and misunderstanding, then what kind of *urbanity* does the Smart City as a *built* ideal city want to establish?

---

[1] For an example, see Corbusier (1986).

[2] See the IBM "Smarter Planet Campaign": www.ibm.com/smarterplanet/uk/en/overview/ideas/ (date accessed: 19 December 2012).

The question we want to ask in this chapter is how we can understand and interpret the conceptual foundations on which this discourse rests. We suggest that one approach is to see it as one example in a longer history of attempts to reform, heal or even salvage a perceived imbalance or illness in the concomitant urban predicament[3] – forming a *response* to an existing urban situation. We attempt here to generate a differentiated critical optic on these issues by discussing the Smart City discourse within a wider cultural-theoretic framework. As a conceptual anchoring point for our argument, we will draw on the distinction between the "disciplinary society", as belonging to modernity's architectures of teaching and control, and "societies of control", as characteristic of the period in which we are now living, as proposed by the philosopher Gilles Deleuze (1992). Of particular significance is the way in which *complexity* (understood as the entanglements of networked societies) is approached – is it reduced and made manageable or is it, in fact, expanded? A discussion of the artworks of the Canadian artist David Rokeby will further help to illuminate the use of automated technologies in the contemporary city. In his works, technologies such as facial recognition software in real-life urban space settings are used to problematise technologies of automatisation, particularly in relation to how they may involve systems of control.[4]

By means of this analytical strategy, which operates across the academic disciplines of architecture, art, urban studies, philosophy and cultural theory, we will argue that the aim of the Smart City is to make the world more manageable without adding further complexity, in fact, even to reduce complexity as such. This will lead us to the inherent paradox in such an aim, entailing as it does a glossing over of the particular form of complexity, which the so-called *smart* technologies embody, thus allowing for a naïvely optimistic discourse reminiscent of the modernist examples mentioned above. First, however, we will take a closer look at the Smart City discourse.

1 The Smart City Discourse and Urban Life

Cities today are faced with significant discrepancies when comparing the way in which they are built and organised locally to the fact that their very structure and functionality is hugely dependent on resources that come from elsewhere. Architecture and urban planning operate through a lavish scheme: the built environment is responsible for something like 40 % of our total energy consumption.[5] Since

---

[3]The modernists talked about the city they were trying to rescue as infected with tuberculosis. In *Towards a New Architecture*, for example, Le Corbusier writes: "the machine we live in is an old coach full of tuberculosis" (1986: 277.)

[4]See www.davidrokeby.com/ (date accessed: 19 December 2012).

[5]See DOE Energy Databook 2003 – http://buildingsdatabook.eren.doe.gov/ (date accessed: 19 December 2012).

the mid-nineteenth century, developments such as suburbanisation and separation of functions have created new, artificial distances. This has had the effect that cities have expanded wildly. This expansion of cities has only been made possible through the plentiful availability of cheap (fossil) fuel during this period – "a paleo-technic paradise" as one critic has called it (Mumford 1961). The problem facing architects and planners today is that if the present resource-distribution and availability falls apart – depending on one's viewpoint, there are many indicators suggesting that we will witness, at the least, a shift in this relation in the years to come – our common urban future might look dramatically different from how we envision it today (Droege 2012). As a response to this uncomfortable situation, in the Smart City, automated computerised technologies are employed in order to optimise resources.

In an online exhibition site, IBM,[6] one of the most significant propagators of the Smart City, tries to explain the beneficial value of this employment of so-called *smart* technologies. The exhibition is entitled "Before Cities Got Smart" and operates by means of a bi-polar juxtaposing of images and diagrams. For each of the 11 examples exhibited on the site, we see a photo or artwork, depicting a well-known problem in cities and urban institutions, such as traffic congestion or human faults, an example being "doctors' incurable bad handwriting". This example is cited as being responsible for errors and even fatal outcomes in treatment plans along with the problem of misplaced paper files, leading to inefficiencies in the hospital sector. In contrast, for each image, a solution is proposed in diagrammatic form, depicting the implementation of smart technologies used to overcome the identified problem. In the example of the doctor's handwriting, the proposed solution is a transition from hand-written to electronic data in hospitals – which leads to increased accuracy and efficiency, better provisions and lower costs.[7]

Another example refers to an unnamed city which has been transformed from representing an energy and public transport disaster to offering a widely used, fast and much more energy-efficient system through the implementation of a smart monitoring system with live feedback mechanisms. Hereby, the provision and routes of buses or trains are optimised. On the main page of the exhibition website, the 11 cases are presented as a matrix on which one can click. Inside the exhibition space itself, visitors find a linear list of the examples where the double pictures are presented in linear progression. The examples can be read as a linear narrative by scrolling down on the page. Each picture is, furthermore, accompanied with a small written statement, designating what the situation was before and how it is now. An example concerns primary education. It is entitled "The Forgery."

---

[6]http://www-07.ibm.com/innovation/my/exhibit/index.html (date accessed: 19 December 2012).

[7]Instead of a situation in which 8 % of paper files are filed erroneously, the image states that the accuracy rate for prescription writing in a so-called smart US hospital is cited as being over 99.997 %. Conversely, whereas average costs per patient in a paper-based hospital are listed as 6.300 USD, in a smart US hospital these are set at 5.000 USD per patient. In this example, however, it is unclear to which extent smart, i.e. self-reflexive and automated technologies, are at all employed. We rather seem to be looking at a situation in which one system of data recording and storage (analogue) is replaced by a new one (digital).

Here, we are confronted with an image of a hand-written note, obviously written by a child, with the following text:

> Dear mrs westre, Sarah was unable to do algebra homework last night as She ~~suddeny~~ suddenly came down with a severe case of chickenpox's. Thank goodness she made a full recovery this morning and is perfectly fit for today's school trip to the zoo. Thank you. Sarah's mum.

The letter is signed with a squiggle that imitates a grown-up's signature. The image of the letter is accompanied with the following text:

> Before smart education systems, homemade notes like "the dog ate Sid's algebra homework" and "Susan's essay was so good, we kept it!" would be quietly penned in the back of school buses. With parents unable to follow their childrens' progress online, 30 % of them failed to graduate each year and the U.S. fell to 25th in math and 24th in science according to global school rankings.

If you scroll further down on the page, the next image is accompanied by the simple message "See how":

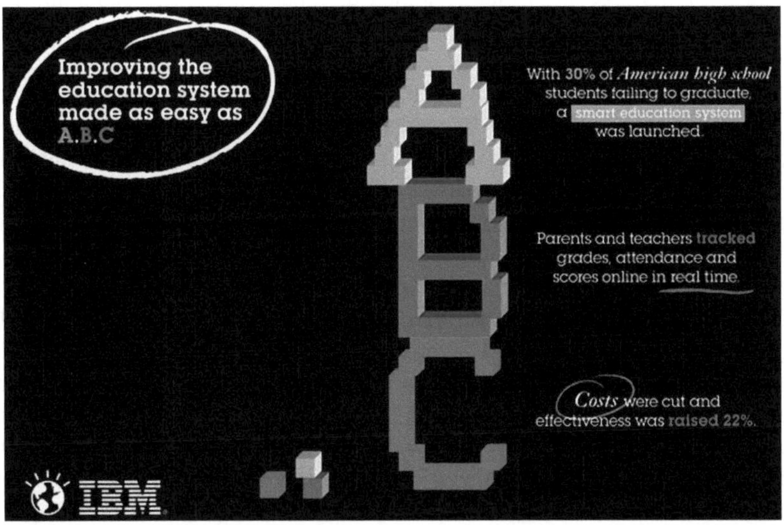

Yet, if we look closer at this image, the "how" of this sentence is unexplained. It compels us to ask a number of questions. The first question that springs to mind is: who would believe in a handwritten note? It is difficult to think of a teacher who is not able to separate the handwriting of a pupil in a class of, say, 9-year-olds from that of an adult. Another obvious question concerns the parents. When parents can monitor their children's performance in school (and that of the teachers) through automated technologies, we are looking at a system that requires a massive human element of interpretation. If one of the suggested benefits of this system is that parents are automatically notified if their children do not appear in class, the system serves as a kind of binary tracking device (my child *is* or *is not* in class). But this, in itself, says nothing about the *ethos* of education or child rearing as a question of a balance between making freedom and independence possible versus practices of monitoring, guiding and restricting. Thus the maximum message of this system, which would allow the parents to make good judgements concerning the information they are given, has to do with the *morality* of the current school system. This is a moral codex that refers to the modern institution of the school as it developed in the late-eighteenth century as a physical container that disciplines and educates all children in a society in a particular way and on equal terms.[8] In the new system, however, it is merely important that the child fulfils all set tasks and, crucially, is present in school, sitting on the designated chair, facing the teacher – probably the most central disciplinary measure of this institution.

Tracking the mere presence of the child inside this container neither tells us about the quality of that education nor of the child's mental presence. It is telling, therefore, that the image gives no information about the desired change in the rankings of US schools – just that "costs were cut and effectiveness raised by 22 %". Furthermore, one may wonder whether or not this increased sense of control on behalf of the parents is mainly aimed at a particular socio-economic group (time- and resource-strong and with IT skills). These are the parents who already possess the kinds of qualities, e.g. socio-economic status, which are among the main parameters determining how well a child will do in school.[9] One may thus also expect that the cost-cutting implied is due to the fact that parents supposedly take over part of the role of the teacher in their function as a control mechanism.

Finally, we may add that the above-mentioned self-manufactured note shows a certain creativity on behalf of the pupil. The so-called smart system of control places parents and teachers in a mutual relationship that bypasses the child, who is subject to a constant double surveillance of their daily performance but who is unable to influence or even eavesdrop on this line of communication. Yet, should the child get access to their parents' account, they can commit much more systematic and perfect forms of forgery than when communication is based on handwritten notes. When it comes to education, the so-called smart school bench is replicated by a *Doppelgänger* at home, one who is mediated by automated digital, mass communication, which in turn means that the state can save money. This seems to involve an extreme form of disciplinary measures which, in fact, belong to an already dated system of education. The school is one of the most typical institutions of the modern society of discipline.[10] At its best, the so-called smart system operates as a kind of warning system for parents and teachers. Yet, at the same time it is reproducing a model and moral code well-known from existing systems.

---

[8] See Foucault 1991.

[9] A recent study has shown that the sheer number of books in a home is an indicator of how well a child will do in school – PIRIS – www.pirls.org, accessed 19 December 2012.

[10] See Foucault 1991.

Similarly, the other examples in the exhibition are, curiously enough, dealing with the core disciplinary institutions of modernity and their associated cultural imaginaries, but without any recognisable move away from the functionality of these institutions. We have mentioned hospitals and schools above, other examples are dikes (disciplining nature, creating more land), the police (the prison, the criminal, the streets of the city as inherently dangerous), the government office (digitalising inefficient bureaucracies but depersonalising the relation between state and citizen). The remaining examples concern the optimisation of traffic and energy infrastructures – here again, no sense of paradigm shift in energy use or politics is suggested, just an optimisation of their systematic aspects, but avoiding the level of ethics or ethos – e.g. what is a *good* education?

It seems certain that these technologies have concrete potential to be useful in the parts of life that operate like systems – places where it makes sense to talk about optimisation of cost and time. However, given that the primary situations of urban, daily life will always revolve around the capacity for spontaneity, creativity, individuality – qualities that go beyond the system and implies significant problems for the Smart City as a vision of betterment. Optimising a system that is already in place may, furthermore, not necessarily make cities fit for the future. When it comes to the city, the question is if a good city by all necessity also is an efficient city?

As an ideal city, rather than proposing new architectural urban forms, the Smart City attempts at optimising urban infrastructures by combining live data collection from different spheres of urban life with different forms of data-processing. In a second step, self-reflexive, automated computer technologies are employed to create concrete responses to these perceived malaises, mainly by reducing complexity, unclogging different forms of congestion and streamlining human processes by an increased use of control mechanisms. By focusing on digital infrastructures, however, the Smart City is a noticeable attempt at dealing with the highly complex, distributed and often invisible technological systems that are crucial to the workings of cities today; structures that often go unnoticed precisely because of their innocuous and elusive character. This makes the Smart City a fascinating case study insofar as it is an attempt at dealing with aspects of the contemporary urban fabric that are often overlooked: it brings to light the complex workings of the invisible and "calm technologies"[11] that underlie contemporary urban life. Yet, as this chapter aims to show, when the Smart City technologies are put forward as a means of complexity reduction, seeking to "save the city from itself", significant slippages in the discourse and the suggested positive effects become apparent.

In the next section we will look to a part of culture where the aspect of "functionality" of technology is no longer in the foreground and where some of the invisible technological processes of automated technologies are made visible: art.

[11]The term "calm technologies" were coined by Mark Weiser in his seminal texts in the early 1990s. For a recent consideration of ubiquitous computing and the reception of Weiser, see Ekman (2012).

2 Unsmarting the City, from Complexity to Control

The Canadian artist David Rokeby explores the potential of ubiquitous computing in his art installations. By bringing forth the ambiguous dynamics at play in these technologies, his pieces allow us to think about how highly automated technologised structures influence the character of urban space and our experience of it. In *Sorting Daemon* for instance, a camera installation and the employment of face-recognition software in the urban space of Toronto are used to problematise technologies of automatisation, particularly in relation to how they may involve systems of control through hidden, that is, calm and often benign, technologies of everyday urban life:

> Like many of my other works, "Sorting Daemon" surveys its environment and uses the resulting images as the primary content of the work. In this specific case, the system looks out onto the street, panning, tilting and zooming, looking for moving things that might be people. When it finds what it thinks might be a person, it removes the person's image from the background. The extracted person is then divided up according to areas of similar colour. The resulting swatches of colour are then placed within the arbitrary context of the composite image projected in the KinoWelt Hall at the Institut. (www.davidrokeby.com/sorting.html)

Sorting Daemon is from 2003 and comments on the increased use of automated systems used for profiling in the aftermath of 9/11. However, it can also be seen in a wider context of human interaction with ubiquitous computing as those we find in the Smart City, which – although developed in other to make urban living smoother and more practical – is also a system of potential surveillance. In this way the piece brings attention to the complexity and the potentially problematic effects of Smart Cities technologies. By dwelling on the complexity that the technology *adds*, not only what it optimises and reduces, Rokeby's artworks make a more ambiguous comment on these technologies, but also one which points to what is in fact new and interesting about them as opposed to the banalisation of the IBM campaign that depicts the technology as providing trivial and common-sense improvements.

Sorting Daemon extracts the individual human being from its urban context and, through algorithmic sorting, divides the human into entities that can be reassembled and grouped in a system of coordinates by hue, size and saturation. In a certain sense, the piece becomes a visual illustration of Gilles Deleuze's "society of control". Deleuze argued in 1992 that we are moving away from Michel Foucault's disciplinary societies of the eighteenth and nineteenth centuries (Foucault 1991, 2007), towards what he terms "societies of control". The disciplinary society involved an individual passing between one closed environment with its own laws after another (the family, the school, the factory, the hospital, the prison). Institutions such as these are embodied in distinct physical and architecturally recognisable settings in which the panoptic principle of a centralised gaze can easily be implemented. Societies of control are, however, less tangible – they are characterised by what Deleuze calls "modulation", phenomenon that is connected to a city pervaded by ubiquitous technology. When virtual and physical spaces are intertwined in new ways through automated technologies, the complexity of the relationship between humans and their spatial surroundings is expanded (Muir 2012). We are no longer only dealing with humans *vis-à-vis* institutions; instead, humans, architecture and digital technology are closely knitted together in a way which, according to Deleuze, renders the disciplinary panopticon model obsolete and replaces it with a system of control:

> Felix Guattari has imagined a city where one would be able to leave one's apartment, one's street, one's neighborhood, thanks to one's (dividual) electronic card that raises a given barrier; but the card could just as easily be rejected on a given day or between certain hours; what counts is not the barrier but the computer that tracks each person's position-licit or illicit – and effects a universal modulation. (7)

This city bears certain resemblances with the city pervaded by smart technologies. Here the city has become something other than a series of enclosed physical spaces with specific architecturally distinguishing characteristics in which the individual moves as an indivisible entity and can be disciplined as such. Rather, control is exerted at levels below the individual:

> In the societies of control, on the other hand, what is important is no longer either a signature or a number, but a code: the code is a password, while on the other hand the disciplinary societies are regulated by watchwords (as much from the point of view of integration as from that of resistance). The numerical language of control is made of codes that mark access to information, or reject it. **We no longer find ourselves dealing with the mass/individual pair. Individuals have become "dividuals," and masses, samples, data, markets, or "banks."** (5 – our emphasis)

The subject is managed not as a whole, but "divided" into parts that can be modulated. *Sorting Daemon* operates precisely at this "dividual" level and makes the process visible by turning algorithmic modulations into visual codes that are projected in the exhibition room.

Through this cross reading of Rokeby's artwork and Deleuze's conceptualisation of contemporary societies we thus seem to see the contours of a more fruitful framework for grasping the complexity that arises in the modern technology-pervaded cities. We also begin to see what is at stake in IBM's marketing campaign, and how its use of technologies is often anything but smart.

In their article "The Surveillant Assemblage", Kevin D. Haggerty and Richard V. Ericson build on Deleuze and Guattari's work, arguing that:

> we are witnessing a convergence of what were once discrete surveillance systems to the point that we can now speak of an emerging 'surveillant assemblage'. This assemblage operates by abstracting human bodies from their territorial settings and separating them into a series of discrete flows. These flows are then reassembled into distinct 'data doubles' which can be scrutinized and targeted for intervention. (606)

Haggerty and Ericson trace a development from the familial and neighbourly relations of the pre-modern rural village to the increased anonymity and possibility for identity-construction prevalent in the urban metropolis – a change which has been regarded as both a freedom (Simmel 2005) and an obligation (Bauman 1997), concluding that the surveillant assemblage involves a "disappearance of disappearance" (Haggerty and Ericson 2006). It is no longer possible to be off the radar: "Instead, knowledge of the population is now manifest in discrete bits of information which break the individual down into flows for purposes of management, profit and entertainment." (ibid.). However, the capture and management of this information has also disappeared, in the sense that it takes place beneath the level of our conscious attention. For the city to be smart, it has to minimise the burden on our cognitive systems, which means that most of its processing takes place invisibly. This is both desirable and practical, yet it also results in complexity being glossed over, which is what IBM's campaign rhetoric purports.

Rokeby addresses the large amounts of invisible data which surrounds us in the piece named *Handheld* (2012). It is an installation that only reveals itself by interaction with the viewer, who makes the sculpture visible by moving her hands in the exhibition space. Here a motion sensor is suspended above the exhibition area, which is connected to two projectors that project images only where they see the body of the visitor: "As you approach an object, it is first blurry, then comes into focus as your hand comes closer. As your hand passes beyond it, the object against loses focus and dissolves." (www.davidrokeby.com/handheld.html). The work thus becomes a comment on the density of information that surrounds us, but is invisible without the right decoding devices. In *Handheld* we can bring forth this information with our hands, thus becoming projecting screens for that which is otherwise invisible. It is especially interesting that the sculpture consists of layers of images of disembodied hands:

> The hands (cut off from their bodies like those of the tormenters in Fra Angelico's fresco of the Mocking of Christ) touch each other, pass objects between them in social, practical, financial and symbolic gestures. It is a sort of network of engagements and transactions haunted by spectres of tactility and intimacy." (ibid.)

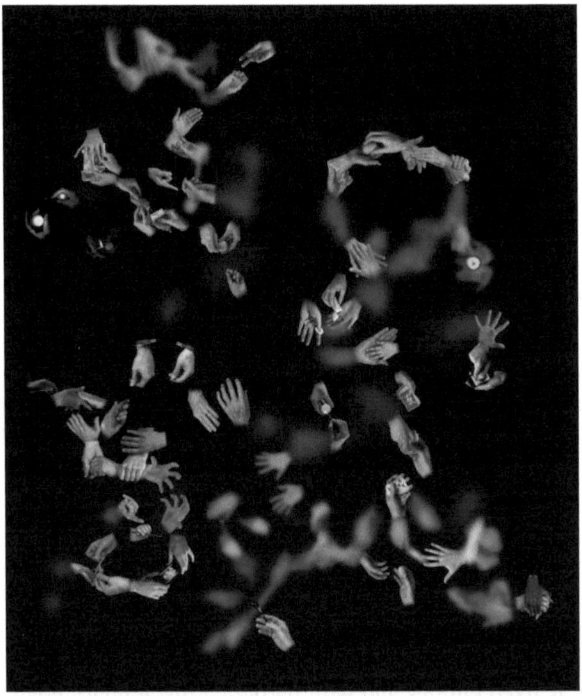

There are thus two sets of hands involved in this piece: firstly, our own – that uncover the sculpture. Rather than minimizing the complexity in the way in which biometric systems such as fingerprint technology or iris recognising do – the interaction between the technology and our bodies makes visible an otherwise invisible world of great complexity. Secondly, there are the 80 layers of images (each is one centimetre thick) of hands, which make up the sculpture. They depict exchanges characteristic of life in contemporary cybercities. By engaging with the sculpture we take the role of the Christ figure in Fra Angelico's fresco with his transparent veil. With our hands we are at the same time the tormenters and the tormented. Veiled yet seeing.

3 Conclusion

What becomes apparent from the juxtaposition of IBM's Smart City exhibition and David Rokeby's work is the way in which the marketing rhetoric of the Smart City seems to propose a notion of the city that still subscribes to a disciplinary society in which a total overview of the system and its potential for control is possible (and desired). However, Rokeby's art pieces, as read through the conceptualisation of Deleuze's notion of societies of control, illuminate the fact that this way of conceptualising the Smart City glosses over the embedded complexity of the contemporary urban environment. As is the case with many of the discussions of surveillance in recent public discourse, nuances are lost and questions are addressed in a way that belong to the disciplinary society in which institutions are contained and controlled as fixed entities and with particular disciplinary goals and morality. The interaction with invisible and often "calm" technologies (Weiser 1991, 1993, 1994), (Weiser and Brown 1996), however, calls for a new understanding of our relation to the city, its built environment and infrastructures.

We thus return to the question of complexity and order in the city with which we began. The rhetoric of the Smart City purports that it aims to reduce complexity – thus placing it in dialogue with age-old conceptualisations of urban planning. This sense of complexity of the unordered city stands in contrast to notions of the city as a smooth machine, one that is efficient and well organised: an ideal city. The Smart City can be seen as an attempt to provide a benign, ideal alternative to a perceived malign, inefficient, unsustainable and humanly disempowering complexity of contemporary urbanity. However, as Rokeby's artworks underline, the integration of smart technology in the city does not reduce complexity, but rather vastly increases it. The vision of the city as a machine that can be controlled and thus reformed for the better is revealed as an out-dated way of approaching the city. It seems to fantasise the city as a closed container of which we can get an overview in the panoptic sense. However, the implementation of automated technologies in fact results in a multiplication of urban and human complexity that needs to be addressed through different interpretative paradigms.[12]

[12]We thus turn to a conception of complexity in the sense that it is addressed in recent theory across disciplines in physics, chemistry, biology, network topology, dynamic systems theory, as well as social and cultural theory (Ekman 2012).

The naïve optimism characterising the Smart City discourse seems to veer our attention away from the consequences of its inherent complexities. This is a technology whose aim it is to make the world more manageable without adding further complexity, in fact, even reducing complexity. In this way, the Smart City discourse fails to acknowledge the new complexity that these so-called smart, automated technologies in fact embody. An acknowledgement of this complexity would entail possibilities for understanding both the potential for a more distributed form of relating to the systematic aspects of urban life and the new regimes of control and micro-management that reach deeply, subtly and unnoticed into our everyday lives.

References

Bauman Z (1997) Postmodernity and its discontents. New York University Press, New York

Corbusier L (1986) Towards a new architecture. Dover, New York

Deleuze G (1992) Postscript on the societies of control. October 59:3–7

Droege P (2012) Beyond sustainability: architecture in the renewable city. In: The sage handbook for architectural theory. Sage, London, pp 590–601

Ekman U (2012) Introduction. In: Ulrik E (ed) Throughout: art and culture emerging with ubiquitous computing. MIT Press, Cambridge, MA, pp 1–59

Foucault M (1991) Discipline and punish: the birth of a prison. Penguin, London

Foucault M (2007) Panopticism. In: Hier SP, Joshua G (eds) The surveillance studies reader. Open University Press, Maidenhead, pp 67–75

Haggerty Kevin D, Ericson R V (2006) The surveillant assemblage. In: Clive N, Dean W, Clive N, Dean W (eds) Surveillance, crime and social control. Ashgate, Burlington

Muir L (2012) Control space? Cinematic representations of surveillance space between discipline and control. Surveill Soc 9(3):263–279

Mumford L (1961) The city in history: its origins, its transformations, and its prospects. Penguin, Harmondsworth

Simmel G (2005) The metropolis and mental life. In: The urban sociology reader. Routledge, London/New York, pp 23–31

Weiser M (1991) The computer for the 21st century. Scientific American. www.ubiq.com/hypertext/weiser/SciAmDraft3.html. Accessed 19 Dec 2012

Weiser M (1993) Ubiquitous computing. Nikeei electronics. December 6: 137–43. http://sandbox.xerox.com/ubicomp/. Accessed 19 Dec 2012

Weiser M (1994) The world is not a desktop. Interactions 1(1):7–8

Weiser M, Brown JS (1996) The coming age of calm technology. http://nano.xerox.com/hypertext/weiser/acmfuture2endnote.htm. Accessed 19 Dec 2012

MIX
Papier aus verantwortungsvollen Quellen
Paper from responsible sources
FSC® C105338
www.fsc.org

If you have any concerns about our products,
you can contact us on
ProductSafety@springernature.com

In case Publisher is established outside the EU,
the EU authorized representative is:
Springer Nature Customer Service Center GmbH
Europaplatz 3, 69115 Heidelberg, Germany

Printed by Libri Plureos GmbH
in Hamburg, Germany